Earthquake Engineering

Earthquake Engineering

Yu-Xian Hu
State Seismological Bureau, Beijing, China

Shih-Chi Liu
National Science Foundation, Washington DC, USA

and

Weimin Dong
Risk Management Solutions Inc., Menlo Park, California, USA

CRC Press
Taylor & Francis Group
Boca Raton London New York

CRC Press is an imprint of the
Taylor & Francis Group, an **informa** business
A TAYLOR & FRANCIS BOOK

CRC Press
Taylor & Francis Group
6000 Broken Sound Parkway NW, Suite 300
Boca Raton, FL 33487-2742

First issued in paperback 2019

Typeset in India by Pure Tech India Ltd., Pondicherry

ISBN-13: 978-0-419-20590-6 (hbk)
ISBN-13: 978-0-367-86590-0 (pbk)

A catalogue record for this book is available from the British Library

Library of Congress Catalog Card Number: 95–74665

Visit the Taylor & Francis Web site at
http://www.taylorandfrancis.com

and the CRC Press Web site at
http://www.crcpress.com

Contents

Preface ix

Acknowledgments xi

PART ONE GENERAL OVERVIEW AND FUNDAMENTALS

1 General Overview 3

1.1 Introduction 3
1.2 Earthquake engineering studies 4
1.3 Earthquake engineering and seismology 7
1.4 Earthquake engineering and socio-economic problems 8

2 Related fundamentals of seismology 9

2.1 Inner structure of the Earth 9
2.2 Plate tectonics 11
2.3 Earthquake distribution and mechanisms 12
2.4 Seismicity 14
2.5 Seismic waves 16
2.6 Seismometry and earthquake magnitude 21

3 Related fundamentals of random vibration 26

3.1 Related concepts of random processes 26
3.2 Spectral parameters and their physical meaning 35
3.3 Distribution of maxima 37
3.4 Random process description of ground motion 38

PART TWO ENGINEERING SEISMOLOGY

4 Earthquake damage and intensity 45

4.1 Earthquake intensity and intensity scale 45
4.2 Earthquake damage and examples 50
4.3 Intensity distribution 55
4.4 Earthquake intensity attenuation 56
4.5 Effect of site conditions on intensity 60

5 Earthquake ground motion 61

5.1 Observation of earthquake ground motion 61
5.2 Characteristics of strong ground motion 67

5.3 Factors influencing ground motion 78
5.4 Earthquake intensity and ground motion 83
5.5 Ground motion evaluation 87
5.6 Simulated earthquake motion 97

6 Earthquakes in China 101

6.1 Seismic belts in China 101
6.2 Earthquake history of China 101
6.3 Seismicity in China 103
6.4 Damage caused by earthquakes in China 105

7 Earthquakes in the USA 117

7.1 Seismicity in the USA 117
7.2 Damage caused by major earthquakes 119

8 Earthquakes in Japan 136

8.1 Seismicity in Japan 136
8.2 Damage caused by some specific earthquakes 138

PART THREE THEORY OF EARTHQUAKE RESPONSE OF STRUCTURES

9 Earthquake response analysis of structures 145

9.1 Evolution of the theory of earthquake response analysis 145
9.2 Earthquake response of SDF systems and response spectrum 146
9.3 Earthquake response of MDF systems 151
9.4 Response analysis in the frequency domain 155
9.5 Vertical earthquake response and P–Δ effect 158
9.6 Theory of damping 160
9.7 Nonlinear earthquake response 164
9.8 Response to multipoint and multicomponent inputs 172
9.9 Soil–structure interaction 177

10 Earthquake response of earthworks and foundations 184

10.1 Field experience on earthworks and foundations 184
10.2 Dynamic properties of soil 186
10.3 Sand liquefaction 192
10.4 Earthquake response analysis of sites 200
10.5 Earthquake stability of earth dams and slopes 205
10.6 Earthquake response of pile foundations 209

11 Vibrational testing and dynamic properties of structures 213

11.1 Purpose of vibrational testing 213
11.2 Development of structural testing 213
11.3 Methods of structural testing 216

11.4 Dynamic properties of materials 217
11.5 Dynamic properties of members 219
11.6 Dynamic properties of structures 224
11.7 Ambient vibration 225
11.8 Failure testing of structures 227
11.9 Structural identification 229

PART FOUR SEISMIC CONSIDERATIONS IN STRUCTURAL ENGINEERING

12 Seismic hazard and risk analysis and zonation 235

12.1 Seismic hazard and risk 235
12.2 Seismicity analysis 237
12.3 Seismic hazard assessment 246
12.4 Structural vulnerability and seismic risk assessment 251
12.5 Earthquake zonation 260
12.6 Microzonation 267

13 Principles of seismic design of structures 273

13.1 Developments and trends in seismic design theory 273
13.2 Principle of selecting the design level 277
13.3 General guidelines for seismic design 278
13.4 Special requirements of various types of engineering works 279
13.5 Seismic constructional measures 300
13.6 Vibrational control 306

14 Seismic design codes 311

14.1 General features 311
14.2 Development of codes in Japan, USA and China 314
14.3 Trends in seismic codes 334

PART FIVE NEW ANALYTICAL TOOLS IN EARTHQUAKE ENGINEERING

15 Stochastic methods 339

15.1 Stochastic models of earthquake occurrence 339
15.2 Stochastic models of earthquake response 343

16 Fuzzy set theory and applications 357

16.1 Basic concepts 358
16.2 Fuzzy information processing 366
16.3 Fuzzy classification 377

List of symbols 383
References 388
Index 403

Preface

Earthquakes are one of the most devastating natural disasters on earth. In China alone, there have been 10 destructive earthquakes in this century which collectively killed more than half a million people and ravaged more than a dozen cities. The most recent Great Hanshin Earthquake in Japan killed more than 5100 people and resulted in heavy economical loss near $200 billion. Earthquakes occurred everywhere, for example in Chile (1985), Mexico (1985), Armenia (1988) and San Francisco (1989) and Northridge, California (1994), causing casualty and damage worldwide.

One major measure to mitigate the earthquake hazards is to design and build structures through better engineering practices, so that these structures possess adequate earthquake-resistant capacity. Considerable research efforts have been carried out in the U.S.A., Japan, China, and other countries in the last two decades to advance earthquake engineering knowledge and design methods. This book summarizes the state-of-the-art knowledge and worldwide experiences, particularly those from China and the U.S.A., and presents them systematically in one volume for possible use by engineers, researchers, students and other professionals in the field of earthquake engineering. Considering the active research and rapid technological advances which have taken place in the field, there has been a surprising shortage of suitable text books for senior level, or graduate level students. This book also attempts to help fill such a gap.

The book consists of two major parts: engineering seismology and earthquake-resistant design. Special attention is placed on bridging the gap between these two disciplines. For the convenience of the reader, fundamentals of seismology, earthquake engineering and random processes, which can be useful tools to describe the ground motions and to analyze the structural or soil response to them, are introduced. This is followed by a chapter describing the earthquake intensity, ground motions and its damaging effects. In the ensuing chapters concerning the earthquake-resistant design, both fundamental theories and new research problems and future directions are presented, including methods for foundation design and structural experiments. Finally, in the chapter on engineering applications, seismic risk and its damaging effects, seismic zonation and features of earthquake-resistant design of different types of structures and design standards are introduced. In this book stochastic methods and fuzzy set theory specially are introduced because of their potential to offer a new dimension in earthquake engineering applications. These two areas have been subjected to intensive research in recent years and there is a potential for them to provide solutions to some special problems which might not be amenable to conventional approaches.

Several additional features of this book are worth of mentioning here. First, we try to provide materials in engineering seismology (Part II) relevant to engineers to understand how the design earthquake motion can be determined or accurately predicted. We also provide examples of engineering applications (Part IV) for seismologists to understand what and how the predicted ground motions will be used by engineers. We feel it is important to discuss earthquake intensity and ground motions in sufficient depth, so that the readers will be able to achieve a comprehensive understanding both of subjects and their inter-relationship. Further, they will also learn how to use both data of historical earthquake events and instrumented strong motion records to analyze field data and design needs, to understand the design requirements and to design a structure and to evaluate or predict losses. For instance, it is emphasized that parameters of design ground accelerations may be approximately specified for low family buildings and peak velocity for tall office buildings, but ground velocity or strain should be specified for underground pipelines in analysis or design considerations.

As a second feature of the book, macroseismic phenomena and instrumental records from strong earthquakes, laboratory experiments and theory are all collated in the in-depth discussions. This is done in an attempt to apply them together in an integrative manner to pursue a solution to an earthquake engineering problem, as illustrated in Part III. Practicing engineers, graduate students, natural hazard mitigation managers, and engineering seismologists should all find useful and interesting materials in the book.

The authors have made a serious effort to address some important yet controversial issues, such as the effects of vertical component relative to the horizontal one, ground motion evaluation for regions with scarce or insufficient strong motion records, alternative ways of testing dynamically full scale structures up to failure, non-linear structural identification, reasons for heavier damage of structures on weaker sites, peak accelerations on soft sites, etc. In these discussions, opinions of pros and cons are both given, together with the authors' own interpretations. It is hoped that these materials will create some room for increased investigations and new discoveries in such subjects by the readers.

It is also hoped that this book will cover a wide and diverse spectrum of interconnecting engineering problems, and that readers will find it a convenient information resource. In Part II field lessons and experiences from strong earthquakes which are the basis of earthquake engineering are presented; in Part III, superstructure, foundation, theories and experiments and strong motion observation are included; and in Part IV, detailed engineering applications including hazard evaluations, design principles of various types of structures and seismic codes are highlighted. The logical path of engineering from static analysis stage, through psuedo-dynamic stage and up to dynamic analysis stage is illustrated, with a hope to show the trend of the development in the earthquake engineering field and to guide the future development to greater benefits. Part V is on new techniques, as mentioned earlier.

Acknowledgements

The authors are grateful to receive valuable comments and suggestions from many experts in the field. Among them Professor George C. Lee, Director of the National Center for Earthquake Engineering Research (NCEER), the State University of New York at Buffalo, is particularly acknowledged.

Y. X. Hu
S. C. Liu
W. M. Dong

PART ONE

General Overview and Fundamentals

1 General Overview

1.1 Introduction

A strong earthquake is a natural disaster which brings sudden fatality, great economic loss and shock to the community. Tables 1.1 and 1.2 list the major earthquakes that have occurred this century together with the resulting casualties.

Table 1.1 Some strong earthquakes that have occurred this century

Date	Earthquake	*M*	Deaths (thousands)
4 April 1905	Kashmir, India	8.6	19
18 April 1906	San Francisco, USA	8.3	>90
16 December 1920	Haiyuan, China	8.6	>200
1 September 1923	Tokyo, Japan	8.3	>100
7 March 1927	Tango, Japan	7.9	3
2 March 1933	Japan	8.9	3
30 May 1935	Pakistan	7.5	30
25 January 1939	Chile	8.3	30
26 December 1939	Turkey	7.9	23
21 July 1952	California, USA	7.7	0.0
28 March 1964	Alaska, USA	7.9	0.13
16 June 1964	Niigata, Japan	7.4	0.026
31 May 1970	Peru	7.7	40
9 February 1971	California, USA	6.6	0.065
10 April 1972	Iran	7.0	17
23 December 1972	Managua	6.5	10
4 February 1975	Haicheng, China	7.0	1.3
28 July 1976	Tangshan, China	7.8	240
3 March 1985	Central Chile	7.8	0.18
19 September 1985	Michoacan, Mexico	8.1	10
7 December 1988	Armenia, (former) USSR	6.9	25
17 October 1989	Loma Prieta, USA	7.1	0.061
16 July 1990	Philippines	7.8	1.6

Table 1.2 Number of earthquakes and deaths in 1949–69

Country	No. of earthquakes	Deaths
Argentina	8	5 100
Chile	32	28 000
China	17	159 000
Ecuador	11	5 500
Greece	47	1 500
India	6	6 500
Iran	29	7 500
Italy	15	1 800
Japan	35	15 800
Pakistan	5	40 000
Turkey	54	46 400
(Former) USSR	2	15 000

In order to reduce the disaster caused by earthquakes, it is necessary to have some means of earthquake prediction and to design structures that will withstand such shocks. Long-term earthquake prediction of 50–100 years provides a foundation for planning and engineering design; short-term prediction of a few days or months may be used as a guide for society to take some emergency action to reduce casualty numbers and to maintain normal functioning. Earthquake prediction is mainly the duty of seismologists; emergency actions are to be taken by society and governmental bodies; but specification and implementation of seismic design and retrofitting of existing engineering works are the duty of earthquake engineers. Earthquake engineers have to participate in seismic hazard evaluation and zonation with seismologists and in damage prediction and emergency planning with governmental bodies.

1.2 Earthquake engineering studies

Figure 1.1 shows the scope of earthquake engineering studies. As an earthquake engineer, working in engineering seismology, structural dynamic analysis, seismic design, earthquake-resistance evaluation and retrofitting, or program management, it is important to understand the purpose of all items in the diagram, the inter-relationship between them, their present status and trends, and also the details of the item with which one is presently involved; by knowing the preceding and following links in such a scientific chain, one can do a much better job.

From the point of view of engineering applications, earthquake engineering may be divided into the following five branches according to scientific content and the sequence of study:

- seismic hazard assessment and zonation;
- earthquake response analysis;
- design of new structures;
- evaluation and retrofitting of existing structures;
- disaster reduction planning.

On the basis of long-term prediction of possible future strong earthquakes, i.e. their time, magnitude, location and probability, it is possible to estimate the value of the selected parameters of ground motion to be expected, with a given probability of exceedance at a given location; this is referred to as seismic hazard assessment (SHA). Then a seismic zonation map can be made according to the result of the SHA of many locations in a region for a common given probability of exceedance. The task of SHA is to convert the result of long-term prediction into a prediction of ground motion required in the design of engineering works. The task of earthquake response analysis is to find the dynamic relationship between structural responses and earthquake input motion to provide a quantitative basis for design, in addition to qualitative experience. Both seismic design and evaluation–retrofitting are based on specifications or guidelines, which specify legal principles of design and construction of structures against earthquakes to ensure a reasonable balance of economy and safety. The principles and measures specified are based on the following three groups of knowledge:

- experiences from field studies of past earthquakes;
- theories from the analysis of strong motion observation, laboratory tests and response computations;
- engineers' experience.

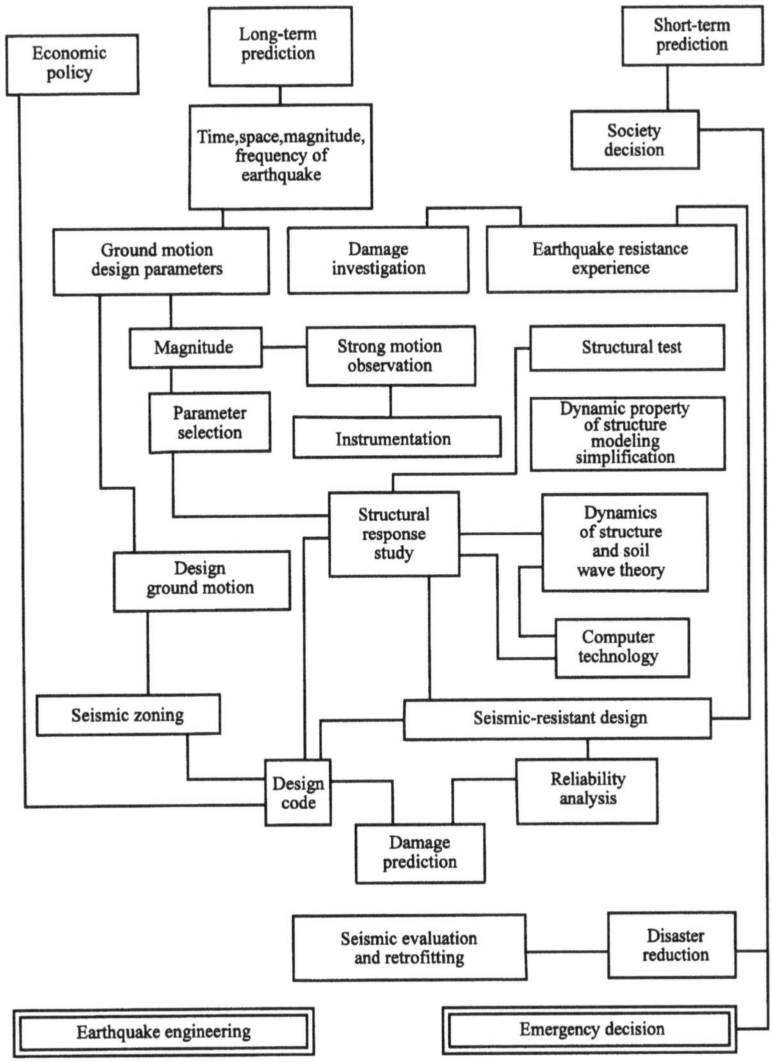

Figure 1.1 Overview of earthquake engineering.

Under the general policy of economy and safety, these technical foundations are converted into specifications.

The boxes in Figure 1.1 show the range of subdivisions under earthquake disaster reduction: earthquake prediction, engineering seismology (SHA and zonation), seismic-resistant structures and seismosociology (damage prevention and emergency decision). Earthquake engineering consists of engineering seismology, seismic-resistant structures and damage prevention. This book discusses primarily the first two subjects.

From a disciplinary point of view, earthquake engineering spans seismology, engineering, geology and sociology. Engineering seismology studies problems of engineering parameters of earthquake ground motion and their evaluation based on seismology and geology. Because large uncertainties exist in earthquake prediction and ground motion evaluation for a given earthquake, the results of ground motion prediction or seismic hazard assessment are usually in terms of probability. The results of seismic hazard assessment for a large region are usually given in a contour line map of ground motion parameters, such as effective peak acceleration or intensity, and this kind of map is called a seismic zonation map. A zonation map for a small region with consideration of the effects of the difference in local geological and geographical conditions on ground motion is called microzonation. Both zonation maps are used for the planning and design of engineering works. Earthquake resistance study includes structural response analysis and design, which cover studies on dynamic properties of foundation soils, structural materials, members and structures, especially those in the inelastic range, and the failure mechanism of the structure through theoretical and experimental methods; these include lessons learned from field results for strong earthquakes, as well as design principles of engineering works, such as reliability theory and engineering decisions to find a suitable balance between safety and economy. Because of the increasing complexity and variety of the types of engineering works, special types of studies are sometimes needed. For example, it is necessary to study interactions of dam, foundation and water for concrete dams, responses under multisupport inputs for long-span bridges, axial deformation analysis for underground pipelines, and highly conservative analysis and design for nuclear power plants.

The following special features make earthquake engineering different from other engineering fields. Firstly, a strong earthquake is a rare phenomenon for a given place; its occurrence is sudden and of short duration and it is very difficult to predict its specific time, location and magnitude; but once it occurs, the damage it produces is very severe. This situation is a challenge to engineering decision: to design all engineering works in a region for the maximum possible earthquake means to increase the constructional cost by 10% or 20% which may be wasted if that earthquake does not happen in the lifetime of the structure, say, 50 years, and that is usually quite probable; but not to design for any earthquake means to gamble the lives and fate of people in a whole city against a rare possibility of disaster. A comparable situation is that of life insurance, but this concerns only one person in a particular case, while a strong earthquake may cost the lives of ten thousand people. The inability to predict a strong earthquake accurately makes the decision very difficult for a very rare and highly uncertain catastrophe at the cost of a very severe loss to society. The same dilemma exists for existing structures when a strong earthquake is predicted for a region: to retrofit or not.

1.3 Earthquake engineering and seismology

Both earthquake engineering and seismology take earthquakes as their primary concern. The relation between seismology and earthquake engineering is quite similar to that between bacteriology and medical science, or hydrology and hydraulic engineering. The latter depends on the results of the former, studies the result on the structures or human beings under the attack of the former and tries to mitigate the risk or damage; and the former finds its application in the latter. Both branches of science need to study their joint link which can hardly be classified to one branch; but each branch has its own major interest and emphasis.

Seismology emphasizes mostly the physics and laws of earthquake occurrence which includes the cause and mechanism of earthquakes, seismic waves and wave propagation. On the basis of damage study of historical earthquakes and instrumental measurements of ground motion, earthquake magnitudes and locations are determined, from which, together with geotectonic data, spatial earthquake distribution and seismicity are then investigated and future earthquakes may be predicted. Earthquake engineering emphasizes the safe and economic design and construction of engineering works on the basis of earthquake prediction and ground motion evaluation. To predict strong earthquakes and to provide adequate strength to engineering works are both necessary for the safety of human beings and the welfare of society.

Ground motion is the joint or overlapping link of seismology and earthquake engineering. Seismologists have to study this problem because it provides the basic data with which to evaluate the location, magnitude and mechanism of an earthquake and to understand the formation of the Earth's interior. Engineers have to study it because it is the most reliable way to understand the characteristics of the earthquake and the ground motion induced and to evaluate ground motion of predicted earthquakes for the design of engineering works. In the past, engineers concentrated on near-field and high-frequency ground motion and seismologists on far-field and low-frequency motion; but their interests are extending towards each other and the gap is closing.

Seismology became a branch of science earlier than earthquake engineering. In the 18th century, scaling of the strength or magnitude of earthquakes and earthquake ground motion was performed on the basis of macroseismic phenomena together with measurements from an early version of the ground motion instrument, while earthquake engineering started at the end of the 19th century when some European engineers suggested designing structures with a few percent of the weight of the structure as the horizontal load. This idea of seismic design was taken up and developed in Japan at the beginning of the 20th century. Intensity and ground motion were an area of common interest, but with different purpose. The seismologist studies the distribution of intensity on the ground surface in order to find out details about the earthquake's source and studies the ground motion in order to find out the arrival time and trace of various types of wave; the final goal is the understanding of the earthquake itself and the earth media. The engineer studies the same subjects, but the final goal is the understanding of the dynamic and engineering properties of ground motion itself. Seismologists used the term intensity to measure the strength of ground shaking; engineers used it to represent some damage-related parameters of ground motion such as peak acceleration or velocity. Recently, many engineers have discarded the term intensity and use the parameters of ground motion directly.

In addition to seismicity and ground motion, seismology also covers geotectonics and other geophysical phenomena, while engineering emphasizes primarily the dynamic properties and response of engineering works and their design to withstand earthquakes.

1.4 Earthquake engineering and socio-economic problems

An earthquake is considered a disaster to mankind. For the mitigation of damage or loss from this disaster, it is necessary to predict the possible loss and then to take countermeasures to reduce the loss to a minimum through some decision policy. One general criterion is to minimize the sum of the investment and possible loss (see section 13.2).

Earthquake loss includes political loss, casualties, loss from secondary disasters (such as fire, flood, tsunami (seismic seawave), disease) and indirect loss to society due to the interruption or disturbance of cultural life and industry, in addition to direct economic loss from damage to engineering works. In indirect loss, a chain action of loss is possible, especially in big cities, which occurs from the reduction of the labor force and means of production and loss in the related links of the chain. Two kinds of countermeasures may be taken to mitigate the earthquake loss: engineering and action by society. The engineering approach tries to mitigate the loss by increasing the earthquake resistance of engineering works, or by decreasing the earthquake input, or both; and the society approach by ensuring the smooth functioning of society during and after a strong earthquake, through public awareness and preparation for its occurrence.

It is important to realize that the level of earthquake protection through engineering means is limited by a city's or nation's economic capacity. A high level of engineering protection requires high economic investment at the cost of reducing investment to other welfare areas in society, while a low level of engineering protection will expose society to risk, though with a small probability. It is a matter of decision as to what chance can be tolerated. For example, it is well known that unreinforced masonry buildings do not resist strong earthquakes well and should not be constructed in regions of strong seismicity, but because of the poor economic situation in developing countries, this kind of building is still common.

2 Related fundamentals of seismology

The purpose of this chapter is to provide engineers with some related fundamentals of seismology, in order to understand the level of reliability and accuracy of the data provided by the seismologist and to design engineering works accordingly. For more details, related books on seismology should be consulted.

2.1 Inner structure of the Earth

The earth is roughly spherical with an average radius of around 6400 km. Its inner structure was determined from the propagation of earthquake waves. One sector through the Earth's center is shown in Figure 2.1. The Earth consists of three spherical shells of quite different physical properties. The outer shell is a thin crust of thickness varying from a few kilometers to a few tens of kilometers; the middle shell is the mantle, about 2900 km thick, and the Moho (Mohorovicic) discontinuity is its interface with the crust; the innermost shell contains the core, of radius approximately 3500 km. The crust is made of various types of rock, differing in composition and thickness in its oceanic and continental parts. The crust in the continental part consists of two layers, granitic in the outer and basaltic in the inner layer, with total thickness about 30–40 km, but reaching 70 km under high mountains, such as in the Qing-Zang plateau of west China. The crust under the oceans is basaltic only, with no granitic deposit, with a thickness of only about 5 km. The mantle consists mainly of comparatively uniform ultrabasic olivine rock; its outer 40–70 km shell together with the crust is usually referred to as the lithosphere, directly under which is a layer of soft viscoelastic asthenosphere a few hundreds of kilometers in thickness. The wave velocity in the asthenosphere is obviously lower than those in its neighboring rocks, perhaps due to its viscoelastic, or creep, property under high temperature and confining pressure. The lithosphere and asthenosphere together form the upper mantle. Below, the lower mantle extends a further 1900 km or so. The core consists of outer and inner cores. Because it is found that no transverse wave can propagate through the outer core, it is agreed that the outer core is in a liquid state.

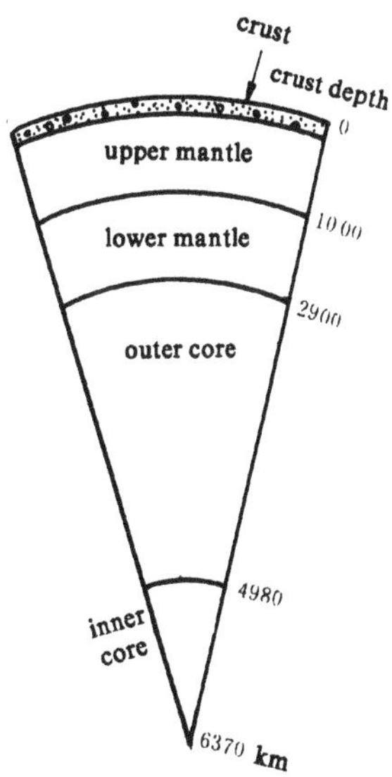

Figure 2.1 Layering of the Earth's interior.

Temperature increases from crust to core. Roughly speaking, the temperature is about 600°C at a depth of 20 km, 1000–1500°C at 100 km, 2000°C at 700 km and 4000–5000°C in the core. The high temperature comes from the heat released from radioactive material inside the earth. The distributions of the radioactive material under ocean or under the continents may be different, and this is considered to be the cause of current movement of materials in the mantle.

Pressure increases also from crust to core, perhaps 9 tf/cm^2 in the upper mantle, 1400 tf/cm^2 in the outer core, and 3700 tf/cm^2 in the inner core.

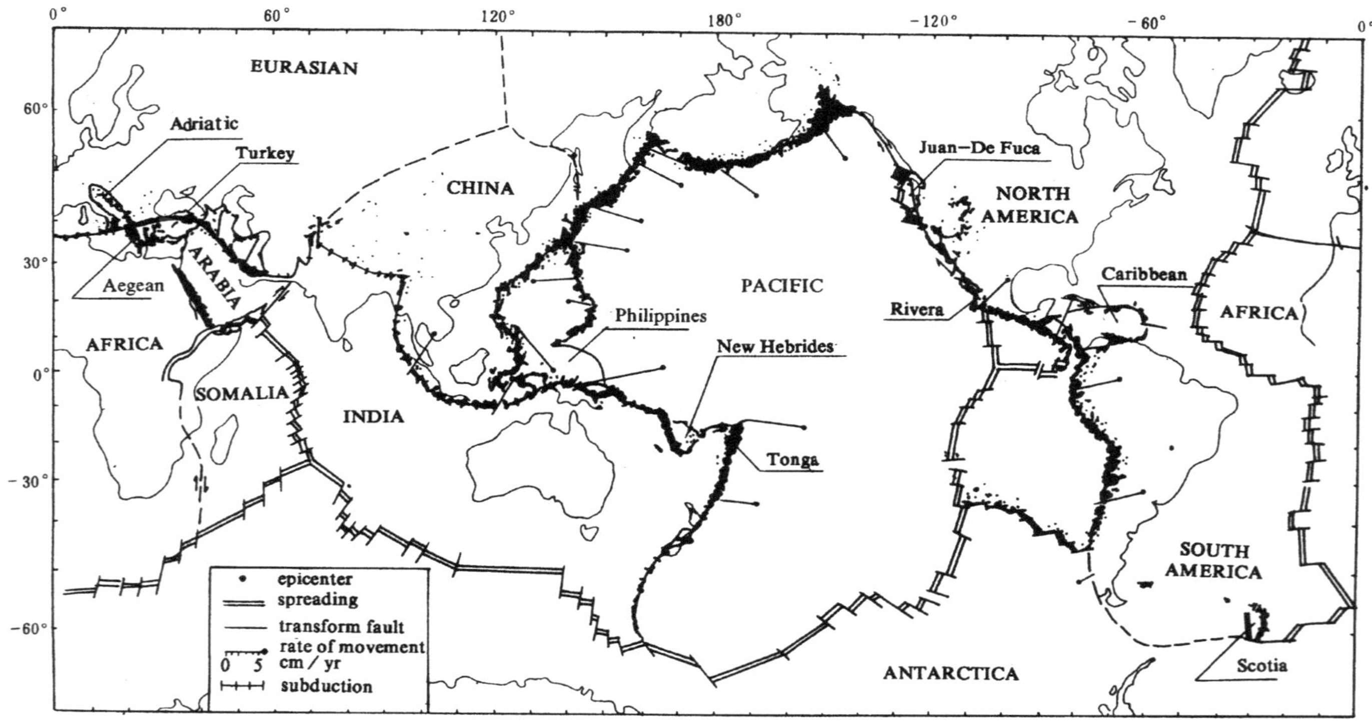

Figure 2.2 Tectonic plates and epicenters of earthquakes.

Under such extreme conditions, the properties of rock are quite different from those under normal temperature and pressure.

2.2 Plate tectonics

Plate tectonics was developed on the hypothesis of sea-floor spreading during the past few decades. According to this concept, the rigid lithosphere, consisting of six major plates, drifts on the rheological asthenosphere, like a ship on the ocean, but with a very slow speed. The six plates are the Eurasian, Pacific, American, African, Indian and Antarctic, as shown in Figure 2.2. Each plate may then be subdivided into smaller plates. The relative movement of the plates is roughly a few centimeters per year, and has continued for at least 200 million years. The theory can be described as follows. Material flows out from the upper mantle through the lithosphere at ocean ridges where the crust is thin, and pushes the lithosphere, whose thickness is a few kilometers, drifting horizontally on the asthenosphere, which shows rheological properties under high temperature, high pressure and permanent horizontal pushing. When two tectonic plates collide, one thrusts under the other and comes back to the lithosphere, which forms a deep ocean trench and subduction zone at the junction of two plates, and volcanos and mountains on the plate which remain on the Earth's surface as shown in Figure 2.3.

The following four facts support this theory:

- matching of the geology, paleontology and geometry of the continents;
- imprinting of historical traces of magnetic fluctuations;

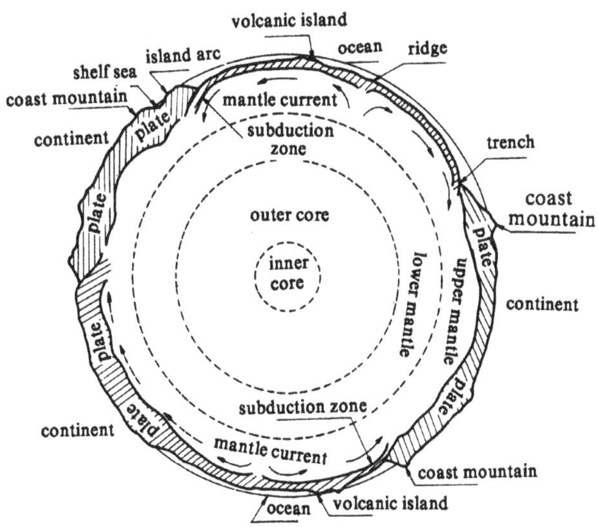

Figure 2.3 Sea-floor spreading.

- characteristics of ocean ridges and trenches;
- instrumental earthquake evidence.

The traces of magnetic fluctuations in the ocean plates are all parallel to the ocean ridge, and can be correlated and dated by radioactive methods and explained only by the sea-floor spreading theory.

In Figure 2.2, the double line shows the ocean ridge where the tectonic plates originate, the jagged line shows the subduction zone where the plates return to the upper mantle, and the thin line shows the transform fault, on which earthquakes occur only in the segment between two offsets of the ocean ridge.

2.3 Earthquake distribution and mechanisms

2.3.1 Earthquake distribution

Various seismological stations now cover almost the whole world, and about 1000 earthquakes of magnitude not smaller than 5.0 are recorded each year. Most of these earthquakes are concentrated along plate boundaries. Earthquake-concentrated zones are called earthquake belts. There are two major earthquake belts on the Earth (Figure 2.2). One is the Circum-Pacific belt, which is very active, with 75% of earthquakes concentrated on it. The other one is the Eurasian or Alpide-Asiatic, with about 22% of earthquakes occurring on it. Earthquakes on these two major belts are interplate earthquakes. The remaining 3% of earthquakes occur inside the plates and are thus intraplate earthquakes.

Intraplate earthquakes are also caused by plate movement. Because the thickness and rigidity of the crust, the Moho discontinuity surface and the driving force of the ocean plates are not uniform, the relative displacements and velocities between plates vary both in space and time. A plate is usually under complex stress condition; one part may be under tension, causing depression, and the other part under compression, causing mountain uplift, or under shear, causing horizontal deformation.

Compared with interplate earthquakes, continental earthquakes have the following three features.

1. They are less frequent and less concentrated.
2. They are more dangerous to humans.
3. The source mechanism varies and is more complicated.

Because the continental plate remains on the Earth's surface and is of varying thickness, it is full of faults and folds under long-term tectonic action, and earthquakes occur with much more scattering and much more varied mechanisms. Figure 2.4 shows strong earthquakes which occurred between 780 BC and 1979 in and around China. Although some rules may be found in their distribution, they are certainly more scattered than those along Japan, where the Pacific plate submerged underneath from the east.

Figure 2.4 Earthquake distribution in and around China.

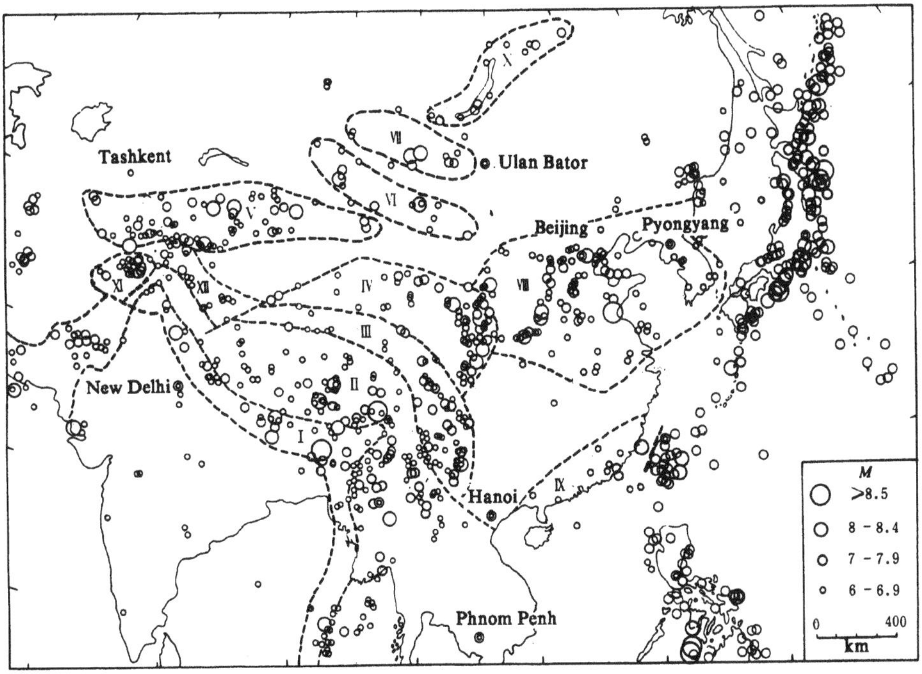

2.3.2 Earthquake mechanisms

Deformation due to plate tectonics is a very slow but persisting process. In a very long time period, deformation accumulates elastic strain in the crust. Crust material may be elastic somewhere but rheologic in other places. If the crust is elastic, strain energy may be accumulated and it may crack suddenly when the strength of rock is overcome and the accumulated energy is transferred into earthquake waves, and an earthquake occurs. This is the elastic rebound theory of earthquakes.

The rigidity and brittleness of rock are higher in the crust, and the rock behaves like an elastic and brittle material. When stress or strain is over the limit capacity, the rock breaks and the stress drops suddenly from its maximum σ_0 at time t_0 instantly to a minimum value σ_1, but it may recover partly to some value σ_2 in-between at time t_1, and the final relative dislocation of the two sides of the fault is u from t_0 to t_1. Both stress σ and dislocation u are averages over the whole fault area A. $\Delta t = t_1 - t_0$ is rise time,

$$\Delta\sigma = \sigma_0 - \sigma_1 \tag{2.1}$$

is stress drop and $\Delta\sigma = \sigma_0 - \sigma_2$ is dynamic stress drop.

$$M_0 = GAu \tag{2.2}$$

is seismic moment, a quantity commonly used to represent the size of an earthquake; G is the shear modulus of rock. The physical meaning of equation (2.2) is the definition of seismic moment as the total work done by a stress equal to G on the faulting area A with a dislocation u. The total energy released is

$$E_T = \sigma A u \tag{2.3}$$

part of which is released in the form of earthquake waves and the remaining part in the form of heat.

Faulting of the whole source of a strong earthquake is not completed at the same instant. The rupture starts usually at one point, the instrumental focus, and propagates outwards with a rupture velocity v_f of 2–3 km/s, somewhat less than shear wave velocity v_s.

In addition to mechanical parameters, the geometrical parameters of the source are also used in earthquake engineering. Since most strong earthquakes occur in the crust, the focal depth of continental earthquakes important in engineering is usually a few or a few tens of kilometers. The linear dimension of the faulting area of one earthquake varies from a few kilometers for small earthquakes to a few hundreds of kilometers for very large earthquakes, and the width in the direction of dip varies from a few to a few tens of kilometers. The magnitude of an earthquake will be discussed in section 2.6.

2.4 Seismicity

Seismicity is a description of the relationship of time, space, strength and frequency of earthquake occurrence within a certain region and its understanding is the foundation of earthquake study. Since there is still no practical way to control earthquakes, we can only try to understand and follow their nature wisely to prepare for possible strong earthquakes through prediction, earthquake engineering and society or governmental efforts of disaster reduction.

The most popular way to study seismicity is empirically or statistically. A seismic belt is defined first to have a similar past earthquake distribution and geological and tectonic background, with dimensions somewhat similar to those shown in Figure 6.1, depending on data available. Past earthquake data in that belt are then analyzed. To describe seismicity, Gutenberg and Richter suggested the following formula (Des Kiureghian and Ang, 1977):

$$\log N = a - bM \tag{2.4}$$

where N is the number of earthquakes occurring within the magnitude interval $(M, M+\Delta M)$, and ΔM should be sufficiently large to have a fairly smooth $\log N - M$ curve; a is related to the total number or annual rate of earthquakes and b to the distribution of earthquakes among magnitudes. The magnitude–frequency relation may be written in terms of accumulated number by integrating N in equation (2.4) as

$$\ln N = \alpha - \beta M \quad \text{or} \quad N = \exp(\alpha - \beta M) \tag{2.5}$$

where N is the number of earthquakes of magnitude $\geqslant M$, and

$$\alpha=\frac{a-\log b}{\log e};\quad \beta=\frac{b}{\log e} \tag{2.6}$$

If the lower limit of magnitude is M_0, the annual number of earthquakes will be

$$N_T=\exp(\alpha-\beta M_0)$$

and the probability of earthquakes not less than M will be

$$1-F(M)=\frac{N}{N_T}=\frac{\exp(\alpha-\beta M)}{\exp(\alpha-\beta M_0)}=\exp(-\beta(M-M_0))$$

The probability of earthquakes not more than M is

$$F(M)=1-\exp(-\beta(M-M_0)) \tag{2.7}$$

The probability density function of M is

$$f(M)=\beta\exp(-\beta(M-M_0)) \tag{2.8}$$

Although the Gutenberg–Richter relation proves good for moderate earthquakes in many regions, especially in large regions, it gives less earthquakes in the small magnitude range and gives no limit for large earthquakes. Cornell and Vanmarcke (1969) suggested an upper limit M_u for magnitude with the probability $F(M)$ of occurrence of earthquake of magnitude not more than M modified as

$$F(M)=\frac{1-\exp(-\beta(M-M_0))}{1-\exp(-\beta-(M_u-M_0))}\quad (M_0\leqslant M\leqslant M_u) \tag{2.9}$$

and the corresponding expressions are

$$N(M)=\exp(\alpha-\beta M_0)-\exp(\alpha-\beta M)$$
$$N_T=\exp(\alpha-\beta M_0)-\exp(\alpha-M_u)$$

When M_u approaches infinity, it reduces to the ordinary case. Figure 6.3 gives the results of some seismic belts.

In practical analysis the lower limit magnitude M_0 is usually taken as 4.0 or more, depending on data completeness. When instrumental data are used, smaller values may be used; when historical noninstrumental data are used, a larger value is preferred. Selection of upper limit magnitude is somewhat difficult; for a large belt of high seismicity, 8.5 is used; for a local belt of low seismicity, smaller ones are used.

Because of the difficulty in selecting the upper limit magnitude, Lomnitz suggested the following formula (Lomnitz and Rosenblueth, 1976):

$$\ln N=a-b\exp(cM) \tag{2.10}$$

Yegulalp and Kuo (1974) suggested the type III extreme value distribution

$$F(M)=\exp\left[-\left(\frac{M_u-M}{M_u-M_0}\right)^k\right] \tag{2.11}$$

where $k > 0$. Because these relations are somewhat complicated, equation 2.4 and 2.9 are preferred.

2.5 Seismic waves

Human understanding of earthquakes, and of other physical sciences, comes first from macroseismic phenomena, but ultimately from instrumental observation. It is the instrumental data for seismic waves that provide quantitative understanding of the earthquake source mechanism and of the interior construction of the Earth. Although the Earth's materials show rheological properties under static forces from plate tectonic movement, they show elastic properties under the dynamic action of seismic waves from earthquakes; the weak viscosity under dynamic forces can be considered in terms of energy-absorbing damping added to the elastic properties.

2.5.1 Wave equation

In isotropic, homogeneous and elastic media, with no damping, the motion of particles should satisfy the stress–strain relation, continuity condition and Newton's second law of motion, and the following wave equation may be derived from elasticity theory:

$$\nabla^2\phi=\left(\frac{1}{c^2}\right)\frac{\partial^2\phi}{\partial t^2} \tag{2.12}$$

where

$$\nabla^2=\frac{\partial^2}{\partial x^2}+\frac{\partial^2}{\partial y^2}+\frac{\partial^2}{\partial z^2}$$

is the Laplace operator and ϕ is a function of time t and coordinates x, y and z. In equation 2.12, for volume strain or dilatation, $c=\alpha$ and

$$\phi=\theta=\frac{\partial u}{\partial x}+\frac{\partial v}{\partial y}+\frac{\partial w}{\partial z}$$

where u, v and w are displacements along axes x, y, and z respectively. For torsional strain or distortion, $c=\beta$ and

$$\phi=\omega_i \qquad (i=x,\ y \text{ or } z)$$

$$\omega_x=\frac{1}{2}\left(\frac{\partial w}{\partial y}-\frac{\partial v}{\partial z}\right),\ \omega_y=\frac{1}{2}\left(\frac{\partial u}{\partial z}-\frac{\partial w}{\partial x}\right),\ \omega_z=\frac{1}{2}\left(\frac{\partial v}{\partial x}-\frac{\partial u}{\partial y}\right)$$

are rotational strains about the x, y and z axes respectively.

$$c=\alpha=v_P=\left(\frac{\lambda+2\mu}{\rho}\right)^{1/2} \tag{2.13}$$

is the longitudinal velocity for dilatation,

$$c=\beta=v_S=\left(\frac{\mu}{p}\right)^{1/2} \tag{2.14}$$

the transverse velocity for distortion; and $\lambda=\nu E/[(1+\nu)(1-2\nu)]$ and $\mu=E/[2(1+\nu)]=G$ are Lamé constants, E and G are respectively Young's and shear modulus, and ν is the Poisson ratio.

2.5.2 *Body waves*

(a) Longitudinal wave

In the case of a plane wave propagating along the x-direction in an infinite space, it is easy to show that

$$\theta=f_1(x-\alpha t)+f_2(x+\alpha t), \quad \omega_i=0 \quad (i=x,\ y \text{ and } z) \tag{2.15}$$

satisfies the wave equation 2.12. Wave f_1 propagates in the direction of x, and wave f_2 in the opposite direction. The plane wave θ takes the same value in the plane $x=$const. and there is dilatation but no distortion. Displacement is only in the x-direction, the same as the direction of wave propagation.

(b) Transverse wave

In the case of a plane wave propagating along the x-direction in an infinite space, it is also easy to show that

$$\omega_y=f_1(x-\beta t)+f_2(x+\beta t) \qquad \omega_x=\omega_z=\theta=0 \tag{2.16}$$

satisfies the wave equation. It is a distortional or transverse wave because $\theta=0$ means no dilatation. This wave has displacement only in the z-direction, perpendicular to its propagating direction x. If the x–z plane is horizontal, it is an SH wave; if the z-direction is vertical, it is an SV wave.

The ratio of v_P to v_S can be obtained from equation 2.13 and 2.14 as

$$v_P/v_S=\left(1+\frac{1}{1+2\nu}\right)^{1/2}$$

Since in the upper mantle, ν may be taken as a quarter, the ratio is then 1.7 or $v_P=1.7v_S$. The difference in arrival times of transverse and longitudinal waves will be

$$\Delta t=t_S-t_P=R\left(\frac{1}{v_S}-\frac{1}{v_P}\right)=\frac{R}{v_0}$$

In the crust, v_P is approximately 7–8 km/s, and v_S 4–5 km; the imaginary wave velocity v_0 is then about 10–11 km/s. The arrival time difference Δt can easily be found from earthquake records; the focal distance is then approximately $R=v_0\Delta t$.

2.5.3 Surface wave

In an infinite space, there are only body waves. If there are boundaries in the space, other types of wave may exist to satisfy the boundary conditions. Rayleigh and Love waves may be generated at free surfaces and are named surface waves. Because the near-surface part of the crust is made of layers of materials of different mechanical properties it is easy to generate surface waves in the process of wave propagation.

Rayleigh waves can be generated at the free surface of a half space. Their particle motion is in the vertical plane only, with a trace of a reverse ellipse having a ratio of vertical to horizontal amplitudes about 3:2. The amplitude attenuates very fast along the vertical axis, reducing to about one-fifth after one wavelength. Rayleigh waves are formed by superposition of body waves at a surface with the reflected waves. They can be generated only at sites of epicentral distance greater than $v_R h/(v_P^2 - v_R^2)^{1/2}$ where h is focal depth and $v_R = 0.92 v_s$ the Rayleigh wave velocity.

Love waves (Q waves) were discovered first in real earthquake records and their existence was proved theoretically by Love. The condition for their formation is a half space with a horizontal layer of soft deposit with a shear wave velocity v_{S1} less than that of the half space v_{S2}. Love waves are SH waves.

The velocity v_L of Love waves is a function of frequency ω or wavelength ℓ, since wave velocity $v = \ell\omega/(2\pi)$. When the wavelength is very small, $\omega \to \infty$, $v_L = v_{S1}$; when ℓ is very large, $\omega = 0$, $v_L = v_{S2}$.

2.5.4 Dispersion of waves

Because the surface of Earth's crust is made of rock layers of different stiffness, increasing with depth, both Love and Rayleigh waves have the dispersion property, i.e. the wave velocity is a function of frequency or wavelength (Okamoto, 1984, pp. 13–14).

First in real earthquake records and later proved by theory, there is a group of waves of very similar frequency and velocity v, forming a regular shape and moving together with a velocity c:

$$c = v - \ell \frac{dv}{d\ell} \tag{2.17}$$

which is called group velocity; v is phase velocity and ℓ is the average wavelength of this group of waves. Near the Earth's surface, $dv/d\ell$ is positive, and group velocity is less than phase velocity.

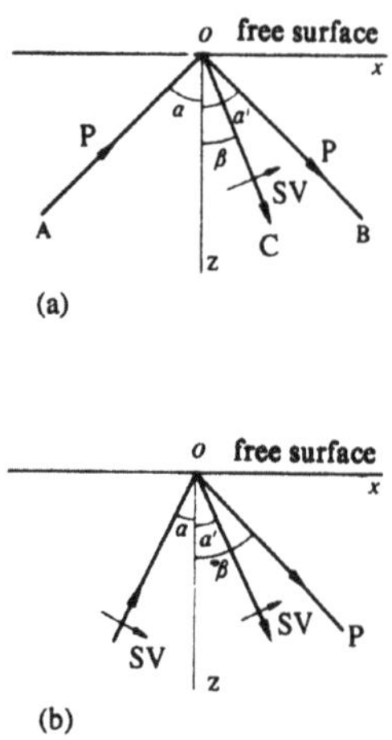

Figure 2.5 Reflection of waves: (a) incident P wave; (b) incident SV wave.

2.5.5 Reflection and refraction of waves

Let us consider a simple case as shown in Figure 2.5. Note here α and β are angles. The y-axis is perpendicular to the paper (not shown), and yox is the free surface of the half space.

(a) Incident P wave

When an incident P wave arrives at the surface with an angle α to the vertical axis z, it is possible to have both P and S waves reflected from the surface.

Their relations may be found from fundamentals of dynamics. Let these waves be

$$\text{incident P: } S(x, z, t) = A\exp[i(k_x x - k_z z - \omega t)] \tag{2.18a}$$

$$\text{reflected P: } S'(x, z, t) = A'\exp[i(k_x' x + k_z' z - \omega' t)] \tag{2.18b}$$

$$\text{reflected S: } S''(x, z, t) = B\exp[i(k_x'' x + k_z'' z - \omega'' t)] \tag{2.18c}$$

where

$$k_x = (\omega/v_P)\sin\alpha,\quad k_x' = (\omega'/v_P)\sin\alpha',\quad k_x'' = (\omega''/v_S)\sin\beta \tag{2.19a}$$

$$k_z = (\omega/v_P)\cos\alpha,\quad k_z' = (\omega'/v_P)\cos\alpha',\quad k_z'' = (\omega''/v_S)\cos\beta \tag{2.19b}$$

It is easy to prove that equation 2.18 satisfies the wave equation. The boundary condition at the free surface is zero shear, i.e.

$$\tau_{xz}|_{z=0} = \sigma_z|_{z=0} = 0$$

from which is obtained

$$\omega = \omega' = \omega'',\quad \alpha = \alpha',\ k_x = k_x' = k_x''$$

$$(1/v_P)\sin\alpha = (1/v_S)\sin\beta \quad \text{(Snell's law)} \tag{2.20}$$

$$A'/A = (v_S^2 \sin 2\alpha \sin 2\beta - v_P^2 \cos^2 2\beta)/D \tag{2.21a}$$

$$B/A = (2 v_P v_S \sin 2\alpha \sin 2\beta)/D \tag{2.21b}$$

$$D = (2 v_S^2 \sin 2\alpha \sin 2\beta + v_P^2 \cos^2 2\beta) \tag{2.21c}$$

The following can be seen from the above equations.

1. When $\alpha = 0$, $B/A = 0$ and $A'/A = -1$, which means that there is no reflected S wave and that the reflected P wave is in antiphase with the incident P wave.
2. When $\alpha = \pi/2$, $A = B = 0$ which means that there is no reflection problem and the wave just propagates.
3. For any α between 0 and $\pi/2$, both reflected P and S waves exist. $\beta < \alpha$ from equation 2.20 because $v_S < v_P$.

(b) Incident SV wave

Similar to the foregoing case of an incident P wave, we have

$$\text{incident P: } S(x, z, t) = B\exp[i(k_x x - k_z z - \omega t)] \tag{2.22a}$$

$$\text{reflected P: } S'(x, z, t) = B'\exp[i(k_x' x + k_z' z - \omega' t)] \tag{2.22b}$$

$$\text{reflected S: } S''(x, z, t) = A\exp[i(k_x'' x + k_z'' z - \omega'' t)] \tag{2.22c}$$

where

$$k_x = k_x' = (\omega/v_S)\sin\alpha,\quad k_x'' = (\omega/v_P)\sin\beta$$

$$k_z = -k_z' = -(\omega/v_S)\cos\alpha,\quad k_z'' = (\omega/v_P)\cos\beta \tag{2.23a}$$

$$B'/B=(-v_S^2\sin 2\alpha\sin 2\beta+v_P^2\cos 2\alpha)/D$$

$$A/B=(v_P v_S \sin 4\alpha)/D \tag{2.23b}$$

and Snell's law. Comparing with the case of incident P wave, the difference is that $\beta>\alpha$; and when $\sin\alpha>v_S/v_P$, angle β does not exist, which means no reflected P wave for large incident angle α.

(c) Incident SH wave

This is the simplest case and there exists only a reflected SH wave with reflected angle equal to the incident angle and with equal amplitude.

More general cases of reflected and refracted waves can be found in standard books on wave theory (Kanai, 1983).

2.5.6 *Attenuation of waves*

In the process of propagation, part of the energy is transmitted as heat into the medium; this is called attenuation or the interior damping action. For small damping, this effect can be considered theoretically by complex damping theory (Section 9.6.1), i.e. by replacing the real Lamé constants or the E or G modulus in the wave equation by corresponding complex constants. In the following, the case of a plane wave is taken to explain it.

Suppose there is a plane P wave propagating along the x-axis. The solution of equation 2.12 for the case of no damping is

$$u(x,\ t)=\exp[\mathrm{i}(kx-\omega t)]$$

where k is real. For the case with damping, use $k(1+\mathrm{i}\delta)$ in place of k, where k and δ are both real and the solution becomes

$$u(x,\ t)=\exp(-k\delta x)\exp[\mathrm{i}(kx-\omega t)] \tag{2.24}$$

where the constant δ is related to damping.

In seismic wave theory, a quality factor Q is used instead of δ. The definition of Q is

$$\frac{1}{Q}=\frac{\Delta E}{2\pi E} \tag{2.25}$$

where ΔE is the difference of energy passing through $x=x_0$ and $x=x_0+\lambda$, or the energy loss between this part of path; E is the energy passing through $x=x_0$ in one cycle of period T of the wave; and $\lambda=2\pi/k$ is wavelength. From the definition,

$$\frac{1}{Q}=\frac{1}{2\pi}\,\frac{\left(\int_0^T |u(x_0)|^2\mathrm{d}t-\int_0^T |u(x_0+\lambda)|^2\mathrm{d}t\right)}{\int_0^T |u-(x_0)|^2\mathrm{d}t}$$

$$= \frac{1 - \exp(-2k\delta\lambda)}{2\pi}$$

$$= \frac{1 - \exp(-4\pi\delta)}{2\pi}$$

When $2\pi\delta$ is sufficiently small,

$$\frac{1}{Q} = 2\delta \tag{2.26}$$

In the Earth's crust Q is of the order of hundreds or thousands; the larger the value of Q, the smaller is the energy loss or damping.

2.6 Seismometry and earthquake magnitude

2.6.1 Seismograph

The first known seismic instrument, the Houfeng seismometer, made in the year AD 132 in the Late Han dynasty by the ancient Chinese scientist Heng Zhang, successfully recorded an earthquake in AD 138. The modern type of seismograph started in the 18th century, and includes three subsystems: sensor, amplifier and recorder. For the special requirements of the seismologist, the seismograph records usually the displacement of the ground motion due to an earthquake.

There are usually three sensors, arranged in two perpendicular horizontal directions and one vertical direction. The most important property of a seismograph is the dynamic amplification spectrum. The pass bandwidth of currently used seismographs is usually not very wide and with maximum amplification not less than a few thousands, mostly at periods not less than 0.7 s; both amplitude (Table 2.1) and phase characteristics may be rather different for different frequency components.

Seismographs are in operation 24 hours a day, and records are produced continuously on a roll of paper. Accuracy of timing is about 0.1 s.

There are several typical seismographs currently in use. One is for local and weak earthquake observation, with natural period around 1 s and

Table 2.1 Amplitude–Frequency characteristics of Wood–Anderson torsional seismograph

Period T(s)	Amplification
0.1	2786
0.2	2745
0.4	2557
0.6	2199
0.7	1972
0.8	1750
0.9	1538
1.0	1350
1.4	807
2.0	424

amplification around 10^4–10^5. The Wood–Anderson torsional seismograph is also used for near earthquake observation, with a natural period 0.8 s and near critical damping (0.8) but with an amplification of only 2800. The long-period seismographs of natural period longer than 10 s are for distant and strong earthquake observation.

2.6.2 Seismic magnitude

In 1935, Richter first introduced the idea of magnitude, to give a quantitative measure of an earthquake through instrumental records. The Wood–Anderson torsional seismograph was in common use in southern California. Richter discovered that the curves of $\log A - \Delta$ for many earthquakes in that region were almost parallel to each other and he suggested the following definition of a local magnitude:

$$M_L = \log A(\Delta) - \log A_0(\Delta) \tag{2.27}$$

where A is the average of the maximum displacements (in μ) of two horizontal components at epicentral distance $\Delta = 100$ km and A_0 is an empirical correction for the effect of local conditions on attenuation of ground motion. As shown in Table 2.1, the amplification at period $T = 1.0$ drops to about half of that at $T = 0.1$ s; this definition is therefore made for magnitude on the basis of short period ground motion and is adequate for local small earthquakes.

Since seismic stations are not normally so densely allocated as in southern California, Gutenberg followed the same idea and defined a surface wave magnitude for shallow and distant earthquakes as follows:

$$M_s = \log A(\Delta) - \log A_0(\Delta) \tag{2.28}$$

At a distance of about 2000 km, the primary wave is a surface wave, of period about 20 s; A is defined as the maximum vector of the horizontal displacement in μ of a 20 s period wave.

Since M_s cannot be used for deep earthquakes, he defined the following body wave magnitude:

$$m_b = \log(A/T)_{max} - Q(\Delta, h) \tag{2.29}$$

where $(A/T)_{max}$ may be the maximum ratio (A/T) of P, PP or S waves, either the vertical or horizontal vector displacement in μ, and T in seconds is the period corresponding to A. But in practice, most seismographs have a natural period of 1 s and A of the vertical component is used to define the body wave.

Empirical conversion relations have been suggested between these magnitudes, such as

$$M_s = 1.59\,m_b - 4.0 = 1.27(M_L - 1.0) - 0.16\,M_L^2 \tag{2.30}$$

In China, when these definitions are used, the correction term $\log A_0$ or $Q(\Delta, h)$ includes also the instrument correction; the conversion relation used is

$$M_s = 1.13M_L - 1.08 \tag{2.31}$$

It is clear that the difference in these magnitudes comes from the difference in frequency band in which the records of some specific seismograph used or the motion are defined. The local magnitude M_L is actually defined in the period band 0.1–1.0 s; the surface magnitude M_s in the period band 3–20 s; and body wave magnitude m_b around 1.0 s. No matter what kind of magnitude is used, the accuracy is certainly no more than 0.1, which is the reason that magnitude is given with only one decimal digit. A difference of 0.5 or even 1.0 of one earthquake between different stations is quite possible. The magnitude of pre-instrumental earthquake is converted from intensity and may have a larger error, say a quarter or a half, which is the reason why this kind of magnitude is indicated not by decimal numbers but by fractions such as 7¾ or 8½, with an accuracy at most of a quarter.

All magnitudes defined above may saturate; the maximum magnitude ever recorded is 9. There are at least two possible reasons for the saturation of magnitude.

1. The strength of the Earth's crust is finite and the stress distribution in the crust is not uniform; the faulting of crust is thus only a small portion of the crust. The energy released in one earthquake is therefore finite.
2. Magnitude is defined by the energy within a narrow frequency band which may not be proportional to the total energy released.

Current data show clearly that the spectra of earthquakes of different sizes are different not only in strength but also in shape. For a source amplitude spectrum, as shown in Figure 2.6, when the magnitude increases,

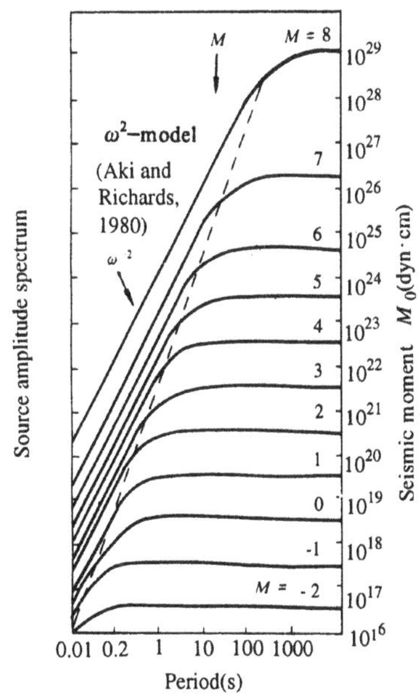

Figure 2.6 Source spectrum.

the energy increases faster in the long-period band than in the short-period band. The source spectrum is controlled by the dynamic properties of the source, such as the dimension and geometry of the source body, stress drop, fault type and the source time function. No narrow-band parameter can reflect changes in other frequency bands of the spectrum; the ordinary magnitude cannot, therefore, be a good representation of changes in all frequency bands. Duda and Nuttli (1974) pointed out this limitation of magnitude and suggested the use of spectral magnitude.

A moment magnitude M_w, based on energy, was suggested by Hanks and Kanamori (1979) as a magnitude that does not saturate. The seismic moment

$$M_0 = E_s \frac{2\mu}{\Delta\sigma} \tag{2.32}$$

is an adequate measure of seismic energy and the empirical relation of Gutenberg and Richter

$$\log E_s = 1.5M + 11.8 \tag{2.33}$$

may be used for a new magnitude. Combining these two equations and introducing suitable numerical values for stress drop $\Delta\sigma$ and Lamé constant μ, Hanks and Kanamori produced

$$\log M_0 = 1.5M + 16.1 \tag{2.34}$$

from which they defined a moment magnitude as

$$M_w = 2/3 \log M_0 - 10.7 \tag{2.35}$$

In the range $M_w = 3$–5, $M_w = M_L$; in the range $M_w = 5$–7½, $M_w = M_s$; when $M_w > 7½$, M_w is greater than M_s and M_L (Figure 5.14).

2.6.3 *Earthquake magnitude and faulting length*

For strong earthquakes of engineering interest, the focus is usually shallow, say 10–40 km in depth. If focal depth does not vary, the fault rupture

Table 2.2 Relation between magnitude and faulting length L (km), $M = a\log L + b$

Reference[a]	a	b	Remarks
Tocher (1958)	0.98	5.65	Data from 10 US earthquakes; $M \geqslant 6$
Iida (1965)	0.76	6.07	Data from 60 earthquakes; $M \geqslant 6$
Price	1.06	5.53	Semi-empirical
Pruce	1.6	4.5	$M \leqslant 6$
Academia Sinica (1965)	2.11	3.3	Data from China and mid- Asia; $M \geqslant 6.5$
Brune-Wyss (1968)	1.91	−6.7	$M = 3$–6
Schebalin (1971)	2.0	3.6	–
	2.2	2.0	L = length of aftershock area
Whitmann (1967)	1.7	3.3	–
Utsu	2.0	3.8	L = length of aftershock area

Source: Earthquake Loss Mitigation Dept (1992).
[a] All citations are given in the source reference.

dimension will be a function of magnitude. Table 2.2 lists some relations of magnitude and faulting length L. This kind of relation is very rough and strongly dependent on regions; and the scatter is very large.

Although of very low accuracy, this kind of relation is very useful in both earthquake engineering and prediction. For example in earthquake engineering, it is necessary in seismic hazard assessment if the fault rupture model (Der Kiureghian and Ang, 1977) is used; it is also necessary in strong motion attenuation, if the extension of fault rupture is considered. In earthquake prediction, the length of a seismic gap, the seismic-active region in which no earthquake has occurred for a long time, may be used to estimate the magnitude of the maximum possible future earthquake.

3 Related fundamentals of random vibration

Early in 1947, earthquake motion was considered for the first time as a random process by Housner (1947), but it was ten years later before many investigators in earthquake engineering started to apply theories of random vibration to the analysis of structural responses. Among them were Rosenblueth (Tajimi, 1960), Thomson (Barstein, 1960), Barstein (1960), Bolotin (1960), Tajimi (1960) and Hu and Zhou (C1962).* In the next decade, many more were interested in this topic, and so much progress was made that random vibration analysis became a powerful method in earthquake engineering.

The application of random vibration theory in earthquake engineering includes the following topics:

- earthquake sequence;
- description of ground motion time history and the evaluation of the parameters of ground motion;
- ambient vibration of ground and structures;
- simulation of ground motion;
- response analysis of structures.

Earthquake sequence analysis is not included in this book; the description of ground motion will be discussed in this chapter as an application example. Other related subjects will be discussed in Chapter 15.

3.1 Related concepts of random processes

3.1.1 Random variables

From the viewpoint of probability theory, physical quantities may be classified into two groups, the deterministic and the random. When observation of a quantity is repeated many times under the same controlled conditions, it is a deterministic quantity if the results of the repeated observations are always the same within the error of measurement, and is a random quantity if the results vary more than the error and with no known rule. Strictly speaking, physical quantities observed in science and engineering, such as strength of material and load, are mostly random. Strength of concrete samples in one mixture, soil samples at one site and ground motion at points within a small area are all random quantities; only when the randomness is small, can they be treated as deterministic for simplicity.

* 'C' before the date in reference citations indicates that the source language is Chinese. These are listed separately in the reference section.

The randomness of a phenomenon comes from the fact that there are numerous factors which affect the phenomenon in unknown ways and their effects are of nearly equal importance. For example, the distribution of energy released depends on the stress distribution and strength distribution of the rock material in the source, which are highly random; therefore the pattern of energy release for each earthquake is random. The particles of rock and soil along the path of wave propagation are arranged with some randomness, and therefore the ground motion in a small area during one earthquake is random. Unfortunately, their scatters are so large that they cannot be taken as deterministic.

Although the outcome of each sample of a repeated observation is not predictable, some averages of the samples are usually found to follow some rules and these averages reflect the effect of the influencing factors. In the case of earthquake ground motion, the specific condition of the area and geometry of the causative fault, the energy distribution, the distance and the soil condition of the site determine these averages. It may be said that there are two groups of factors which control a random phenomenon: one group is known which controls the averages of the quantity; another group is not entirely known which adds the random property to it; the deterministic school considers the average or some safe estimate, while the probabilistic school considers the average and the scatter.

In probability theory, the property of a random variable may be described by its probability density distribution function or its integration, the probability function. Let x be a random variable. The probability density function $p(x)$ is the probability, or the possibility, that x takes a value in the range $(x, x+\mathrm{d}x)$; and the probability function

$$F(x)=\int_{-\infty}^{x} p(x)\,\mathrm{d}x \qquad (-\infty<x<\infty) \tag{3.1}$$

is the probability that the variable takes a value smaller than x. The function must satisfy the conditions $F(-\infty)=0$ and $F(\infty)=1$ from the definition of probability. $P(x)=1-F(x)$ is the probability that the variable takes a value larger than x.

If the probability density function $p(x)$ of a variable is known, the expectation of a function $f(x)$ is defined as

$$E[f(x)]=\int_{-\infty}^{\infty} f(x)\,p(x)\,\mathrm{d}x \tag{3.2}$$

and the mean and variance of x are respectively

$$m=\bar{x}=E[x]=\int_{-\infty}^{\infty} xp(x)\,\mathrm{d}x \tag{3.3}$$

$$\sigma^2=E(x-m)^2=\mathrm{Var}[x-m]=\int_{-\infty}^{\infty}(x-m)^2p(x)\,\mathrm{d}x \tag{3.4}$$

The mean m represents the center of gravity of the area under the curve $p(x)$ and the variance σ^2 the width of the peak of $p(x)$ or the scatter of the variables. The coefficient of variance is $c_v=\sigma/m$. Given a set of observation data for a variable, it is not easy to determine the probability function, but it is easy to determine the mean and variance. Using a small number of parameters such as mean and standard deviation to describe a probability function is the most popular way in the engineering field because it is both simple and practical. If only the first parameter m is used, a random variable is reduced to a deterministic one.

(a) Normal distribution

The most commonly used probability distribution function is the normal or Gaussian distribution, whose probability density function is

$$p(x)=\frac{1}{\sigma(2\pi)^{1/2}}\exp\left[-\frac{1}{2}\left(\frac{x-m}{\sigma}\right)^2\right] \qquad (-\infty<x<\infty) \tag{3.5}$$

which is determined entirely by two parameters m and σ.

(b) Log-normal distribution

Because in many engineering problems, a physical quantity cannot be negative, a log-normal distribution is then used instead of normal:

$$p(y)=\frac{1}{y\sigma(2\pi)^{1/2}}\exp\left[-\frac{1}{2}\left(\frac{\ln y-m}{\sigma}\right)^2\right] \qquad (0\leqslant y<\infty) \tag{3.6}$$

where m and σ are respectively the mean and standard deviation of $\ln y$. Because $x=\ln y$ is normal $p(y)$ is called log-normal.

The parameters of the distribution function are obtained by statistical analysis as follows. Let $x_i (i=1,\ldots, N)$ be the observed data for x. The parameters are usually calculated from central moments:

$$\mu_j=E[(x_i-m)^j]=\frac{1}{N}\sum n_j\cdot(x_j-m)^j \qquad (j=1, 2, \ldots) \tag{3.7}$$

For the ordinary case, only the first two central moments are used to determine the mean and variance:

$$m=\sum\frac{x_i}{N} \tag{3.8}$$

$$\sigma^2=\sum\frac{(x_i-m)^2}{N-1} \tag{3.9}$$

where in the variance, $(N-1)$ is used for N to obtain an unbiased estimation of variance.

3.1.2 Random process

If a random variable x is a function of a deterministic variable t, the random function $x(t)$ is then a random process when t is time. The ambient

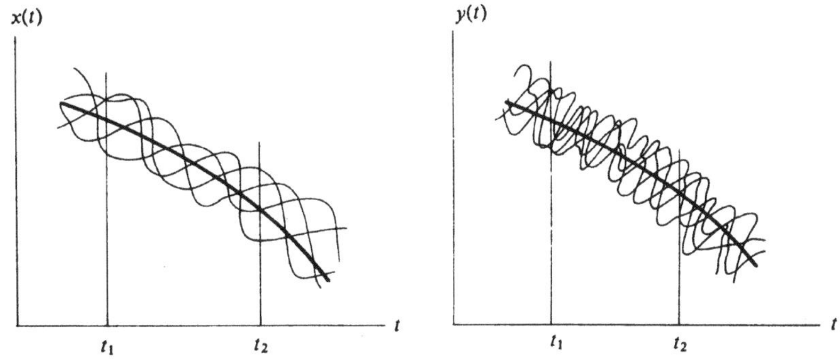

Figure 3.1 Repeated observation of processes $x(t)$ and $y(t)$.

vibration of ground or structure is a typical random process. For a random process $x(t)$, when it is observed repeatedly under controlled conditions, the results $x_i(t)$ $(i=1, 2, \ldots)$ of the observation can never be identical, no matter how one shifts the records along the time axis.

There are two ways to describe a random process, a probabilistic way and a statistical way. The statistical one is more intuitive and will be given here.

The most important properties of a random variable are the mean and variance. It is quite similar for a random process, although more complicated. In Figure 3.1, repeated observations of two random processes $x(t)$ and $y(t)$ are plotted. At any given time t_1, the mean and variance of $x(t_1)$ will be same as equations 3.8 and 3.9:

$$\bar{x}(t_1)=\frac{1}{N}\sum x_i(t_1) \tag{3.10a}$$

$$\sigma_x^2(t_1)=\frac{1}{N-1}\sum[x_i(t_1)-\bar{x}(t_1)]^2 \tag{3.10b}$$

Let t vary from $t=0$ to $t=T$ by some finite step; then functions of mean $m_x(t)$ and variance $\sigma^2(t)$ at all steps will be obtained. If only linear problems are considered, the mean curve $x_x(t)$ may be separated from the random process and considered by a deterministic method; the remaining random portion $x'(t)=x(t)-\bar{x}(t)$ will be of zero mean, and is the one considered later in this book, but with the superscript omitted for simplicity. For a random process with zero mean, the random property of the process will be described by a single parameter, the variance σ^2 or the standard deviation σ.

The discussion above is almost the same as for a random variable. In a random process, however, there is another important quantity, i.e. the autocorrelation function $R(t_1, t_2)$:

$$R_x(t_1, t_2)=E[x(t_1)\cdot x(t_2)]=\frac{\sum[x_i(t_1)\cdot x_i(t_2)]}{N-1} \tag{3.11}$$

which reduces to variance when $t_1=t_2$. A correlation function can also be defined between two random processes $x(t)$ and $y(t)$ as

$$R_{xy}(t_1, t_2)=E[x(t_1)\cdot y(t_2)]=\frac{1}{N-1}\sum_i\sum_j[x_i(t_1)\cdot y_j(t_2)] \tag{3.12}$$

which is called the cross-correlation function. Both these correlation functions are defined in terms of assembly average, i.e. by taking an average over the repetitions or assembly.

The correlation function has very important physical meaning. It is seen from equation 3.12 that $R_{xy}(t_1, t_2)$ represents the correlation between $x(t_1)$ and $y(t_2)$ in a probabilistic or statistical sense. It is only an average or an expectation, not a deterministic relation between a particular pair $x(t_1)$ and $y(t_2)$. When $R_{xy}(t_1, t_2)$ is negative, it means that, on average, $x(t_1)$ and $y(t_2)$ are of opposite signs; when it is zero, it means that $x(t_1)$ and $y(t_2)$ are independent. Other important physical meaning can easily be seen in the frequency domain (next section).

3.1.3 Stationarity of random processes and stationarized processes

(a) Stationary processes

If all statistical properties of a random process are independent of time t, this process is stationary; if only the mean $E[x(t)]$ and the correlation function $E[x(t_1)\cdot x(t_2)]$ are, it is then a weak stationary process. For them, we have

$$m_x=E[x(t)]=\text{const.}$$

$$E[x(t_1)\cdot x(t_2)]=R_x(t_2-t_1)=R_x(\tau)$$

where $\tau=t_2-t_1$. The nondimensional form is

$$\rho_x(\tau)=\frac{R_x(\tau)}{R_x(0)}$$

(b) Ergodic stationary random process

An ergodic stationary process is the stationary process commonly used, for which one observation $x_1(t)$ includes all properties of the process and averages in the time domain may be used for averages in assembly. The proof is as follows.

$$E[x(t)]=E\left[\frac{1}{T}\int_0^T x(t)\,\mathrm{d}t\right]=\frac{1}{T}\int_0^T E[x(t)]\,\mathrm{d}t$$

$$=E[x(t)]\,\frac{1}{T}\int_0^T \mathrm{d}t=m_x$$

$$E[x(t)\cdot x(t+\tau)]=E\left[\frac{1}{T}\int_0^T x(t)\cdot x(t+\tau)\,\mathrm{d}t\right]=\frac{1}{T}\int_0^T E[x(t)\cdot x(t+\tau)]\,\mathrm{d}t$$

$$=E[x(t)\cdot x(t+\tau)]\,\frac{1}{T}\int_0^T \mathrm{d}t=R_x(\tau)$$

which shows that the expectations of the average of $x(t)$ and the average of the product $x(t) \cdot x(t+\tau)$ in the time domain over the whole range T or $T-\tau$ are respectively the mean and correlation in assembly, i.e.

$$m_x = \frac{1}{T} \int_0^T x(t)\,\mathrm{d}t \tag{3.13}$$

$$R_x(\tau) = \frac{1}{T-\tau} \int_0^{T-\tau} x(t) \cdot x(T+\tau)\,\mathrm{d}t \tag{3.14}$$

In earthquake engineering, because it is not possible to obtain many ground motion records under similar conditions to estimate averages in the assembly, one has to use the ergodic assumption to estimate them in time averages, although it is obviously not ergodic.

(c) Stationary random process

In addition to the ergodic assumption, stationarity is also often used in earthquake engineering, and is later generalized into a stationarized assumption for ground motion. Assuming that only the amplitude of the ground motion is nonstationary and its variation with time can be described by a deterministic function $f(t)$, the ground motion can then be written in the following form:

$$x(t) = f(t) \cdot y(t) \tag{3.15}$$

where $y(t)$ is an ergodic stationary random process, the autocorrelation function of $x(t)$ will be

$$R_x(t_1, t_2) = f(t_1) \cdot f(t_2) \cdot R_y(\tau) \tag{3.16}$$

(d) Spectral density function

The power spectral density function $S(\omega)$, or just spectral density function, is equivalent to the correlation function. They form a Fourier transformation pair:

$$S(\omega) = \int_{-\infty}^{\infty} R(\tau) \exp(-\mathrm{i}\omega\tau)\,\mathrm{d}\tau \tag{3.17a}$$

$$R(\tau) = \frac{1}{2\pi} \int_{-\infty}^{\infty} S(\omega) \exp(\mathrm{i}\omega\tau)\,\mathrm{d}\omega \tag{3.17b}$$

This is the well-known Wiener–Khintchine relation. Autospectrum and autocorrelation functions are both even, i.e.

$$R(\tau) = R(-\tau)$$

$$S(\omega) = S(-\omega)$$

The cross-spectrum and cross-correlation function are also a Fourier transformation pair but they are not even; the cross-spectrum is a complex function.

It is easy now to show another important property of a random process and the difference between the two random processes $x(t)$ and $y(t)$ shown in Figure 3.1. Figure 3.2 shows the autocorrelations and autospectra of $x(t)$ and $y(t)$, if they are of zero mean and are stationary. It can be seen from the figure that the slower changing phenomenon of $x(t)$ in the time domain is shown here as a spectrum of lower predominant frequency in the frequency domain.

The power density spectrum is closely related to the Fourier spectrum. The Fourier spectrum and the time history of a random process $x(t)$ are a Fourier pair by definition:

$$X(\mathrm{i}\omega)=\frac{1}{2\pi}\int_{-\infty}^{\infty} x(t)\exp(-\mathrm{i}\omega t)\,\mathrm{d}t \tag{3.18a}$$

$$x(t)=\frac{1}{2\pi}\int_{-\infty}^{\infty} X(\mathrm{i}\omega)\exp(\mathrm{i}\omega t)\,\mathrm{d}\omega \tag{3.18b}$$

Let T be the duration of this process. It can be derived that

$$R(0)=E[x^2(t)]=\lim_{T\to\infty}\frac{1}{2\pi T}\int_{-\infty}^{\infty}|X(\mathrm{i}\omega)|^2\mathrm{d}\omega \tag{a}$$

From equation (3.17b) for $\tau=0$,

$$R(0)=\frac{1}{2\pi}\int_{-\infty}^{\infty}S(\omega)\mathrm{d}\omega \tag{b}$$

Comparing equations (a) and (b), a relation between the power spectrum and the Fourier amplitude spectrum is obtained as follows:

$$S(\omega)=\lim_{T\to\infty}\frac{1}{T}|X(\mathrm{i}\omega)|^2 \tag{3.19}$$

which means that power density function is the expectation of the square of the Fourier amplitude spectrum if the right-hand term is considered as the average per unit time of a sufficiently long time T. It is seen also that these two spectra have two important differences. First, the Fourier spectrum $X(\mathrm{i}\omega)$ is random because $x(t)$ is, but the power density function is deterministic. Second, the Fourier spectrum includes phase information in its imaginary part but the power density is real and loses the phase information.

(e) Power density function of the derivative process

From the correlation function $R_x(\tau)$ and power density function $S_x(\omega)$, it can be derived that the autocorrelation and power density of the nth derivative process $x^{(n)}(t)$ are respectively

(b)

Figure 3.2 (a) Correlation function; (b) spectra.

$$R_n(\tau)=(-1)^n\left(\frac{d^{2n}}{dt^{2n}}\right)R_x(\tau)$$

$$S_n(\omega)=\omega^{2n}S_x(\omega) \tag{3.20}$$

(f) Cross-correlation and cross-spectrum

The case of n random processes $x_i(t)$ ($i=1, \ldots, n$) (note that these are not n repeated observations of one process), is similar to that of two processes $x(t)$ and $y(t)$ discussed above, but the cross-correlation and cross-spectrum are matrices:

$$\boldsymbol{R}_n(\tau)=\begin{bmatrix} R_{11}(\tau) & R_{12}(\tau) & \ldots \\ R_{21}(\tau) & R_{22}(\tau) & \ldots \\ \vdots & \vdots & \ldots \end{bmatrix}$$

$$\boldsymbol{S}_0(\omega)=\begin{bmatrix} S_{11}(\omega) & S_{12}(\omega) & \ldots \\ S_{21}(\omega) & S_{22}(\omega) & \ldots \\ \vdots & \vdots & \ldots \end{bmatrix}$$

The cross-correlation function $R_{ij}(\tau)$ and cross-spectrum $S_{ij}(\omega)$ are also a Fourier pair; the terms on the diagonals of the matrices are the autocorrelation and autospectral functions.

Based on the definition of cross-correlation function, it can be proved that the cross-correlation function is odd and

$$R_{ij}(\tau)=R_{ij}(-\tau)$$

but the autocorrelation function $R_{ii}(\tau)$ is even. On the basis of the relation of correlation function and spectrum it is obtained that

$$S_{ij}(\omega)=\int_{-\infty}^{\infty}R_{ij}(\tau)\exp(-i\omega\tau)\,d\tau=\int_{-\infty}^{\infty}R_{ij}(\tau)\cos\omega\tau d\tau-i\int_{-\infty}^{\infty}R_{ij}(\tau)\sin\omega\tau d\tau$$

Because $R_{ij}(\tau)$ is not even, the second term on the right-hand side is not zero and the cross-spectrum is thus complex.

(g) Coherence function

The coherence function $\gamma_{ij}(\omega)$ is defined as follows:

$$\gamma_{ij}^2(\omega)=\frac{|S_{ij}(i\omega)|^2}{S_i(\omega)\,S_j(\omega)} \qquad (0\leqslant\gamma_{ij}^2(\omega)\leqslant 1) \tag{3.21}$$

Its physical meaning may be given as follows. If $x_1(t)$ and $x_2(t)$ represent respectively the input and output of a linear system, and $z(t)$ is noise or the effect of other input on $x_2(t)$, it can be proved that

$$S_z(\omega)=S_2(\omega)\,[1-\gamma_{12}^2(\omega)] \tag{3.22}$$

It is seen here that $\gamma_{12}^2(\omega)=1$ means the noise $S_z(\omega)$ at this frequency is zero and the output $S_2(\omega)$ comes entirely from the input $S_1(\omega)$; on the other hand, $\gamma_{12}^2(\omega)=0$, means that the output at this frequency is entirely from noise.

3.2 Spectral parameters and their physical meaning

3.2.1 Spectral parameters (Vanmarcke, 1972, 1976)

Let $x(t)$ be a stationary random process of zero mean, variance σ^2, correlation function $R(\tau)$ and spectrum density $S(\omega)$. The one-sided spectrum is $G(\omega)=2S(\omega)$ for $0\leqslant\omega<\infty$. The ith moment of the one-sided spectrum is defined as

$$\lambda_i=\frac{1}{2\pi}\int_0^\infty \omega^i G(\omega)\,d\omega \qquad (i=1, 2, \ldots) \tag{3.23}$$

The most often used parameters are the first three, λ_1, λ_2 and λ_3, from which the parameters σ, ω_2 and q can be obtained.
From equation 3.17b, it is obtained that

$$\sigma^2=R(0)=\frac{1}{2\pi}\int_0^\infty G(\omega)\,d\omega=\lambda_0 \tag{3.24}$$

The following parameters are then defined:

$$\omega_i=\left(\frac{\lambda_i}{\lambda_0}\right)^{1/i} \qquad (i=1, 2) \tag{3.25}$$

$$q=\left(1-\frac{\lambda_1^2}{\lambda_0\lambda_2}\right)^{1/2} \tag{3.26}$$

The physical meaning of these parameters will be explained first and followed by their application.

3.2.2 Meaning of spectral parameters in the frequency domain

From equation 3.24, it is known that $\lambda_0=\sigma^2$ is $1/(2\pi)$ times the area under the one-sided spectral density function $G(\omega)$ as shown in Figure 3.3, and

$$\omega_1=\frac{\lambda_1}{\lambda_0}=\frac{\int_0^\infty \omega G(\omega)\,d\omega}{\int_0^\infty G(\omega)\,d\omega}$$

is the frequency at the center of gravity of the area of $G(\omega)$ and shows in general where the spectrum is concentrated. Similarly, it may be shown that ω_2 is the radius of inertia and indicates the width of the spectrum, and

$$q=\left(\frac{\omega_2^2-\omega_1^2}{\omega_2^2}\right)^{1/2} \qquad (0\leqslant q\leqslant 1) \tag{3.27}$$

represents also the shape of the spectrum, spreading away from the center of gravity. The smaller the q value, the narrower is the spectrum. When

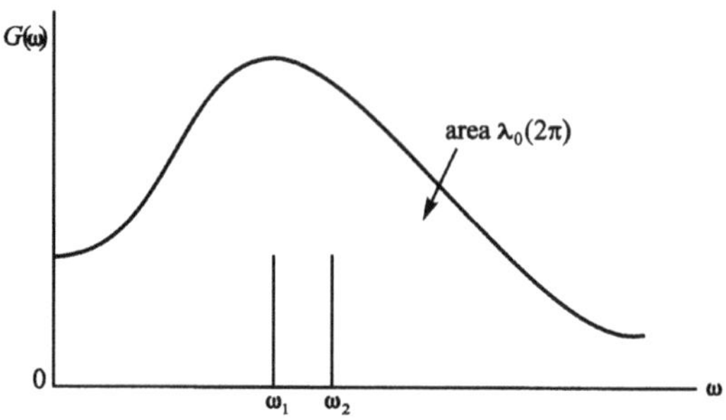

Figure 3.3 Spectral parameters.

$q=0$, the spectrum $G(\omega)=\delta(\omega-\omega_1)$ is a line at the center of gravity and the process is deterministic; when $q=1$, $G(\omega)$ is constant and the spectrum is white. A white spectrum means statistically equal amplitudes for all frequencies, like white light.

3.2.3 Meaning of spectral parameters in the time domain

Let velocity $v(t)=\dot{x}(t)$ be a derivative process of time history $x(t)$. From equation 3.20, the density function of $v(t)$ is

$$G_v(\omega)=\omega^2 G(\omega) \tag{3.28}$$

It follows that

$$\lambda_2=\frac{1}{2\pi}\int_0^\infty \omega^2 G_X(\omega)\,d\omega=\frac{1}{2\pi}\int_0^\infty G_v(\omega)\,d\omega=\sigma_v^2 \tag{3.29}$$

$$\omega_2=\left(\frac{\lambda_2}{\lambda_0}\right)^{1/2}=\frac{\sigma_v}{\sigma_x} \tag{3.30}$$

Figure 3.4 shows a process $x(t)$ and its envelopes $\pm r(t)$. If the process $x(t)$ has a narrow frequency band, its envelopes vary slowly. On the basis of the definition of the envelope, $|r(t)|\geqslant|x(t)|$ for any value of t. It may be proved that

$$q=\frac{\sigma_{\dot{r}}}{\sigma_v} \tag{3.31}$$

where σ_v is the standard deviation of the slope of $x(t)$ at t, and $\sigma_{\dot{r}}$ the standard deviation of the slope of the envelope $r(t)$.

For the Gaussian process $x(t)$, the positive or negative rate of crossing a is

$$\begin{aligned}\nu_a^-=\nu_a^+&=\frac{1}{2\pi}\left(\frac{\lambda_2}{\lambda_0}\right)^{1/2}\exp\left(-\frac{a^2}{2\lambda_0}\right)\\&=\nu_0^+\exp\left(-\frac{a^2}{2\lambda_0}\right)\end{aligned} \tag{3.32}$$

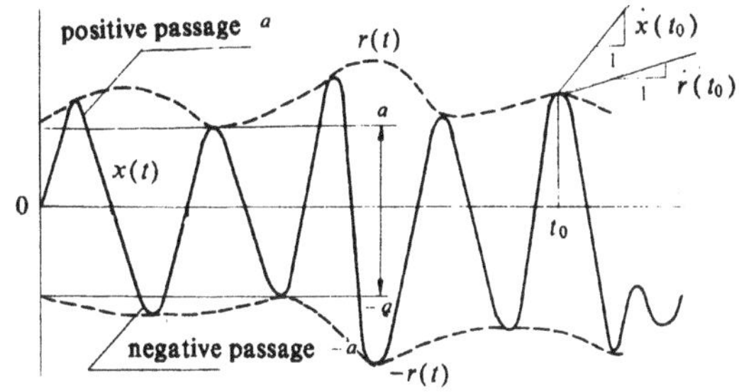

Figure 3.4 Process $x(t)$ and its envelopes $r(t)$.

where ν_a^+ is the number of times per second of the positive a crossing when $x(t)=a$ is positive. When $a=0$, the average zero crossing rate ν_0 is

$$\nu_0=2\nu_0^+=2\nu_0^-=\frac{\omega_2}{\pi} \tag{3.33}$$

which means that ω_2 is the apparent frequency. The average positive a crossing rate of the envelope process $r(t)$ n_a^+ is

$$n_a^+=\frac{\sigma_{\dot{r}}}{\sqrt{2\pi}}\cdot\frac{a}{\lambda_0}\cdot\exp\left[-\frac{a^2}{2\lambda_0}\right] \tag{3.34}$$

For a Gaussian process $x(t)$, the envelope process $r(t)$ has a Rayleigh distribution, its derivative process $r(t)$ is normal, and they are independent. The related parameters are

$$\sigma_r^2=\frac{2\lambda_0}{\pi};\qquad m_{\dot{r}}=0;\qquad \sigma_{\dot{r}}^2=\frac{q^2\omega_2^2\lambda_0}{\pi} \tag{3.35}$$

3.3 Distribution of maxima

The accumulative probability distribution $F_{x_T}(a)$ of the maximum $x_T=\max|x(t)|\leq|a|$ between $t=0$ and $t=T$ is of important practical meaning, but there is still no method to make an exact analytical estimation. In a special case of response $x(t)$ of a linear system with small damping under normal white noise excitation, the following assumptions are usually made: for very large values of $a=r\sigma_x$ (here r is a constant), events $|x(t)|>a$ are independent and have a Poisson distribution, with average rate $2\nu_a$. In this case, from equation 12.2b and 3.32, it is obtained that

$$F_{x_T}(a)=\exp(-2\nu_a T)=\exp\left[-2\nu_0 T\exp\left(-\frac{r^2}{2}\right)\right] \tag{3.36}$$

When $r\rightarrow\infty$, this result is exact; but for a not very large value of a, this equation may have considerable error, and the result will be very conservative for narrow frequency band process.

A better approximation (Vanmarcke, 1976) is

$$F_{x_T}(a) = A\exp\left[-2\nu_0^+ T\exp\left(-\frac{r^2}{2}\right)\right]g(r)$$

$$g(r) = \frac{1-\exp[-(2\pi)^{1/2}qr]}{1-\exp(-r^2/2)} \tag{3.37}$$

where A is the initial value of $F_{x_T}(a)$ at $T=0$, which is related to the initial values of the system and may often be taken as 1.

Let $x_{T,p}$ be the value of x with probability of nonexceedance of $x_T = \max|x(t)|$ in the duration T. Vanmarcke obtained the following semi-empirical approximation:

$$x_{T,p} = \sigma_x\left\{2\ln\left[\frac{2\nu_0^+ T\{1-\exp[-(\pi/2)^{1/2}qr_e]\}}{\ln(1/p)}\right]\right\}^{1/2} \tag{3.38}$$

where $r_e = \{2\ln[2\nu_0^+ T/\ln(1/p)]\}^{1/2}$. When q is very large,

$$x_{T,p} = \sigma_x\left\{2\ln\left[\frac{2\nu_0^+ T}{\ln(1/p)}\right]\right\}^{1/2}$$

$$p = \exp\left[-2\nu_0^+ T\exp\left(-\frac{x_{T,p}^2}{2\sigma_x^2}\right)\right]$$

which is just equation 3.36 because $r = x_{T,p}/\sigma_x$.

It is clear now from the relation

$$x_{T,p} = r(p,\ T)\,\sigma_x \tag{3.39}$$

that the standard deviation σ_x is a statistical constant but the maximum $x_{T,p}$ is random, because the peak factor $r(p, T)$ is a function of probability p of $x \geqslant x_{T,p}$ and the duration T of the process.

3.4 Random process description of ground motion

The stationary random process was the first one applied in the random process theory of the analysis of ground motion. Several popular models of ground motion are listed in Figure 3.5, in which white noise (4) was the earliest one adopted, for example by Housner (1947), and is still often used because of its simplicity and because earthquake ground motions are usually of wide spectrum. It is a good first order approximation. Band-limited white noise (3) is a modification of the white noise model, which avoids an infinity of the energy

$$\int_{-\infty}^{\infty} S(\omega)\mathrm{d}\omega$$

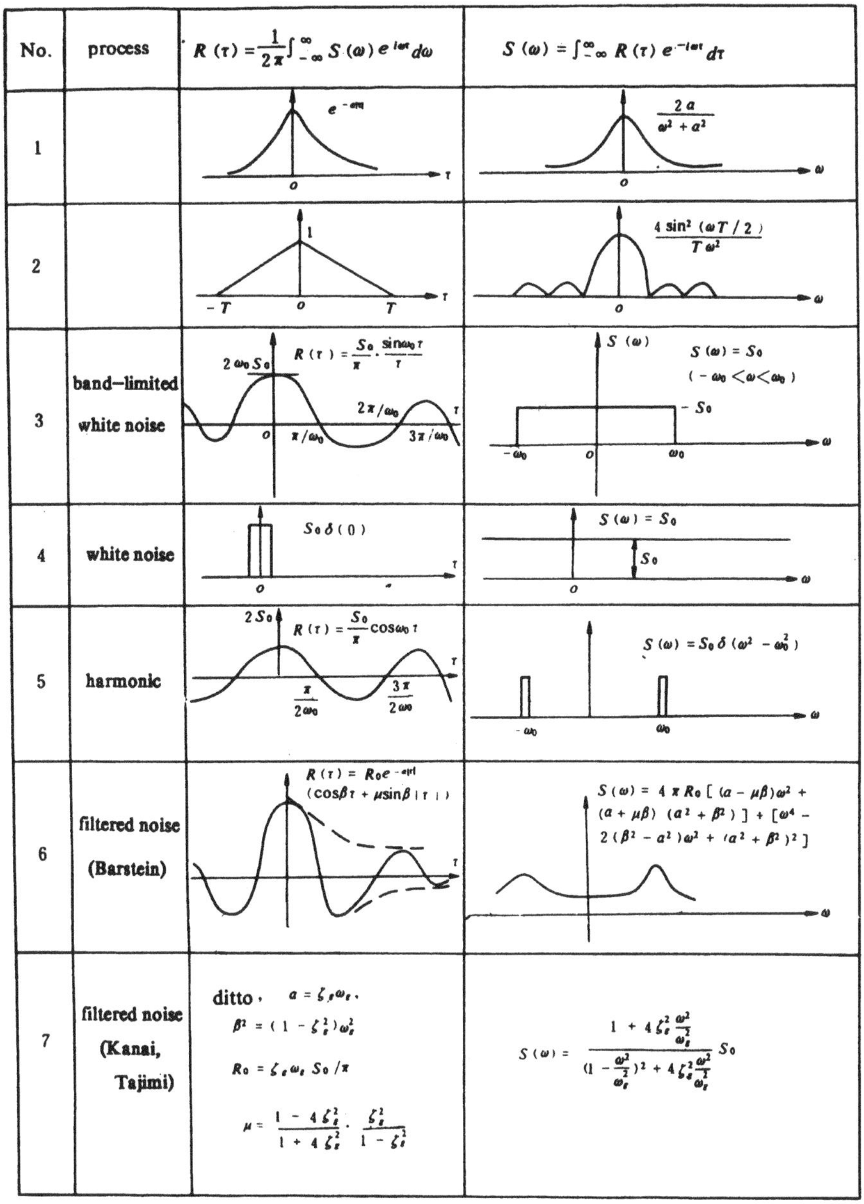

Figure 3.5 Some typical random representations of ground motion.

The models used by Barstein (1960), Kanai and Tajimi (Tajimi, 1960) (6 and 7 in Figure 3.5) may be considered as filtered white noise of a single peak spectrum with the peak near the natural frequency ω_g of the filter.

The duration of strong earthquake motion usually lasts only from a few to a few tens of seconds, and at most one or two minutes. Several typical earthquake acceleration time histories are shown in sections 5.1 and 5.2. The common feature of nonstationarity of motion shows three stages:

1. The motion increases rapidly from weak to strong.
2. The motion maintains its average strength.
3. The motion gradually decreases.

The model considering this feature may be generally written as the following stationarized process:

$$x(t)=\sum f_i(t)\,y_i(t) \qquad (i=1,2,3) \tag{3.40}$$

Three terms in this formula are usually taken with $f_1(t)$, $f_2(t)$ and $f_3(t)$ for the nonstationarity of the three stages (Figure 3.6); random motions $y_i(t)$ ($i=1$, 2 or 3) are stationary within each stage. For example, Amin and Ang (1968) suggested

$$f_1(t)=\frac{t^2}{t_1^2} \qquad 0\leqslant t\leqslant t_1 \tag{3.41a}$$

$$f_2(t)=1 \qquad t_1\leqslant t\leqslant t_2 \tag{3.41b}$$

$$f_3(t)=\exp[-c(t-t_2)] \qquad t_2\leqslant t\leqslant t_3 \tag{3.41c}$$

$$f_4(t)=0 \qquad t_3\leqslant t \tag{3.41d}$$

and

$$y_1(t)\colon \frac{R_1(\tau)}{R_1(0)}=\exp(-\alpha\tau) \tag{3.42a}$$

$$y_2(t)\colon \frac{R_1(\tau)}{R_1(0)}=\exp(-\gamma_2\omega_2\tau)\left[\cos\omega_{2D}\tau+\frac{\gamma_2}{(1-\gamma_2^2)^{\frac{1}{2}}}\sin\omega_{2D}\tau\right] \tag{3.42b}$$

$$y_3(t)\colon \frac{R_3(\tau)}{R_3(0)}=\exp(-\gamma_3\omega_3\tau)\left[\cos\omega_{3D}\tau+\frac{\gamma_3}{(1-\gamma_3^2)^{\frac{1}{2}}}\,\frac{1-4\gamma_3^2}{1+4\gamma_3^2}\sin\omega_{3D}\tau\right] \tag{3.42c}$$

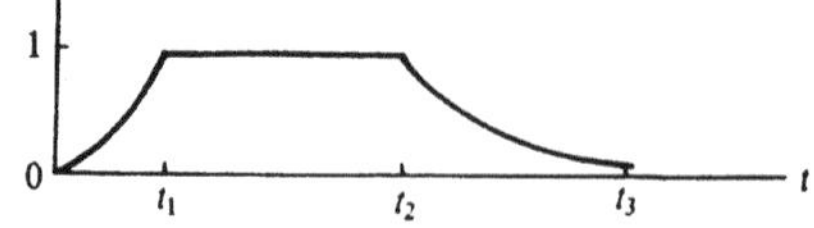

Figure 3.6 Stationarizing or strength envelope function.

where $\alpha = 10\,\text{Hz}$, $0.5 < \gamma_2 < 0.6$, $4 \leqslant \omega_2/(2\pi) \leqslant 5\,\text{Hz}$, $0.5 < \gamma_3 < 0.8$, $2 \leqslant \omega_3/(2\pi) \leqslant 3\,\text{Hz}$; $\omega_{iD} = (1-\gamma_i^2)^{\frac{1}{2}}\omega_i$, $(i = 2$ or $3)$. But, for simplicity, most investigators prefer to take only one term in equation 3.40, i.e.

$$x(t) = f(t)y(t)$$

and the commonly used stationarizing or strength envelope function is one of the following:

$$f(t) = \alpha\left(\frac{t}{t_p}\right)\exp\left(1 - \frac{t}{t_p}\right) \quad \text{(Goto and Toki, 1969)}$$

$$= \exp(-\alpha t) - \exp(-\beta t) \quad \text{(Shinozuka and Sato, 1967; Hu and Zhou, C1962; Shimizu, J1972*)}$$

$$= (a + bt)\exp(-\alpha t) \quad \text{(Iyengar and Iyengar, 1969)} \qquad (3.43)$$

In addition to the models mentioned above, shot noise has also been used for the random ground motion model, which may be written as

$$x(t) = \sum x_k \delta(t - t_k) \qquad (3.44)$$

where x_k $(k = 1, 2, \ldots, m)$ are independent random impulses with zero mean and variance $\text{Var}[x_k] = f(\tau_k)\,dt$, $f(\tau_k)$ is the strength envelope function and m is the total number of impulses in the period $(0, t)$. The interval between impulses is dt. For a sufficiently small interval dt,

$$R_x(t_1, t_2) = f(t_1)\,\delta(t_2 - t_1) \qquad (3.45)$$

If $f(t) \equiv 1$, $x(t)$ will be white noise.

* 'J' before the date in a reference citation indicates that the source language is Japanese. These are listed separately in the reference section.

Part Two

Engineering Seismology

4 Earthquake damage and intensity

4.1 Earthquake intensity and intensity scale

4.1.1 Earthquake intensity and its usage

Earthquake intensity is an index of the local destructive effect of an earthquake in a small area. Although this term is very old and popular, its interpretation may vary widely.

Earthquake intensity is measured through four categories of macroseismic phenomena:

- human feeling;
- damage to artificial structures;
- response of objects;
- change in natural conditions.

The word 'intensity' may be defined qualitatively as 'the quality or condition of being intense' or quantitatively as 'magnitude, as of energy or a force per unit of area or time'. The term earthquake intensity was introduced to be a physical quantity, but through qualitative or fuzzy definitions. In the earthquake engineering field, some consider it a qualitative rating through the intensity scale, but many engineers consider it an equivalent of ground peak acceleration. For example, in Japan, the term 'intensity' (or *shinto* in Japanese) is used for the ratio of peak ground acceleration to that of gravity, or a/g, and at the same time in the term 'intensity (*shinto*) scale'. In this book it will be used only for its qualitative definition, as accepted by most seismologists and engineers, and Roman numerals are used; if it has to be used as a quantitative term, quotation marks will be added.

There are three properties of earthquake intensity defined by macroseismic phenomena.

1. A composite measure of multiple scales: intensity is defined through the levels of human response, structural damage, changes in natural phenomena and behavior of objects during or after an earthquake. Each of these four responses is a scale, and intensity ratings are specified in each scale. When the measurements of each individual scale differ from each other as usually happens, a subjective decision is made, such as in China, by individuals who rate the intensity, but the maximum rating is taken in Europe.
2. A ranking and fuzzy measure through macroseismic description: on each individual scale, the scaling is not quantitative but qualitative in terms of fuzzy ranks, such as a 'few', 'some', 'many', 'most' for the

number of items considered; 'slight', 'moderate', 'heavy', 'serious' for strength of response. These are typical fuzzy set terms, i.e. they are defined through qualitative description, and the boundaries between ranks are not clear but fuzzy.

3. Indirectness of defining a cause through the results if intensity means ground motion: many building design codes still define intensity as ground motion. It is quite clear that ground motion is the cause of its effects on objects. The relation between cause and effects must be rough or fuzzy if the objects, as well as the cause and effects, are not clearly defined as in the case of an intensity scale.

The term earthquake intensity has been used for more than 170 years, before the term earthquake magnitude, and is still widely used. Its usage includes the following three aspects.

1. As a simple measure of earthquake damage: after a strong earthquake, government and society want to know the general level of damage and its distribution in order to take emergency action. It is relatively easy to estimate after an earthquake and to be understood by ordinary people, just like wind intensity.
2. As a simple macroseismic scale for the strength of earthquake motion for seismologists: historical earthquake data are damage data, which have been used to study past seismicity. Without damage data, the history of earthquakes would cover only a few decades, too short to know the spatial and temporal distribution of earthquakes or their seismicity.
3. As a rough but convenient index to sum up experience on earthquake engineering construction and to depict zones of seismic hazard (Hu, C1983): many countries use intensity as index for zonation, such as Russia, China and several European countries, and many countries have previously used it, such as the United States, Canada and India.

4.1.2 Earthquake intensity scale

The macroseismic idea of earthquake intensity was a natural product 170 years ago when no instruments were available to measure any physical quantity of ground motion and the requirement of earthquake description was also simple, just a scale or a single number or grade to describe its strength or magnitude (Xie, C1977). The earliest record of earthquake intensity may be traced back to 1564 in Europe. When discussing the effect of an earthquake, J. Gataldi used different colors to show different effect on a map. Up to the 19th century, there appeared many earthquake intensity scales, mostly used to grade the local damage or effect. During the early period, macroseismic phenomena were the only means possible to define the indices of the intensity scale. As an example, the commonly used abridged Modified Mercalli (MM) intensity scale, from which a general and brief idea of intensity may be obtained, is described below; this is followed by the 1957 China intensity scale.

(a) Abridged MM scale

I. Not felt. Marginal and long-period of large earthquakes.

II. Felt by persons at rest, on upper floors, or favorably placed.

III. Felt indoors. Hanging objects swing. Vibration like passing of light trucks. Duration estimated. May not be recognized as an earthquake.

IV. Hanging objects swing. Vibration like passing heavy trucks; or sensation of a jolt like a heavy ball striking the walls. Standing motor cars rock. Windows and doors rattle; glasses clink. Crockery clashes. In the upper range of IV, wooden walls and frames crack.

V. Felt outdoors; direction estimated. Sleepers wakened. Liquids disturbed, some spilled. Small unstable objects displaced or upset. Doors swing, close, open. Shutters, pictures move. Pendulum clocks stop, start, change rate.

VI. Felt by all. Many frightened and run outdoors. Persons walk unsteadily. Windows, dishes, glassware broken. Knick-knacks, books and so on, off shelves. Pictures off walls. Furniture moved or overturned. Weak plaster and masonry D cracked.* Small bells ring (church, school). Trees, bushes shaken visibly, or heard to rustle.

VII. Difficult to stand. Noticed by drivers of motor cars. Hanging objects quiver. Furniture broken. Damage to masonry D including cracks. Weak chimneys broken at roof line. Fall of plaster, loose bricks, stones, tiles, cornices, unbraced parapets and architectural ornaments. Some cracks in masonry C. Waves on ponds; water turbid with mud. Small slides and caving in along sand or gravel banks. Large bells ring. Concrete irrigation ditches damaged.

VIII. Steering of motor cars affected. Damage to masonry C; partial collapse. Some damage to masonry B; none to masonry A. Fall of stucco and some masonry walls. Twisting and falling of chimneys, factory stacks, monuments, towers, elevated tanks. Shifting of frame houses on foundations if not bolted down; loose panel walls thrown out. Decayed piling broken off. Branches broken from trees. Changes in flow or temperature of springs and wells. Cracks in wet ground and on steep slopes.

IX. General panic. Masonry D destroyed; masonry C heavily damaged, sometimes with complete collapse; masonry B seriously damaged. General damage to foundations. Frame structures, if not bolted down, shifted off foundations. Frames wrecked. Conspicuous cracks in ground. In alluviated areas sand and mud ejected, earthquake fountains, sand craters.

* Masonry strength classifications are: A = good workmanship, mortar and design, reinforced, especially laterally, and bound together using steel, concrete etc., designed to resist lateral forces; B = good workmanship and mortar, reinforced, but not designed in detail to resist lateral forces; C = ordinary workmanship and mortar, no extreme weaknesses such as failing to tie in at corners, but neither reinforced nor designed to resist horizontal forces; D = weak materials such as adobe, poor mortar, low standards of workmanship, weak horizontally.

X. Most masonry and frame structures destroyed with their foundations. Some well-built wooden structures and bridges destroyed. Serious damage to dams, dikes, embankments. Large landslides. Water thrown onto banks of canals, rivers, lakes etc. Sand and mud shifted horizontally on beaches and flat land. Rails bent slightly.
XI. Rails bent greatly. Underground pipelines completely out of service.
XII. Damage nearly total. Large rock masses displaced. Lines of sight and level distorted. Objects thrown into the air.

(b) Chain intensity scale (1957)

I. Not felt. Detected only through instrument.
II. Felt by individuals who are very sensitive and at rest.
III. Felt by a few indoors at rest. Vibration feels like fast passing trucks. Careful observers note slight swing of hanging objects.
IV. Felt by most indoors and a few outdoors, some woken up.
V. Felt by all indoors and most outdoors; many woken up; livestock restless. Doors, windows, floors, ceilings and timber joints rattle; open doors and windows swing. Dust and plaster drop; wall surfaces may show fine fissures. Hanging objects swing. Pendulum clocks stop. Liquids spilt from containers when full. Unstable objects on shelves overturn or drop.
VI. Many run outdoors; it is hard to stand still. Most livestock run out of stables. Water sloshes strongly and sometimes spills. Books on shelves sometimes overturn or drop. Light furniture may move.

Many class I buildings damaged, some seriously damaged, a few collapse; most class II and III slightly damaged; some class II more significantly.* Memorial archways, brick and stone pagodas, and fence walls slightly damaged.

Small fissures on wet and loose ground; occasionally small slope slides, rolling rocks and pitfalls may be found.
VII. All frightened and run outdoors; tremor felt by drivers. Hanging objects swing violently; may be damaged and drop. Light furniture moves; books and knick-knacks shaken off shelves.

A great majority of class I buildings damaged, many heavily; a few collapse. Most class II damaged, a few heavily. A great majority of class III slightly damaged, and many significantly. Fences damaged, some collapse, weak city walls flake, bricks drop from some weak battlements. Memorial archways, brick and stone pagodas, tombs and chimneys may be damaged. Some loess caves collapse.

Fined fissures and occasional slides on road surfaces, foundations or earth embankments. More and larger cracks on wet and loose ground,

* Class I building: simple shed made of adobe, pebble and rubble without mortar, with a straw–mud top. Class II building: ordinary earth, adobe, pebble and rubble dwelling, low standards of workmanship; old and rotten timber frames. Class III building: timber-framed building such as palace, temple and gate tower; new masonry building.

sandblows and waterspouts from a few of them. Slides and rockfalls possible on steep mountain slopes. Changes of spring flow and ground water level possible.

VIII. Difficult to walk. Death and injuries occur to people and cattle due to building failures. Furniture moves and overturns. A great majority of class I buildings heavily damaged; many collapse. Many class II heavily damaged, a few collapse. A great majority of class III damaged, a few heavily damaged, collapse possible. Fences are heavily damaged, partially collapsed; most weak city walls badly damaged, some collapse, many battlements collapsed. Memorial archways, tombs, pagodas and chimneys are damaged, weak ones heavily damaged or even collapsed. Most stone tablets and other memorials shift, twist or overturn. Individual joints of underground pipelines are damaged.

Width of ground surface cracks several centimeters, wider for those on flood lands of loose and wet soils; often with sand and mud ejected on ground with a high water table. Moderate slides on banks of roads, embankments and steep slopes. Conspicuous debris flows, rockfalls, slides often on areas of loose soils, some rivers blocked, forming new lakes. Springs sometimes stop or are created.

IX. Furniture overturned and damaged.

A great majority of class I and many class II buildings collapse; many class III heavily damaged, a few collapse. Most fences collapse, including relatively solid ones. Many parts of relatively solid city walls heavily damaged, many battlements collapse. Memorial archways, tombs, pagodas and chimneys often heavily damaged or even collapsed. Many relatively stable tablets and monuments also overturn. Some underground pipelines crack.

Some road foundations damaged, cracks on roads. Rails locally bend. Many cracks on ground, more than 10 cm wide; sometimes crossing each other on slopes and on alluvial river deposits with width up to a few tens of centimeters and very great length. Many debris flows, rockfalls and slides in mountain areas.

X. Most furniture and other articles damaged. Many class III buildings collapse. Many memorial archways heavily damaged; almost all tombs, pagodas and chimneys collapse; solid stone tablets and other monuments collapse. Rails bend slightly; underground pipelines crack. Cracks on ground tens of centimeters wide, some more than 1 m; large and wide bands of cracks, extending a few kilometers. Cracks sometimes on rocky ground. Huge slides on mountain cliffs block rivers to form lakes. Strong lake waves (seiche) hit banks.

XI. Buildings usually badly damaged and most collapse, causing many deaths of people and cattle and loss of property. Road foundations and embankments destroyed on a large scale; rails bent over a great length; underground pipelines completely destroyed and out of service.

Many widely open cracks, muddy water with ancient buried items ejected. Large scale landslides and slope slides, new faulting with considerable horizontal and vertical offsets. Water levels above and below ground greatly changed.

XII. All buildings collapse. Topography greatly changed over a large region; overground and underground water systems destroyed, flood everywhere, plains submerged; animals and plants in mountain areas destroyed by slides and pitfalls.

Summary

To summarize, earthquake intensity is a very simple idea to give a rough grading of local damage level. Simplicity is its advantage. It is very useful for rough but simple indication to guide emergency disaster relief planning or national investment. For any specific case or by specific person, individual ratings may give quite different grades at one location, sometimes differing by four grades (Ambraseys, 1969).

4.2 Earthquake damage and examples

4.2.1 Two types of earthquake damage

Earthquake damage can be classified into direct and indirect damage. Direct damage includes those directly involved with casualties and property losses. Property losses include only the monetary value of the damaged parts of engineering works, such as buildings, roads etc., as well as of natural resources, such as rivers, groundwater and oil. Indirect damage includes damage due to loss or malfunctioning caused by the direct earthquake damage, such as casualties and economic losses from earthquake-induced fires, flood, disease, loss of workforce and capital, interruption of communication and energy supply. The last is very important for major cities. But only direct damage will be referred to further in this book; and it will be discussed in two categories, ground and structural failure.

Ground failure refers to damage induced by loss of bearing capacity of the foundations, which includes relative surface displacement or underground fault rupture, landslides and slope slides from fault rupture and vibration, differential settlement and deformation, and partial or complete loss of bearing capacity.

Many field examples of damage will be given in Chapters 6, 7 and 8. Landslides from fault rupture may block rivers and form earthquake lakes, burying villages and roads; although the damage may be serious, it is limited to the epicentral area. Sand liquefaction may cover farmland; induce slope slide; damage dams, bridges, buildings, underground works and roads; sink or float structures; and differential deformation is particularly harmful to lifeline systems. This type of damage occurs mostly on soft and weak site soils, such as saturated fine sand, silt and very soft clay sites. Structural damage resulting from ground failure has a pattern of static deformation of differential displacement.

Structural damage from vibration includes all kinds of damage from inertial forces, such as inclined cracks and X-type cracks from torsion or repeated push–pull horizontal or vertical vibration, brittle or ductile, partial or global, and is usually different from ground failure.

There are two reasons to distinguish these two kinds of damage. Firstly, structural damage indirectly from ground failure comes from relative displacement of the foundation and is of a static nature; but structural damage directly from structural vibration is from inertial forces and is of a dynamic nature. Secondly, it is easier from the design point of view to take different measures to reduce the damage.

4.2.2 Examples from earthquake field experience

A few examples are given below for some of the strong earthquakes that have occurred worldwide; examples of earthquakes in China, USA and Japan will be given in Chapters 6 to 8.

(a) Gediz earthquake in Turkey

An earthquake of magnitude 7 occurred on 28 March 1970 in the northwest of Gediz in Turkey, with isoseismals shown in Figure 4.1. The first special phenomenon was the intensity in the epicentral area, IX according to ground damage, but VIII according to buildings; the second was many abnormal intensity areas; the third was serious damage of long-period structures far away from the epicenter. For example, the Tofas factory in Bursa City was 135 km away, in an area of intensity V, but the structures were badly damaged, with a local intensity of VIII. The ground deposit was very soft sandy clay with a depth of a few hundred meters. Aftershock records showed a ground predominant period of 1.2 s and the natural

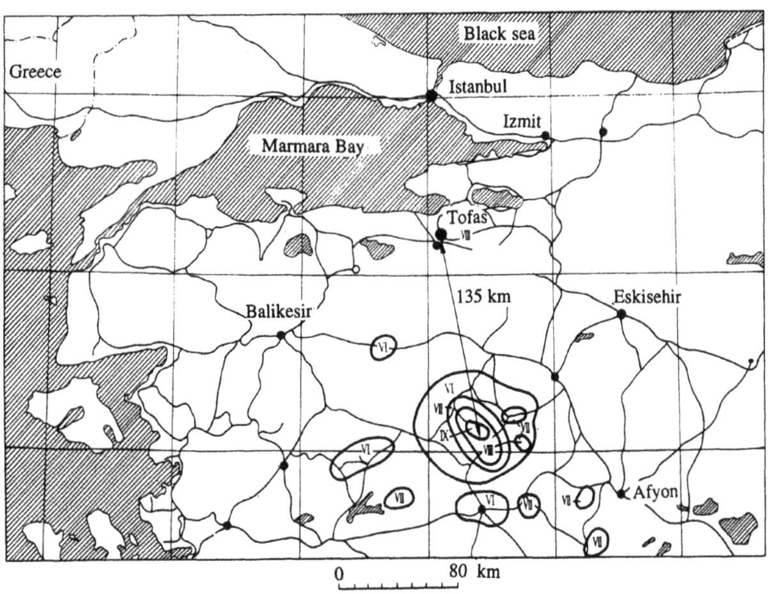

Figure 4.1 Isoseismals of Gediz earthquake in Turkey.

period of collapsed structures about 1.25 s. The intense damage came from resonance effects.

(b) Mexican coastal earthquakes (Cassaro and Martinez-Romero, 1987)

As shown in Figure 4.2, a series of strong earthquakes of magnitude greater than 7 along the Pacific Coast of Mexico were caused by the subduction of the Cocos plate beneath the North American plate with an average relative velocity of 5 cm/year to the northeast.

The first feature of the Mexican earthquakes is perhaps the clearly defined seismic gaps. Dividing the subduction zone into segments where the observed return period in each segment varies from 30 to 75 years, Kelleher *et al.* pointed out some seismic gaps for future strong earthquakes (Cassaro and Martinez-Romero, 1987, p. 10) and recent earthquakes have occurred in some of them, but both the Tehuantepec and Michoacan gaps have no strong earthquake records. Because of the relatively short return periods, an array of digital strong-motion stations was sited along the coast of Michoacan and Guerrero to wait for large earthquakes. In 1985, earthquakes did occur in the Michoacan gap and resulted a series of fine accelerograph records.

The second feature is the damage and ground motion in Mexico City, roughly 300–400 km away from the epicenter. In several recent strong earthquakes since 1957, Mexico City showed abnormally high damage to tall buildings in the old city, built on very thick and soft deposits. This combination of very soft ground, strong distant earthquake and tall buildings, was considered the cause.

1985 GUERRERO ACCELEROGRAPH ARRAY

Figure 4.2 Tectonic setting and strong earthquakes of Mexico.

On the basis of site conditions, Mexico City may be classified into three zones, A, B and C. Zone A is located on an old lake bed where the foundation soil is highly compressible volcanic clay, 10–230 m thick, with average water content 300%, mean compressive strength of upper layer 0.6 kg/cm^2, shear wave velocity as low as 40 m/s and nearly linear behavior under large dynamic strains. The bedrock is about 1000 m below surface. Zone C is the hilly slopes of the mountain, with subsoil of volcanic tuffs, conglomerates, cemented or dense sands, or similar relatively incompressible soils. The unconfined compressive strength is usually more than 5 kg/cm^2. Zone B is the transition zone between zones A and C.

In the 1957 earthquake, for example, the damage in Mexico City was heavy. Of the 1500 buildings recorded as damaged, 1000 were comparably heavily damaged, of which less than 0.3% were in zone C, about 4% in zone B and the remaining 95% were in zone A; the collapsed buildings were all in zone A.

The latest example is the 1985 earthquake, about 400 km away from Mexico City, which consisted of two events, 30 s apart. The zone of maximum damage (primarily due to vibration and liquefaction) was not found in the city, but was located entirely in the ancient lake bed (zone A), where the intensity rating reached a city record high of IX or X, while the damage close to the epicenter on hard ground was fairly minor. The relatively rigid buildings, with masonry bearing walls, in general suffered little damage. The framed buildings, mainly made of reinforced concrete, were the most heavily damaged. Those with shear walls or diagonal bracing were not damaged.

The third feature is the unusually long duration of strong shaking in Mexico City which resulted from the occurrence of two consecutive earthquakes on the same causative fault. In Mexico City, because of its long distance away, the dispersion of waves from two events overlapped into one continuous motion and lengthened the duration. This phenomenon may repeat for a city on thick and soft deposits when a large and distant earthquake attacks, and flexible structures should be designed for it.

The fourth feature is the low peak acceleration in the epicentral area. In 1985 when the earthquakes occurred, there were about 150 accelerographs installed, the II array inland and the Guerrero array by the coast. Among them, 49 instruments triggered. This is one of the best records both near to and away from the epicentral area of a strong earthquake of magnitude around 8. The maximum peak acceleration reached 0.65g at the rock site in the epicentral region and 0.18g at the soft site in Mexico City. The comparatively low acceleration in the epicentral area for an earthquake of $M_s = 8.1$ was also not expected.

(c) 1977 Vrancea earthquake in Romania

An earthquake of magnitude 7.3 occurred in Vrancea, Romania, on 4 March 1977 with focal depth about 100 km, unusually deep for a highly damaging earthquake, which caused 1570 casualties; total losses exceeded two billion dollars. Because of the deep focus, the epicentral intensity of this earthquake of $M=7.3$ was only VIII but the attenuation was rather slow.

Clear evidence of long-period effects was the first feature of this earthquake. For example, tall buildings in Sofia, 450 km away from the epicenter, suffered significant damage: many multistory reinforced concrete buildings with infill masonry walls needed repairs to cracks; a few tall buildings with reinforced concrete shear walls located on alluvial cracked. At Svishtov in Bulgaria, 240 km from the epicenter, rigid one-story brick buildings were not damaged, but buildings, tall or with a flexible first story, were damaged or even collapsed. Strong acceleration records showed that the ground motion was similar to a sinusoidal motion with long predominant periods of 1.0–2.4 s (Brankov and Sachanski, 1980).

(d) 1985 Chile earthquake

A major earthquake of magnitude 7.8 occurred near the coast of central Chile, focal depth 33 km; the isoseismals are shown in Figure 4.3 (Wallace and Moehle, 1989).

The earthquake consisted of two events: the main event with $M_s=7.8$ and a smaller one with $M_b=5.2$ preceded the main one by about 10 s. The maximum intensity observed was VIII, with 180 deaths and 10 000 injured. It affected a large populated area with many modern engineered structures. Most modern reinforced concrete structures performed well, but adobe houses were severely damaged.

The overall feature of this earthquake experience is the good behaviour of the modern structures, including highrise buildings, and equipment, which made some engineers from the United States say: ‘it was our hope and dream that a similar earthquake in the United States would result in a

Figure 4.3 Intensity distribution in Chile earthquake.

similar satisfaction with the performance of modern engineered structures' (Wallace and Moehle, 1989).

Although structures performed well, ground failures due to liquefaction and slope slides caused heavy losses to seawalls, bridges, tailing dams, roads and embankments.

More than 30 accelerographs recorded the event. Perhaps because the rupture propagated from north end of the source to south, those stations near the north end, though closer, recorded peak accelerations of $0.2g$ with predominant frequency of 1–2 Hz, but those near the south end, though farther, recorded $0.6g$ with more high-frequency components (>4 Hz). The response spectra are similar to the El Centro record.

4.3 Intensity distribution

4.3.1 Isoseismal map

The intensity contour or isoseismal map is a simple way to show the general damage of a strong earthquake on a map. After field investigations an isoseismal map is drawn in the following two stages.

1. The original intensity ratings of all sites investigated are indicated on a detailed map.
2. Contour lines of equal intensities are drawn which group sites of the same intensity in the same zone.

Because of the wide scattering of data, judgement is necessary for the second stage. The routine compromise is to have one set of isoseismals starting from the epicentral area with the highest intensity and decreasing outwards by an decrement of one grade with fairly smooth contours, and some individual contours scattered almost randomly. The former is considered to represent the general intensity distribution, which reflects attenuation of damage away from the epicenter, and the latter as abnormal intensity areas, which reflect the local variations of site conditions and other effects.

If intensity is rated by the highest indication of an intensity scale, the isoseismals will be the envelope of intensity region; if intensity is rated by weighting all indications of the intensity scale, as the Chinese do, the isoseismals will be some average.

4.3.2 Abnormal intensity region

Since site conditions over a wide region change quite often, abnormal intensity areas are actually quite common, and can be found in almost any strong earthquake. In mountain regions, local topography and soil condition are mainly responsible; in alluvial plains, soft and weak soils are the major cause of abnormality. There are areas where abnormal intensities have been reported for several historical earthquakes. For example, the Yutian area, about 50 km northwest of Tangshan City, was an area of abnormally low intensity in both the 1679 Sanhe-Pinggu and the 1976 Tangshan earthquakes.

4.3.3 Factors controlling intensity distribution

(a) Source effect

The term 'earthquake source' is used here for the whole region where strain energy is released in the form of earthquake waves during one earthquake. Energy density, geometry of the source and its position in space are the controlling factors of intensity distribution. The overall shape of isoseismals is controlled by the source. For example, small earthquakes have small sources and the isoseismals are usually circular; but very large earthquakes have in general extended sources, i.e. with one horizontal dimension much longer than the other, and the isoseismals of higher intensities are usually elongated or elliptical. For example, the surface faulting of the 1970 Tonghai earthquake was about 60 km long and the isoseismals of intensities VII, VIII, IX, and X were very narrow, with a short-axis/long-axis ratio less than one-third.

(b) Distance effect

The effect of distance is clear and it is the main factor of attenuation. There are several definitions of distance and their distinction is important for the near field, i.e. the epicenter and the neighboring areas. The epicentral distance depends on the definition of epicenter. The instrumental epicenter is the easiest to use with instrumental records, but it has the lowest correlation with intensity distribution. The macroseismic epicenter or the damage center can be found only after field investigation, but it has the closest correlation with intensity distribution. The fault distance or the shortest distance to the fault is even better when surface faulting can be found.

Site effects are discussed in section 4.5.

4.4 Earthquake intensity attenuation

The relationship of intensity variation with magnitude and distance is usually referred to as the attenuation law.

4.4.1 Epicentral intensity and magnitude

The epicentral intensity I_0 was used to represent the size of an earthquake before the earthquake magnitude was suggested. Although epicentral intensity also depends on focal depth h, it can still be used for a certain specified narrow range of focal depth, say 10–30 km, which is the rough average of most strong earthquakes. To estimate the magnitude of historical earthquakes, the relation between epicentral intensity I_0 and magnitude M is useful. Gutenberg and Richter (1956) obtained for the southern California region

$$M = 2/3\, I_0 + 1 \qquad (h = 16\,\text{km}) \tag{4.1}$$

In China, the following relation was suggested in the 1970s:

$$M = 0.66 I_0 + 0.98 \qquad (h = 15\text{–}45\,\text{km},\ \sigma = 0.33) \tag{4.2}$$

based on data of 152 earthquakes since 1900 (Deng *et al.*, C1980), very close to equation 4.1. The earlier result by Li (C1980), derived from earlier earthquakes, is

$$M = 0.58 I_0 + 1.5 \tag{4.3}$$

Other results from studies considering the effect of focal depth h (km) are

$$M = 0.68 I_0 + 1.39 \log h - 1.4$$

by Fu and Liu (C1960) from Chinese data;

$$I_0 = 1.5M - 3.5 \log h + 3.0 \tag{4.5}$$

by Schebalin (Medvedev, 1962) from Russian data;

$$I_0 = 1.5M - 1.2 \log h + 3.0 \tag{4.6}$$

from Chinese data, $h = 10$–40 km.

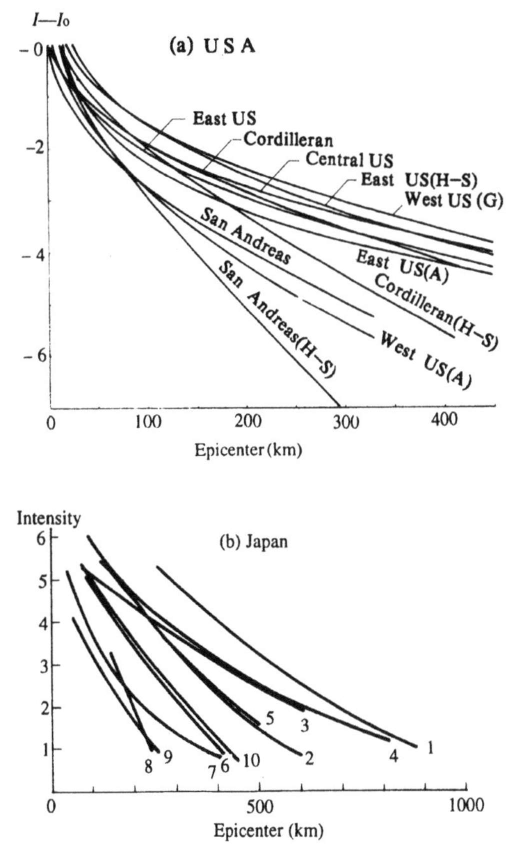

Figure 4.4 Attenuation relations of intensity. In (a): A = Anderson (1978); B = Gupta (1976); H–S = Howell and Schultz (1978). (Redrawn from Chandra, 1979; all citations given in source.) In (b): 1 = Sanrikuoki, 1933; 2 = Tonankai, 1944; 3 = Nankai, 1946; 4 = Tokachioki, 1952; 5 = Kanto, 1923; 6 = Tottori, September 1943; 7 = Fukui, 1948; 8 = Tottori, March 1943; 9 = Tojiki, December 1943; 10 = Niigata, 1964. In (c)–(e): 1 = Whole of China; 2 = east; 3 = west; 4 = Taiwan; 5 = southeast; 6 = north; 7 = southwest; 8 = Xinjiang.

4.4.2 Circular attenuation

Most available attenuation relations consider circular isoseismals, which is good for smaller earthquakes ($M \leqslant 6.5$). In early studies, the following formula was used:

$$I = I_0 + a - b\Delta - c\log(\Delta + R_0) \tag{4.7}$$

where coefficients a, b and c are obtained from regression, and constant R_0 was taken to be zero in the early days. Recently, with magnitude data available, the following formula is often used:

$$I = a + bM - c\Delta - d\log(\Delta + R_0) \tag{4.8}$$

Magnitude can also be obtained for historical earthquakes through areas of isoseismals instead of from epicentral intensity. For large earthquakes, the area inside the intensity VIII curve, i.e. for intensity VIII and above, was found to correlate better with magnitude. Some results of attenuation are given for the United States in Figure 4.4(a), for Japan in Figure 4.4(b) and for China in Figure 4.4(e).

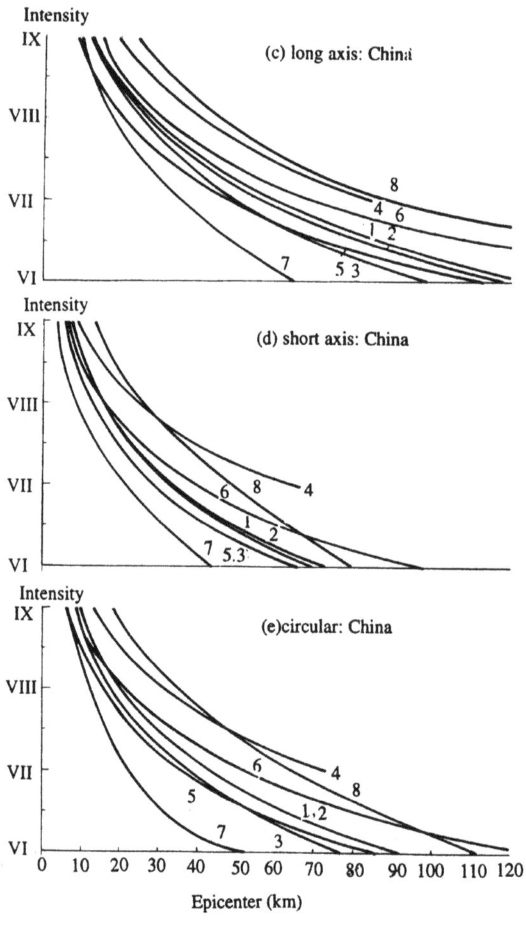

Figure 4.4 (c)–(e)

4.4.3 Elliptical attenuation

For very strong earthquakes with enough data, an elliptical attenuation may be assumed by defining its long and short axes of attenuation. Historical earthquakes in western China are typical examples; and elliptical attenuation has been used for the second and third zoning maps of China. Figure 4.4(c) and (d) show the data used for the second zoning map.

Chandra *et al.* (1979) gave an elliptical attenuation for Iran:

$$\text{long axis: } I = I_0 + 4.824 - 0.00548\Delta - 3.708\log(\Delta + 20) \quad \Delta < 160\,\text{km} \tag{4.9}$$

$$\text{short axis: } I = I_0 + 8.729 - 0.01158\Delta - 6.709\log(\Delta + 20) \quad \Delta < 110\,\text{km} \tag{4.10}$$

4.4.4 Selection of epicentral intensity

Selection of epicentral intensity used in equation 4.7 is not an easy problem for the following two reasons. Firstly, the epicentral intensity of a large earthquake may be rated too high if only according to surface faulting or some other individual index, and may therefore be higher than surrounding next high intensity by two grades or more. Secondly, because of different focal depth and the integer property of intensity, the maximum difference in epicenter intensity may be one grade.

When studying the intensity attenuation, Chandra gave the following result on the basis of earthquakes listed in Table 4.1, where I_1 is the intensity following I_0:

$$I - I_0 = 6.453 - 0.00121R - 4.96\log(R + 20)$$

where R is focal distance in kilometers. He suggested a method of iteration as follows to select I_0.

1. Plot attenuation curves of all earthquakes to be analyzed on one graph.
2. Draw an approximate average curve visually with special attention to the slope or the general shape of the attenuation curves.

Table 4.1 Epicentral intensities used by Chandra (1979) (Cordillera, western USA)

No.	Date	Earthquake	M	I_0	I_0'	I_1
1	18 October 1935	Helena, MT	$M_S = 6\frac{1}{4}$	VIII	7.3	–
2	14 February 1946	Puget Sound, WA	$M_S = 5\frac{3}{4}$	VII	7.3	–
3	15 December 1950	Herlog, CA	$M_S = 5.6$	VII	6.8	–
4	15 May 1954	Puget Sound, WA	–	VII	6.6	–
5	6 July 1954	Fallon, NV	$M_L = 6.8$	IX	8.1	VII–VIII
6	23 August 1954	Fallon, NV	$M_L = 6.8$	IX	8.2	VII–VIII
7	16 December 1954	Dixie Valley, NV	$M_L = 7.2$	X	8.7	–
8	21 December 1954	Eureka, CA	$M_S = 6.6$	VII	7.5	–
9	17 September 1959	Hebgen Lake, MT	$M_S = 7.1$	X	9.1	VIII
10	30 August 1962	Logan, UT	$M_S = 5.7$	VII	7.2	–
11	29 April 1965	Puget Sound, WA	$M_b = 6.5$	VIII	8.0	–
12	12 September 1966	Boca Dam, CA	$M_L = 6\frac{1}{4}$–$6\frac{1}{2}$	VII	7.4	–
13	4 October 1967	Marysvale, UT	$M_b = 5.2$	VII	7.0	–

Note: I_0' is the epicentral intensity finally obtained.

3. Determine the value of I_0 for each earthquake by shifting the average curve parallelly to a best fit to each curve.
4. Perform the regression to find out the coefficients a, b and c from attenuation relations with the I_0 values found from step 3 and to obtain an attenuation relation.
5. Use this regressed attenuation curve in place of the visual average, and repeat the forgoing steps until all I_0 values are stable.

Chandra's approach of using regressed epicentral intensity instead of the maximum given in the epicentral area has at least one advantage of avoiding an unreasonably high epicentral intensity.

4.5 Effect of site conditions on intensity

'Site conditions' refers usually to local engineering geological conditions, such as soils within tens or hundreds of meters, and the water table and local topography, as well as faulting. Worldwide experience repeatedly shows the important effect of site conditions on intensity and ground motion. As far back as 1906, during the great San Francisco earthquake it was realized that damage or intensity was much heavier on downtown soft ground. In 1932 when K. Suyehiro gave his famous lecture on engineering seismology in the United States, he mentioned the harmful effect of soft ground on wooden houses and the beneficial effect on rigid houses during the 1923 Kanto earthquakes. S.V. Medvedev in the former Soviet Union was perhaps the first to summarize systematically the effect of site conditions on intensity. On the basis of field data, he suggested the following formula in 1952 (Medvedev, 1962)

$$\Delta I = 1.67 \log\left(\frac{\rho_0 v_0}{\rho v}\right) + \alpha \exp(-0.04h^2) \tag{4.11}$$

where v was firstly longitudinal wave velocity, but later changed to shear wave velocity; subscript 0 refers to the reference soil and no subscript for the soil under consideration; h is the depth of water table in meters; the constant $\alpha = 0.5$ (gravel) or 1.0 (sandy clay, clayey sand and fine sand). He defined the intensity corrected by this formula for a specific site as site intensity. The 1957 Seismic Design Code for the Soviet Union adopted this suggestion and defined the moderate site as the reference site in a three-type classification of hard (rock), moderate and soft, and the intensity given in the earthquake zoning map was taken as that on moderate sites, while site intensity was one grade lower for hard sites and one grade higher for soft sites.

It is important to realize that, especially in this section, intensity means damage, which includes both structural and ground failure, but not ground motion. Weak sites will intensify the ground failure and thus the total damage, the velocity and displacement of ground motion, but not acceleration. It is wise to think of ground motion in terms of spectrum and not as a single quantity (section 5.3.3).

5 Earthquake ground motion

Earthquake ground motion is a special kind of load. Other loads ordinarily acting on structures are usually in terms of force, but an earthquake load is in terms of ground motion; other loads are mostly static, not changing within a short period of time, but an earthquake load is dynamic, changing rapidly within a very short period of time, say in one second; other loads act mostly in the vertical direction, but an earthquake load acts simultaneously in both the vertical and horizontal directions, and also with torsion.

The terms 'earthquake motion', 'earthquake ground motion', or 'ground motion' have been used to describe the vibrational motion of the ground near the ground surface, caused by earthquake waves propagated away from the earthquake source. On the one hand, this is the main interest of seismological study in engineering subjects; for example, 'strong motion seismology' is a term coined in the 1960s to indicate this branch of study. On the other hand, structural engineers have to design their structures on the basis of earthquake ground motion. This is the reason that earthquake motion is generally an important subject discussed in all recent books on earthquake engineering.

Earthquake ground motion is a complicated phenomenon and the complication comes from several factors, critical but not accurately known, such as the dynamics of source breakage and the details of the transmitting media; these vary either from source to source, from path to path or from site to site and therefore introduce large scattering in numerical values of ground motion. It is the important development in strong motion observation in the past few decades that makes a scientific seismic design of structures possible. This chapter will introduce some important ground motion characteristics and the methodology of their evaluation; others may also be important but not so well-known (Bertero *et al.*, 1980).

5.1 Observation of earthquake ground motion

5.1.1 Instruments of observation

There are two categories of earthquake observation instruments currently available. One is that used by seismologists to determine the location, time and mechanism of the earthquake, with the aim of understanding the nature of the earthquake and the medium. This is the seismometer or seismograph. The other category of instrument is that used by engineers, with the purpose of understanding the vibrational characteristics of the site and the structures built on it, in order to evaluate the earthquake input to the structure, the structural response and finally how to design structures economically that will withstand earthquakes safely. This is the accelerometer or accelerograph.

Table 5.1 gives a comparison of the different features of these two categories of instruments. But owing to the development of both branches

Table 5.1 Comparison of seismograph and accelerograph

Instrument	Earthquake type	Operation	Record speed	Amplification	Parameter	Features observed	Site	Frequency band
Seismograph	Weak	Continuous	Slow	High	$u(t)$	Arrival of waves	Rock	Low & narrow
Accelerograph	Strong	Trigger	Fast	Low	$a(t)$	Whole process	Soil & rock	High & wide

of science, seismologists and the engineers are coming closer in their approach to studying strong motion, the types of instruments employed and the scope of measurement. Their observations of ground motion parameters, choice of frequency band, and measurement locations, either near to the source or epicentral region or in the far field, tend therefore to combine or be supplementary to each other.

Since accelerographs are triggered by the ground motion of the earthquake itself, the records obtained from this category of instrument usually show a phenomenon of losing the initial portion of the motion. The lost initial portion, say lasting Δt seconds, consists of the first peak of amplitude over $0.01g$ or lower and the preceding weaker motion. And because there is a delay before the mechanical parts of the accelerograph operate after the triggering signal comes from the trigger (at least 0.1s), the total duration of the lost initial portion will be at least $\Delta t + 0.1$s. For an earthquake of small magnitude occurring at a close distance, the duration of strong shaking is usually about a few seconds; if the strongest peak is the same one which triggers the accelerograph, it may be lost in the 0.1s delay before the instrument starts to work, so that only the later weak portion of motion is recorded, as shown in Figure 5.1. It is, therefore, necessary to be particularly careful when the maximum peak acceleration of a recorded time history of ground motion shows a value near $0.01g$, especially at the beginning of the record, if the stronger motion is lost. For strong earthquake motion, a good acceleration record is usually triggered by the beginning part of the P waves and shows the strong P wave portion together with the later S wave and sometimes the surface wave, such as shown in Figures 5.1 (top) and 5.2.

It is then obvious that, for an accelerograph, if the natural period $T_0 = 2\pi/\omega$ of the pick-up is 0.05 s, as commonly used, an accurate measurement of ground motion with period $T > 0.05$ s can be obtained, but not those of shorter periods. Fortunately, the suppressed components may be partly recovered because the dynamic property of the accelerograph is known and may be described by differential equations. Different accelerographs may have different features in addition to their common characteristics. For example, the accelerograph RDZ-1 currently used in China has the following special features.

1. There are 12 channels, instead of the usual three, with common recording paper and a common time signal, which is convenient for the observation of structural vibration;

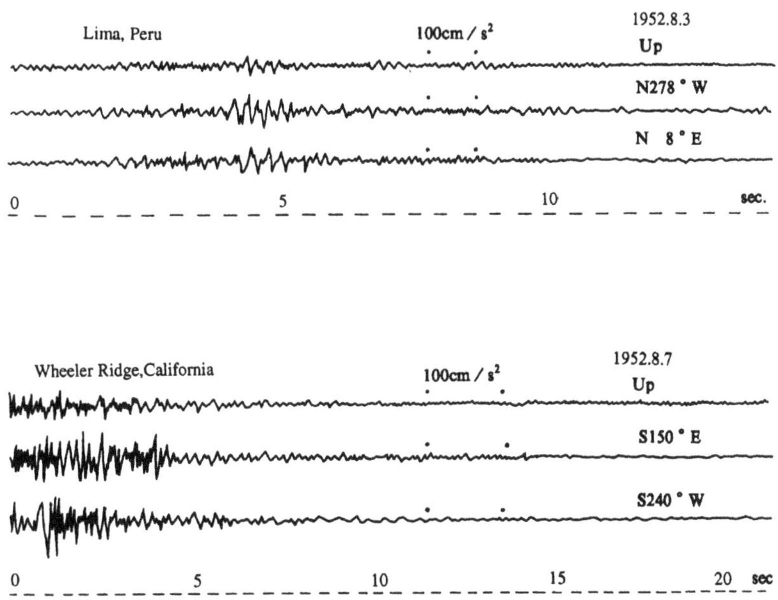

Figure 5.1 Typical acceleration records.

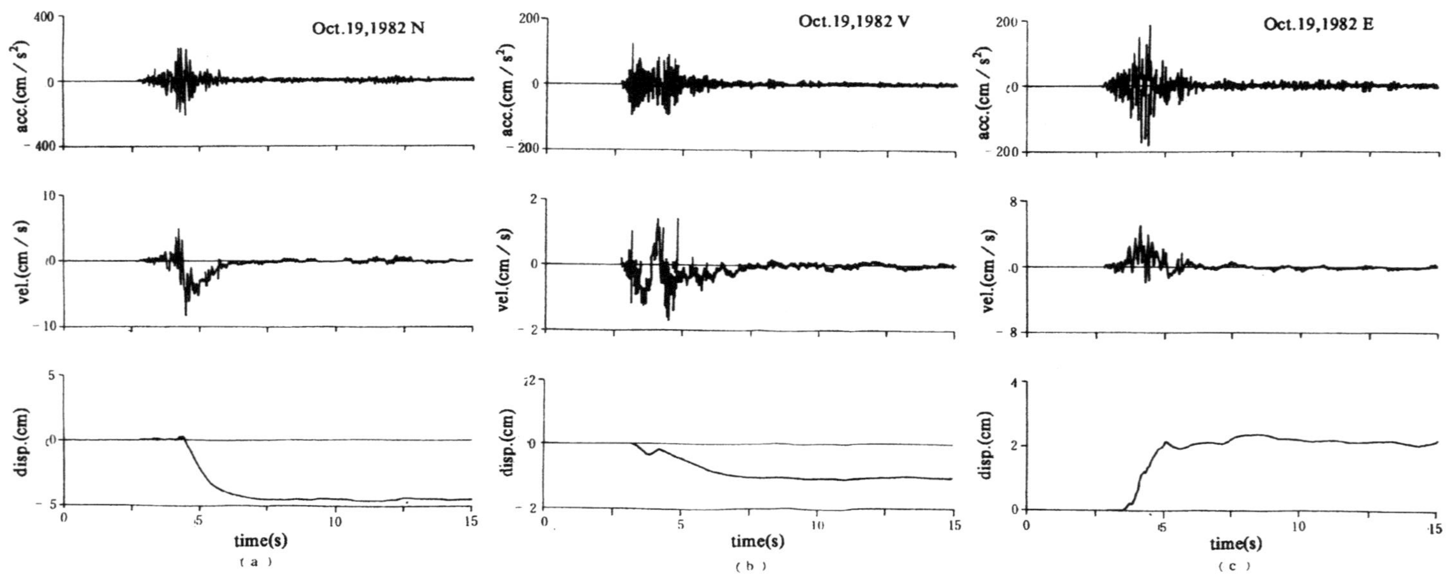

Figure 5.2 Records of a digital accelerograph at Lulong.

2. It is an accelerograph with galvanometer and the lower frequency band is wide.

5.1.2 Current status of strong motion observation

Measurement of strong motion during an earthquake with the accelerographs described above is termed strong motion observation. One of the purposes of strong motion observation is to understand, through the analysis of the records obtained, the characteristics of the ground motion and the structural response in order to study further some problems of earthquake engineering such as selection of the important parameters related closely to the damage of structures; the relation of ground motion to the earthquake, path of propagation of earthquake waves and site conditions; and the failure mechanism of structures.

There are at least six types of array for strong motion observation systems, each with a different purpose:

1. array for ground motion attenuation;
2. array for ground motion variation in a region;
3. array for ground motion on the fault;
4. array for structural response;
5. dense or differential array;
6. underground and three-dimensional array.

A network of strong motion observation has developed quickly worldwide. There were 1500 stations around the world up to 1970, with approximately 500 each in the United States and Japan; the total number of stations nearly doubled up to 1978 and reached roughly 5000 in 1980, with 2000 each in the United States and Japan; the total number increased to about 7000 up to 1985, with 3000 each in these two countries.

5.1.3 Application of strong motion observation

Internationally speaking, strong motion observation has almost a 60-year history, with about 7000 traces of records of engineering relevance. It is useful then to look at its role in engineering applications. Generally speaking, it has an important function in the following areas.

1. It supplied to earthquake engineers the necessary data for the introduction of response spectra to facilitate the transition of seismic design from static to dynamic theory – a milestone in earthquake engineering.
2. The strong motion records provided a strong impetus to the introduction of random vibration theory to earthquake analysis of structural responses. Early in the 1960s, random vibration theory was the theoretical basis for the SRSS (square root of sum of squares) method for the simplified consideration of modal contributions to the total response of a multi-degree-of-freedom (MDF) system. This method is still the one most commonly used in seismic design codes.
3. It provided the necessary quantitative data for the study of earthquake ground motion characteristics. In earthquake engineering, although a

qualitative understanding of structural behavior during an earthquake is very important, it would be very incomplete without quantitative measurement; only when qualitative understanding is supported with quantitative measures, can experience become science. Results from strong motion observation can supply us with not only quantitative but also qualitative understanding.
4. It supplied ground motion inputs to the response analysis in the time domain and earthquake-simulating test of structure. Harmonic waves were used in the 1940s; the 1940 El Centro record has been widely used since the 1950s; and some typical records obtained in the country where a structure is to be built have also been used, such as the Qian-an and Tianjin records of the Tangshan aftershocks in China.
5. It helped to understand quantitatively the structural behavior during an earthquake. The responses calculated from the recorded input motion compared well with the records on structures.

5.1.4 Digital accelerograph

(a) Defects of the analog accelerograph

Although the analog accelerographs used widely have made an important contribution to the progress of earthquake engineering science, it is necessary to realize that they do have the following defects:

1. missing initial portion of wave;
2. low accuracy in recording waves of low frequency and very large accelerations of high frequency;
3. time-consuming pre-processing of records before analysis.

(b) Requirements for a new type of accelerograph

For the reasons mentioned above, requirements for a new type of accelerograph were suggested in the late 1970s and early 1980s. For example, a 'Strong Motion Observation Instrument Study Committee' had been appointed in 1980 in Japan by the Strong Motion Observation Promotion Council, and specific requirements were agreed upon in 1983 (Architectural Institute of Japan, 1983, Section 5.3), of which the main items are as follows:

1. period range of 0.05–20 s (it is necessary to have an A/D converter with a sampling rate of 200/s);
2. amplitude range of $0.0001g$–$2.0g$ (it is necessary to have an A/D converter of at least 12–14 bits and a dynamic range of 86 dB);
3. arrival time with an accuracy of 1/100 s;
4. temporary memory of at least 2–5 s.

In 1988, new suggestions were given, such as to widen the period range to 0.02–20 s and the dynamic range to 114 dB and some requirements were added to assure a high signal-to-noise ratio (Omote *et al.*, 1988).

(c) Digital accelerograph (Iwan, 1984)

The first digital accelerograph appeared in the USA in 1974, in Japan in 1978 and in China in 1982, and developed rapidly because it could supply more and better information than those currently used.

The recorder of a digital accelerograph consists of an amplifier, mixed frequency filter, floating-point amplifier, A/D converter, memory device, trigger, power and control equipment, in addition to a cassette recorder. The recorder has to be used with sensors, of which the force-balancing sensor is the one commonly used. Sensors convert the ground motion signal into an electric signal which is then converted into a digital signal through an A/D converter.

When compared with typical analog accelerographs, the advantages of the digital accelerograph are as follows.

1. It generally does not miss the initial portion of motion.
2. The amplitude may vary widely and the accuracy will still be high.
3. The frequency band is wide to include long period motion with good accuracy, which is controlled primarily by the signal/noise ratio.
4. Playback is easy and can be used directly for computer processing, which reduces pre-processing work from a few months to a few hours.
5. Further improvements can be made easily.

The main disadvantage of the digital accelerograph is the possibility of malfunctioning of the tape cassette when used for a long time in severe conditions of high humidity and strong shaking.

5.2 Characteristics of strong ground motion

It is generally agreed that, for the safety of structures, characteristics of the strong ground motion may be represented by three basic elements: amplitude, spectrum and duration. In earlier days, strong ground motion was treated from a static point of view and only the importance of the amplitude was emphasized. The importance of the spectrum of ground motion was clear and was considered in design codes in the 1950s. After the 1971 San Fernando earthquake, the importance of duration on the safety of engineering works was then accepted but has still not been included in the code.

5.2.1 Some typical strong ground motion

Figures 5.3 to 5.5 show some typical earthquake accelerograms $a(t)$ and their integrated time histories of velocity $v(t)$ and displacement $u(t)$, with their relevant information given in Table 5.2. The following facts may easily be seen from these traces of time histories.

1. High-frequency components are rich in accelerograms and long-period components in displacement time histories. A very simple reason is the much shorter apparent periods of the accelerograms than those of the displacements.

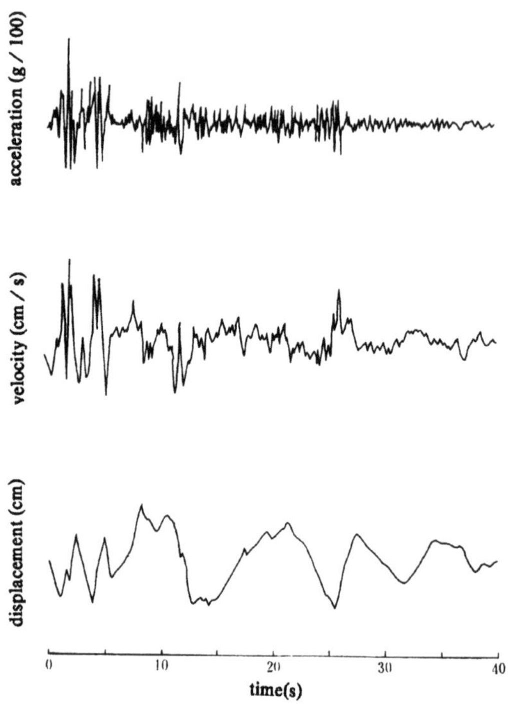

Figure 5.3 El Centro record (18 May 1940).

2. The duration of motion is very long for some accelerograms, e.g. 40s, but some are very short, as shown in Figure 5.5.
3. There is sometimes only one high peak and sometimes several in one accelerogram.
4. The predominant period is short for some but long for others.

The component motions at one point are correlated, and there are three principal axes along which the cross-correlation between components is zero, similar to the principal axes of the moment of inertia of a section (Penzien and Watabe, 1975; Kubo and Penzien, 1979).

The above-mentioned features of the accelerograms are not entirely random but are related to the source mechanism and size of the earthquake, distance from the earthquake and the local site conditions, which will be discussed in the following sections of this chapter.

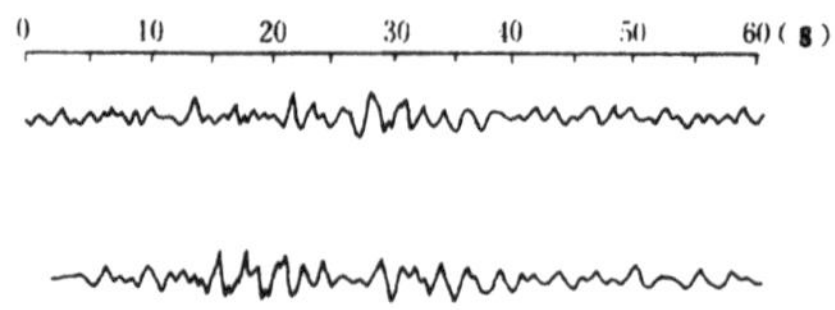

Figure 5.4 Mexico City record (11 May 1962).

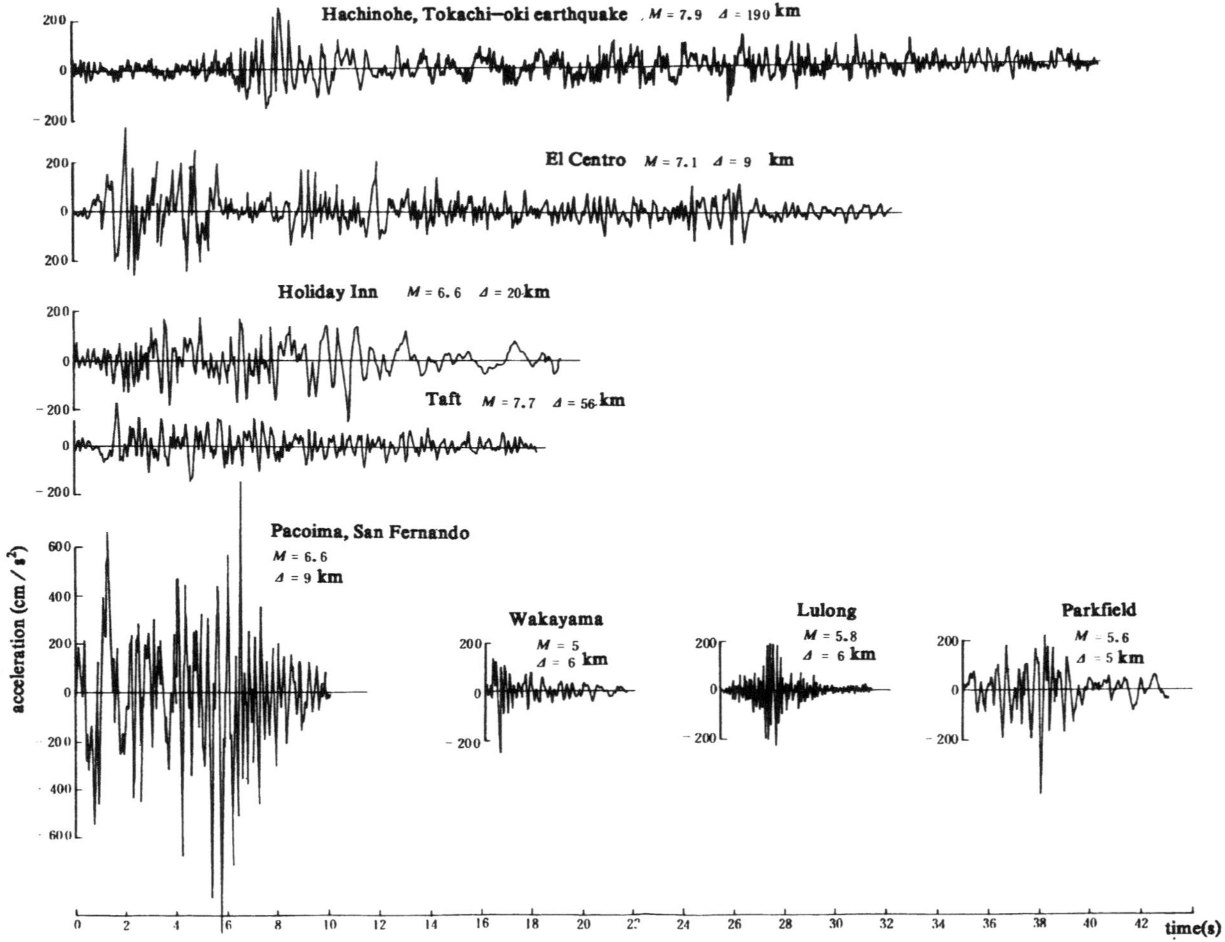

Figure 5.5 Comparison of several typical accelerograms.

Table 5.2 Relevant data for some accelerograms

Date	M	Station	Δ (km)	I	a/g	v (cm/s)	u (cm)	T_d (s)	Site
18 May 1940	7.1	El Centro	9	VIII	0.32	42	28	30	Average
11 May 1962	7.0	Mexico City	280	VI	0.05	–	–	40	Soft
27 June 1966	5.6	Chalome No. 2	0.08	VII–VIII	0.50	71	22	8	Average
17 October 1966	7.5	Lima	205[a]	VII	0.40	20	15	10	Hard
10 May 1968	7.9	Hachinohe	165	VIII	–	–	–	–	Average
15 November 1976	6.9	Tianjin	90	VI	0.15	–	–	10	Soft

[a] Fault distance may be very short.

5.2.2 Amplitude

The amplitude of ground motion may mean all or one of the peak values or effective peak values of the acceleration, velocity and displacement of ground motion.

As mentioned above, acceleration has long been considered as an index of the strength of ground motion, and is still the most thoroughly studied quantity after a large set of accelerograms has been accumulated.

Peak acceleration is usually controlled by very-high-frequency components, say 20–100 Hz, which are too high to affect ordinary structures. For this and other reasons, effective peak accelerations are used. The commonly used ones are as follows.

(a) Effective peak acceleration (EPA) and velocity (EPV) in ATC-3 (ATC, 1978)

If the average value of the 5% damping acceleration response spectrum in the period range 0.1–0.5 s is S_a and the average of the velocity response spectrum around the period 1.0 s is S_v (McGuire, 1974), then the effective values are defined in ATC-3 as

$$\begin{aligned} \mathrm{EPA} &= S_a/2.5 \\ \mathrm{EPV} &= S_v/2.5 \end{aligned} \tag{5.1}$$

which are statistically correlated to the real peak values.

(b) Static effective acceleration

On the basis of the overturning and shifting of a simple rigid body and Newton's second law, an effective peak acceleration a_{eff} of the rigid body may be derived by inverse computation, considering a static force of ma_{eff}, acting on the body of mass m. The accelerations relating to the earthquake intensity scales in the early days of earthquake engineering were obtained in this manner.

(c) Root-mean-square acceleration (a_{rms})

From the point of view of random vibration, the maximum peak of an accelerogram is random and not a suitable quantity for the index of strong motion characteristics, but the variance or the standard deviation (section 3.3) is. Arias (1970) suggested using the energy dissipated in an elastic single-degree-of-freedom (SDF) system per unit mass

$$I_A = \frac{\pi}{2g} \int_0^{T_d} a^2(t)\, dt \tag{5.2}$$

as an index of the strength of ground motion; this is referred to as the Arias intensity. In this formula, T_d is the total duration. If the motion is non-stationary, T_d may be taken as the duration of the strong portion where the motion is nearly stationary, and the energy per unit time will be proportional to the variance:

$$a_{rms}^2 = \frac{1}{T_d} \int_0^{T_d} a^2(t)\, dt \tag{5.3}$$

which is the definition of root-mean-square acceleration.

Joyner and Boore (1981) have obtained the following relation:

$$a_{max}^2 = a_{rms}^2 \left[2 \ln \left(\frac{2T_d}{T_p} \right) \right] \tag{5.4}$$

where T_p is the predominant period of ground motion.

(d) Spectral intensity (SI)

Housner (1952, also in Wiegel, 1970) defined the spectral intensity as

$$SI_\zeta = \int_{0.1}^{2.5} S_v(T\zeta)\, dT \tag{5.5}$$

where S_v is the velocity response spectrum of damping ratio ζ which takes a value here of 0 or 0.2; T is the period. Both the Arias and the spectral intensity are not defined in relation to macroseismic phenomena and are therefore quite different from the macroseismic intensity.

The peak values of ground motion mentioned above are not independent, and statistical relations may be found between them, but with very large scattering. Figure 5.6 shows an example.

5.2.3 Spectrum

Theoretically speaking, it is easy to accept the important influence of the ground motion spectrum on structural response; but, in practice, it was not until the 1950s, when there were enough strong motion data, that the average spectrum of the ground motion was obtained and accepted in the engineering design code.

Though the standard or average design spectrum is still popularly used in most seismic design codes, it is well known that the shape of the spectrum of ground motion varies with earthquake and site. There are three reasons for accepting a variable spectrum.

1. It is repeatedly shown by damage phenomena that earthquake spectra might be different for different cases. Even in the 1920s, people realized

Figure 5.6 Correlation of a_h and a_v.

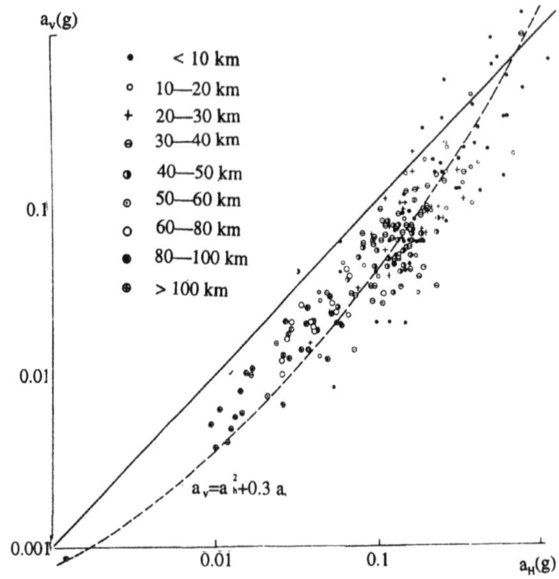

from the damage data of buildings in the 1923 Kanto earthquake that rigid structures suffered heavier damage on rigid ground and flexible structures on flexible ground, as shown in Fig. 5.7, where the site on the right is harder than that on the left.

2. Quantitative changes of the spectrum of ground motion for different magnitudes, distances and site conditions were known from the strong motion records which can, at many instances, explain specific damage phenomena as a direct effect of the change of spectrum.
3. According to the theory of wave propagation, waves of higher frequencies attenuate faster than waves of lower frequencies.

There are three kinds of spectra commonly used in earthquake engineering, namely Fourier, response and power spectra.

Figure 5.7 Effect of site on damage of buildings in 1923 Kanto earthquake.

(a) Fourier spectrum

An accelerogram $a(t)$ and its Fourier spectrum $A(i\omega)$ are a pair of Fourier transformations

$$A(i\omega) = \int_{-\infty}^{\infty} a(t) \exp(-i\omega t)\, dt \tag{5.6a}$$

$$a(t) = \frac{1}{2\pi} \int_{-\infty}^{\infty} A(i\omega) \exp(i\omega t)\, d\omega \tag{5.6b}$$

From the point of view of random theory, an accelerogram is but a sample from the assembly and it should be random, together with its Fourier spectrum (sections 3.1.2–3.1.3). If sufficient sample accelerograms could be obtained under similar conditions, the mathematical method described in section 3.1.3 might be used to find the expected Fourier spectrum $E[A(i\omega)]$ which could then be used as the statistical characteristics of this group of ground motions. Unfortunately, this approach is almost impossible and one has to compromise with some rough assumptions.

(b) Response spectrum

The following formulas are obtained from section 9.2:

$$S_d(\omega, \zeta) = S_v(\omega, \zeta)/\omega \tag{5.7a}$$

$$S_v(\omega, \zeta) = \left| \int_0^t a(\tau) \exp[-\zeta\omega(t-\tau)] \sin[\omega(t-\tau)] \, d\tau \right|_{max} \tag{5.7b}$$

$$S_a(\omega, \zeta) = \omega S_v(\omega, \zeta) \tag{5.7c}$$

The velocity spectrum thus computed will be named the pseudo-velocity response spectrum PSV(T, ζ) or PSV(ω, ζ), where $T = 2\pi/\omega$.

The response spectrum is defined through the response of an SDF system, just like a moving window filter. It is right to consider the response spectrum as a representation of the ground motion characteristics, not the structure.

(c) Power spectral density function

The power spectral density function, or power spectral density, is a quantity used in random process analysis to describe the characteristics of the process in the frequency domain. As stated above, power spectral density may be defined as the expectation of the square of the Fourier amplitude,

$$S(\omega) = \frac{E[A^2(\omega)]}{2\pi T_d} \tag{5.8}$$

where T_d is the duration of motion.

Spectra of typical accelerograms are given in Figures 5.8 to 5.10.

From the statistical point of view, a maximum $y_{max} = r\sigma_y$ of the response y, with variance σ_y^2, of an SDF system is of random nature and is a function of the probability of exceedance p. Vanmarcke (1976) has obtained a relation between the power spectrum and response spectrum as follows:

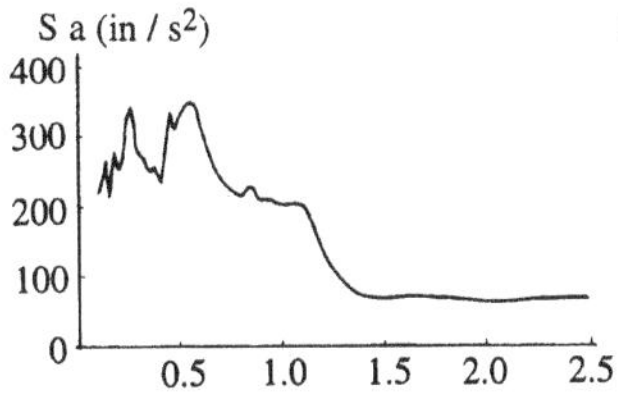

Figure 5.8 Spectrum of 1940 El Centro record.

$$S_a^2(\omega_0, \zeta) = r^2 \cdot \left[\omega_0 G(\omega_0) \cdot \left(\frac{\pi}{4\zeta} - 1 \right) + \int_0^{\omega_0} G(\omega) \, d\omega \right] \tag{5.9}$$

If the frequency is not very high, the last term may be neglected. Kaul (1978) obtained the same relation, i.e.

$$S_a^2(\omega_0, \zeta) = \frac{\pi\omega_0}{\zeta} S(\omega_0) ln \left[\frac{\omega_0 T_d}{-\pi \ln p} \right] \tag{5.10}$$

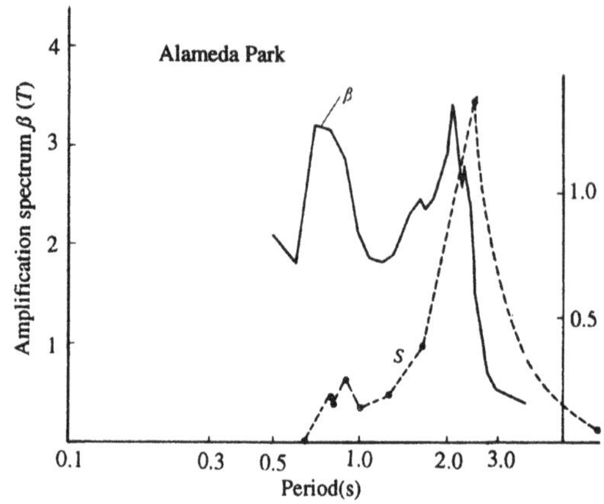

Figure 5.9 Spectra of 1962 Mexico City record.

by taking

$$r^2 = 2\ln\left(\frac{\omega_0 T_d}{-\pi \ln p}\right) \tag{5.11}$$

(d) Indirect expression of spectrum

As shown in Figure 5.11, the response spectrum can be simplified by a few straight line segments. The popularly used one consists of three segments which can be defined respectively by the peak values of ground acceleration a, velocity v and displacement d. It can be derived from Figure 5.11, that

$$\frac{2\pi v}{a} = T_1 \tag{5.12}$$

$$\frac{2\pi d}{v} = T_2 \tag{5.13}$$

$$\frac{ad}{v^2} = \frac{T_2}{T_1} \tag{5.14}$$

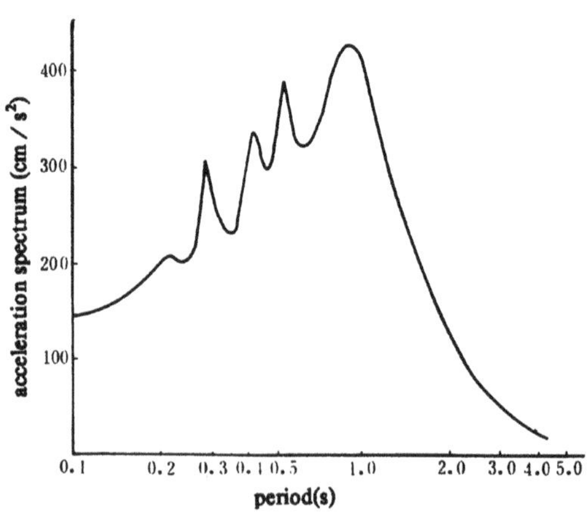

Figure 5.10 Spectrum of 1976 Tianjing City record.

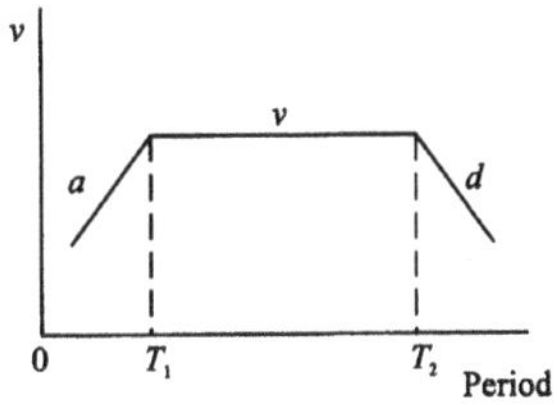

Figure 5.11 Simplified spectrum.

where T_1 and T_2 are corner periods. The ratio ad/v^2 is an indication of the chaotic nature of the motion, $ad/v^2=1$ means pure harmonic and the greater the ratio ad/v^2, the more chaotic the ground motion. For earthquake ground motion it is usually in the range of 3–7, but in some cases it may take an even greater value.

Equation (5.12) has been specified in some seismic design codes, such as the current Chinese Code GBJ 11–89 as follows:

$$T_1 = 0.2, 0.3, \ldots, 0.85\,\mathrm{s} \tag{5.15}$$

depending on site conditions and epicentral distance.

5.2.4 Duration

Most earthquake engineers believe that duration is one of the three important characteristics of engineering importance and their belief is based on their understanding from field data of structural damage, phenomena of low-cycle fatigue of structural members and materials, the effect of accumulative damage and other tests and analyses (McGuire and Barnhard, 1979; Nagahashi, 1980; Hu and Tao, 1984).

Table 5.3 compares two sets of data. The maximum accelerations of the ground motions in the second set are generally 0.05g–0.2g, lower than those in the first set, but the durations are much longer, usually longer than 30 s and the intensity (VI–VIII) is higher than those in the first set. The fact that lower acceleration with longer duration corresponds to higher intensity reveals the important effect of duration on structural safety.

Ambraseys (1969), among many others, pointed out that accumulated damage had considerable influence, and, in many cases, those structures that sustained the first strong earthquake collapsed in the following. For

Table 5.3 Comparison of ground motion and damage

Date	Earthquake	Station	M	Δ (km)	I	a/g	T_d (s)
14 November 1964	Ibaragi (offshore)	Tokaimura	5.1	–	V[a]	0.22	1
11 November 1967	Hokkaido (east)	Kushiro	–	–	VI[a]	0.23	1
19 November 1967	Ibaragi Bay	Tokai?	6.0	–	V[a]	0.48	1
4 September 1972	Central California	Melendy Ranch	4.7	10	VI	0.69	2
4 November 1973	Leukas	Leukas	6	25	VII	0.54	3
4 August 1966	Matsushiro	Matsushiro	4.7	2	V	0.39	3
5 April 1966	Matsushiro	Matsushiro	5.1	4	VI	0.42	4
18 March 1957	Pt. Hueneme	Hueneme	4.7	7	VI	0.17	1–2
28 June 1966	Parkfield	No. 2	5.6	0.1	VII	0.50	2–8
9 February 1971	San Fernando	Pacoima	6.6	10	IX	1.25	7
9 December 1965	Mexico	Acapulco	6.8	60	VI	0.27	8
13 April 1949	Olympia	Seattle	7.1	40	VIII	0.07	20–45
16 May 1968	Takachioki	Hachinohe	7.9	179	VII–VIII[a]	0.23	20–30
		Mororan	7.9	310	VI[a]	0.21	20–30
12 June 1964	Niigata	Niigata	7.7	38	VII–VIII	0.16	30
18 May 1940	Imperial Valley	El Centro	7.0	20	VIII	0.32	30
11 May 1962	Mexico	Alameda Park	7.0	260	VI–VII	0.05	60
19 May 1962	Mexico	Alameda Park	7.1	260	VI–VII	0.05	60

[a] Converted from Japanese intensity scale.

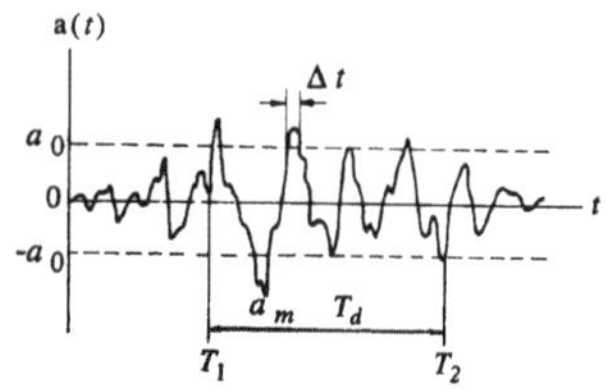

Figure 5.12 Definition of duration.

example, a highway bridge of 35 simple spans of reinforced concrete T-beams over the Luan River collapsed during one aftershock although it had sustained the main Tangshan shock, and the intensity was estimated there as VIII for both earthquakes.

Although the effect of duration on structural safety is recognized, a definition has not yet been agreed upon. The definitions available may be classified into two categories. Relative duration refers to the relative amplitude of motion, and absolute duration to the absolute.

Consider the time history of a motion $a(t)$ as shown in Figure 5.12, with maximum peak a_{max}, or simply a_m. The most popular definition is

$$T_d = T_2 - T_1 \tag{5.16}$$

or the time interval between the first and the last crossings of the time history $a(t)$ and the two horizontal lines of constant acceleration $\pm a_0$. If a_0 is an absolute acceleration, the definition is called a_0 duration. In the USA, a_0 is usually taken as $0.05g$ or $0.1g$, because smaller acceleration will have little effect on structural safety. It is called the fractional or bracket duration, if a_0 is specified as a fraction k of the maximum a_m, i.e. $a_0 = ka_m$. The value k is usually taken as one-third, or in the range one-fifth to a half.

Husid (1969) suggested that

$$I(t) = \frac{\int_0^t a^2(t)\,dt}{\int_0^{T_d} a^2(t)\,dt} \tag{5.17}$$

might be used to indicate an increase in the ground motion energy with time. In the equation, T_d is the total duration and $I(t)$ varies from 0 to 1. Trifunac and Brady (1975) suggested the following definition of strong motion duration through Eq. (5.17)

$$I(T_1) = 0.05 \text{ or } 0.15 \tag{5.18a}$$

$$I(T_2) = 0.95 \text{ or } 0.85 \tag{5.18b}$$

The first set of T_1 and T_2 defines a 90% duration and the second a 70% duration. There are two reasons for the wide use of the definition of relative duration.

1. Duration is only one of the several parameters of motion and is used usually together with other parameters such as peak acceleration a_m or spectrum $S_a(T)$. In this case, amplitude is already included in a_m and there is no reason to include it again in duration.
2. If absolute duration is used, for example $a_0 = 0.1$ g, the ground motion at a long distance may be smaller than $0.1g$ and the duration of $0.1g$ will be zero. Zero duration might imply no vibration and introduce confusion, especially when it is known that ground motion with a peak

acceleration less than $0.1g$ but of long duration has caused considerable damage to buildings, such as those shown in Table 5.3. It is puzzling to consider the damage made by ground motion of zero duration.

A subjective decision is introduced in the above-mentioned definitions of duration. The values assigned to a_0, a_0/a_m, $I(T_1)$ and $I(T_2)$ are subjective definitions of the fuzzy term 'strong'. One way to avoid this subjective decision is to use a function $g(t)$ showing the variation of acceleration strength with time, such as

$$g^2(t)=\frac{1}{\Delta t}\int_{t-\frac{\Delta t}{2}}^{t+\frac{\Delta t}{2}} a^2(t)\,\mathrm{d}t \tag{5.19}$$

On the basis of the definitions in equation (3.10b), $g^2(t)$ is just the variance of $a(t)$, as shown in Figure 5.13.

The definitions of duration given above, relative or absolute, can all be made in terms of the amplitude standard deviation function $g(t)$.

5.2.6 Rotational components of ground motion

Rosenblueth was perhaps the first to point out that there are all together six independent components of a ground motion, namely three translational displacements $d_i(t)$ along the axes i ($i=x$, y and z) and three rotations about these axes. Each component has also velocity and acceleration, or angular velocity and angular acceleration. For each component, there are three independent characteristics, i.e. amplitude, spectrum and duration and other properties.

While there are almost no data on direct observation of rotational ground motion (Shibata *et al.*, 1978), rotational motion may be estimated from the translational motion $x(t)$. Suppose that y is the direction of wave propagation, which is perpendicular to the recorded direction x of motion. From wave propagation theory,

$$x(t)=x\left(t+\frac{y}{v_c}\right) \tag{5.20}$$

where v_c is the wave velocity and y is the distance along the direction of propagation. From equation 5.20 it can be obtained that

$$\frac{\mathrm{d}x}{\mathrm{d}y}=\frac{(\mathrm{d}x/\mathrm{d}t)}{v_c}=\frac{v}{v_c}$$

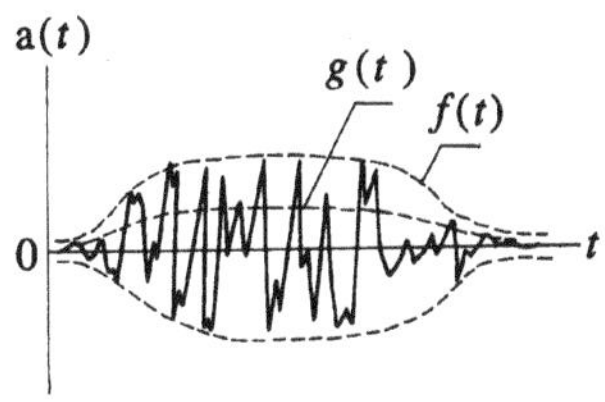

Figure 5.13 Standard deviation and envelope function.

where $v=\mathrm{d}x/\mathrm{d}t$ is the particle velocity and $\theta=\mathrm{d}x/\mathrm{d}y$ is the angle of rotation. The foregoing relation is finally

$$\theta=\frac{v}{v_c} \tag{5.21}$$

If there are strong motion records obtained at very closely spaced points, finite differences may be used in place of derivatives. For example, the rotation is

$$\theta = \frac{\Delta x}{\Delta y} \tag{5.22}$$

if the wave propagates along the y-axis. If not, the following general equation may be used instead of equation (5.22)

$$\theta = \frac{1}{2}\left(\frac{\Delta x}{\Delta y} - \frac{\Delta y}{\Delta x}\right) \tag{5.23}$$

There are two components of rotational motion on ground surface considered in earthquake engineering, i.e. the rocking and the torsional. If z is the vertical displacement, and x and y are horizontal, the rocking motion about the x-axis will be

$$\theta x = \frac{\Delta z}{\Delta y} \tag{5.24}$$

and the torsional motion about the z-axis will be

$$\theta z = \frac{1}{2}\left(\frac{\Delta y}{\Delta x} - \frac{\Delta x}{\Delta y}\right) \tag{5.25}$$

Information on rocking and torsional motion has been studied on the basis of accelerograms of translational motion obtained from dense arrays, such as the SMART-1 array in Taiwan, China (Loh *et al.*, 1982; Bolt *et al.*, 1982; Wang and Hu, C1991).

Since there are too few rotational motion data available to be of practical use, ground motion mentioned later in the book refers only to translational.

5.3 Factors influencing ground motion

There are three categories of factors which have important influence on ground motion, namely earthquake source, propagating media and local site conditions, as for intensity (sections 4.3.3 and 4.4). Each category consists of many factors, but only those having important influence and which are possible to estimate will be considered in detail in this section.

5.3.1 Earthquake source

(a) Linear dimension of source

For large earthquakes, the shape of the source or energy release is elongated; it is therefore reasonable to expect an elongated shape of distribution of the equal peak amplitudes of ground motion in the near field; however, the data from accelerograms obtained mostly from small and moderate earthquakes are too few to prove this, and ground motion attenuation laws are given more often for a point source assumption than for an elongated source.

(b) Source spectrum

Duda and Nuttli (1974) pointed out that magnitude of earthquake is actually a spectrum and the widely accepted source models, such as Brune's (Hanks, 1979b), also considered the source spectrum. Huo (C1989) obtained from engineering data that γ in the ω^{γ} model varies from 2 for small earthquakes to 2.5 for very large ones.

(c) Saturation of high-frequency ground motion

There are three reasons for the saturation of high-frequency motion. Firstly, the magnitude of the earthquake saturates when M_s or M_L is used, as shown in Figure 5.14.

Secondly, the acceleration will saturate when M increases because the released energy is distributed over an extended source. Hanks (1979a, b; Hanks and Kanamori, 1979) pointed out that high-frequency component motion near a causative fault depends on the local stress drop closest to the site, but not the magnitude of the earthquake. He obtained the RMS acceleration

$$a_{RMS} = C \cdot \frac{\Delta\sigma}{R} \cdot \left(\frac{f_{max}}{f_0}\right)^{1/2} \tag{5.26}$$

where R is the focal distance, $f_0 = 1/T_d$ and f_{max} are respectively the minimum and maximum frequencies considered; C is a constant and T_d the duration of fault breakage.

Thirdly, the upper limit value of ground motion may be derived from the limit strength of the fault. Housner (1965, also in Wiegel, 1970) and Ambraseys (1969) have tried separately. Ambraseys considered the fault under simple shear with shear modulus of the fault material G and average residual shearing strength $rP/2$, and obtained the maximum velocity

$$V = Pv_s \cdot \frac{1-r}{2G} \tag{5.27}$$

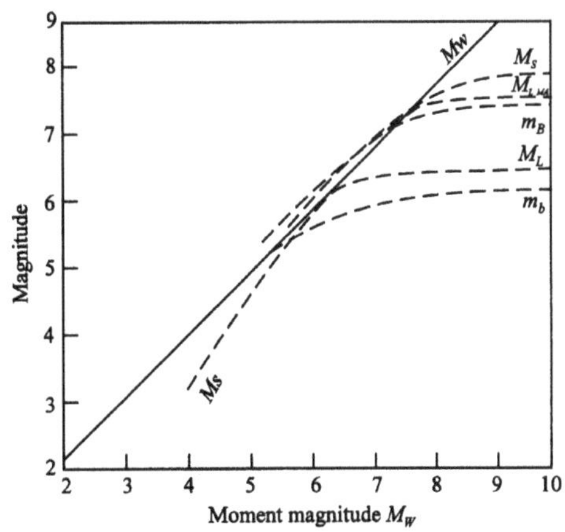

Figure 5.14 Magnitude situation.

If $r = 0$, this formula gives the Housner's result; he obtained a rough estimate of the velocity on the fault

$$V = (3\times10^{-3})\cdot(3\times10^{5})\cdot\frac{0.25}{2}\approx 110\,\text{cm/s}$$

It is not wise to put an upper limit on peak acceleration. When the frequency is very high, peak acceleration may be very high, several times gravity. It has been estimated as $0.5g$ in the 1960s and $2g$ in the 1980s, but both have been exceeded by actual records.

(d) Wider spectrum for larger earthquake

Either from theoretical considerations of seismologists or from strong motion records obtained by engineers, the spectra of ground motion are different for earthquakes of different magnitudes, with wider spectra for larger earthquakes. Under certain assumptions of the fault rupture model, seismologists obtained theoretical displacement spectra for the far field, such as the ω^{γ}-model,

$$\Omega(\omega) = \frac{\Omega(0)}{[1+(\omega/\omega_0)^2]^{\gamma}} \tag{5.28}$$

where ω_0 is the corner frequency. No matter what the model used, the spectrum at a given distance increases with magnitude much faster in the low-frequency band than in the high-frequency band, which changes the shape of the spectrum.

Engineers have a similar understanding, firstly from damage data, and confirmed later by the accelerograms they obtain. For example, in the 1957 and 1985 Mexico earthquakes, long-period structures suffered very heavy damage in cities far away from the epicenter compared to neighboring rigid structures. This selective damage is usually considered as the consequence of an earthquake of large magnitude at a long distance and at a soft site.

On the basis of accelerograms of the main shock and aftershocks, a set of spectra for different magnitudes and distance combinations was obtained separately for the Tangshan and Mexico regions (Figure 5.15(a), (b)) (Anderson and Quaas, 1988). It is clear that the greater the magnitude, the longer is the predominant period of the spectrum.

5.3.2 Propagation path

Distance from the epicenter or source is the only factor of propagation path and media used in seismology and earthquake engineering. The crust is rather complicated because of its stratified structure and faults; the path of waves after many reflections and refractions is certainly not a straight line. Actually, different waves travel along different paths before arriving at a site. The straight line from site to epicenter or source is but a rough parameter used in place of the complicated path of propagation.

The definition of distance is very important for ground motion in the near field. Epicentral distance Δ, focal distance R and fault distance D have all been used. The last one is perhaps the best for engineers to use in areas very

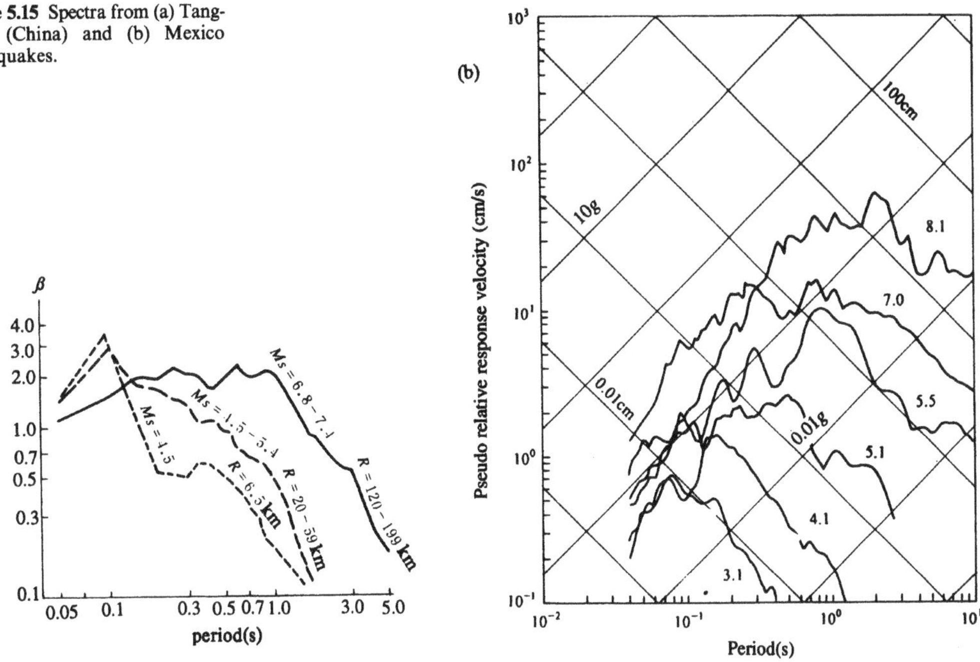

Figure 5.15 Spectra from (a) Tangshan (China) and (b) Mexico earthquakes.

close to the causative fault of a large earthquake. With fault distance, an attenuation different from a circle may be easier to formulate.

The azimuth effect is emphasized by seismologists, and a theoretical formula is given for displacement, but engineers did not widely accept it because there are not enough data to give such a modification of the effect of azimuth on acceleration and the direction of fault rupture is not easy to predict.

5.3.3 Local site conditions

Local site conditions may include soil and topography. Damage or intensity is much higher on average at soft sites than at hard sites (section 4.5). This fact has been repeatedly confirmed by many strong earthquakes.

Contrary to intensity, the effect of local soil conditions on acceleration was not universally acknowledged. In the 1950s and early 1960s, many believed that the peak acceleration at a soft soil site was greater than at a hard site because damage was greater, and acceleration was considered equivalent to intensity; some believed it smaller because high-frequency motion cannot pass through soft soil, and some in the USA thought that it was greater at a soft site for weak motion but smaller for strong motion because of the nonlinear properties and the low strength of the soft soil. Current understanding from available records of ground motion is that the peak acceleration for hard sites is greater at close distances but smaller at far distances, as shown in Figure 5.20 (section 5.5.2). As for low-frequency motion such as peak displacement, it is universally accepted that it is greater for soft sites.

The effect of soil conditions on spectrum was accepted by earthquake engineers in the 1960s and adopted in their design codes. The first code to accept this effect and to specify a wide spectrum for soft sites was the Mexican code because of the special soil conditions and damage experienced in Mexico City.

On the basis of records in the USA available at that time, the Chinese obtained (Chen, C1965) different spectra for different soil conditions as shown in Figure 14.5 which were adopted in the 1964 seismic design code (draft) (Liu, C1965) and used in the later official codes with minor modification (section 14.2.3). It is now popularly adopted in the seismic design codes all over the world.

5.3.4 Combined effect of magnitude, distance and site conditions

Effects of earthquake magnitude, distance and site conditions appear in real records usually in a combined form and care is needed to separate them because most records of accelerograms for soft sites are from large earthquakes at long distances. The results are sometimes given in a combined form, such as Figure 5.15.

5.3.5 Effect of earthquake magnitude on duration

As far as the total duration is concerned, the larger the magnitude, the longer the total duration. But for strong motion duration (section 5.2.4), this is not so at sites near the fault for very strong earthquakes (Huo, Hu and Feng, C1991), because those parts far away from the site close to the fault will add almost nothing to the maximum peak acceleration from the nearby part of the fault due to attenuation.

An earthquake starts at the first break on the causative fault and ends when the breakage stops. The fault rupture velocity is v_f, and the time of fault breakage will be in the range $(0.5\,L/v_f)$–(L/v_f), where L is the length of fault. The lower boundary is for bilateral rupture and the upper boundary for one-dimensional rupture.

Housner and Jennings (1964) suggested the parameters given in Table 5.4 for the strength envelope of strong ground motion in equations 3.41.

In addition to magnitude and distance, site soil conditions also affect duration. The softer the site, the longer the duration because a soft layer retains the energy.

Table 5.4 Effect of magnitude M on duration parameters

	Parameter (s)		
M	t_1	t_2	t_3
5	2	2.5	3.5
6	2	4	10
7	4	15	30
8	4	30	80
$f(t)$	1	1	0.2

5.4 Earthquake intensity and ground motion

The relation between earthquake intensity and ground motion is a very old problem. It is still of interest because there are many regions in the world where few acceleration data have been recorded.

5.4.1 Relation of intensity and a single parameter of motion

In addition to intensity scales, Gutenberg and Richter (1942, 1956) were perhaps the first to correlate intensity with recorded ground motion parameters. With many weak motions, their result

$$\log a(\mathrm{cm/s^2}) = \frac{I}{3} - \frac{1}{2} \tag{5.29}$$

gives naturally much lower acceleration. Neumann (1952, 1954, 1958) used only data from the epicentral area and obtained a relation between intensity and acceleration of approximately 0.3 Hz, as shown in Figure 5.16, which gives a-values about twice those of equation 5.29. Neumann's result was the best at that time for the following two reasons. Firstly, only data for the epicentral area were used which reduced the range of variation of the other ground motion parameters. Secondly, the relation of peak acceleration to frequency was clearly mentioned. He gave in the same figure the relation of acceleration and intensity for other distances. Hershberger (1956) obtained the following relation:

$$\log a = \frac{3}{7} I - \frac{9}{10} \tag{5.30}$$

but at the same time he concluded: 'About the only conclusion that I can draw from this investigation so far is that this is not the way to arrive at

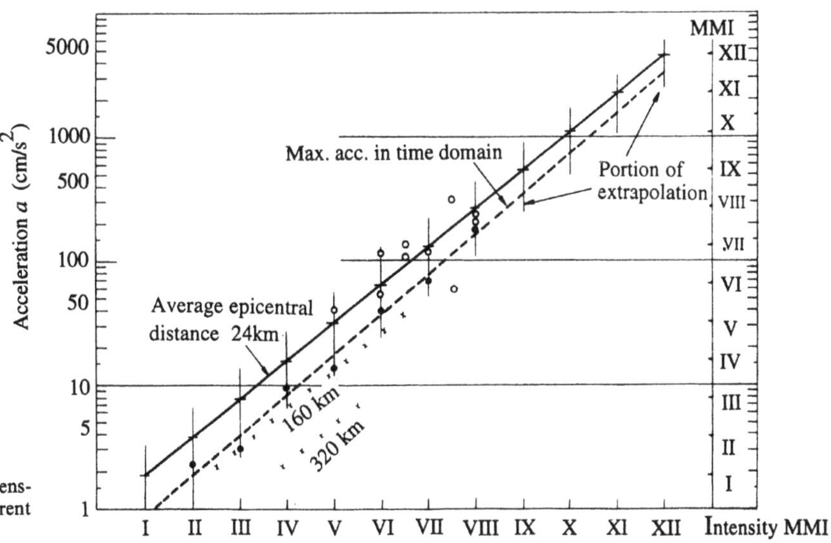

Figure 5.16 Relation of intensity and acceleration at different distances.

Table 5.5 Correlation of intensity to ground motion

Intensity	a (cm/s^2)	v (cm/s)	d(cm)
V	12–25	1–2	0.5–1
VI	25–50	2.1–4	1.1–2
VII	50–100	4.1–8	2.1–4
VIII	100–200	8.1–16	4.1–8
IX	200–400	16.1–32	8.1–16
X	400–800	32.1–64	16.1–32
Related period (s)	0.1–0.5	0.5–2.0	0.25

any conclusion. The scatter in these results is too great to permit the assumption of any definite quantitative correlation of intensity and peak acceleration alone.' Medvedev (1962) analyzed the strong motion records for the USA and obtained a rough correlation of intensity to peak acceleration, velocity and displacement of different periods, as shown in Table 5.5.

Several results commonly referred to are plotted in Figure 5.17 together with a set of data from the US records. It is seen that scattering is typically very large so that acceleration corresponding to one intensity may vary by several tens or hundreds of times, but the difference of one grade of intensity changes the acceleration only by a factor of two. Many attempts have been made to correlate intensity with other factors, such as peak velocity, root-mean-square acceleration, spectral intensity and duration, but the results are not improved. Figure 5.18 shows one of them. Ambraseys (1973) gave a good review of this subject, which is convincing and represents the main current trend. Several other experts, such as Housner and Newmark have also clearly pointed out that it is not possible to find

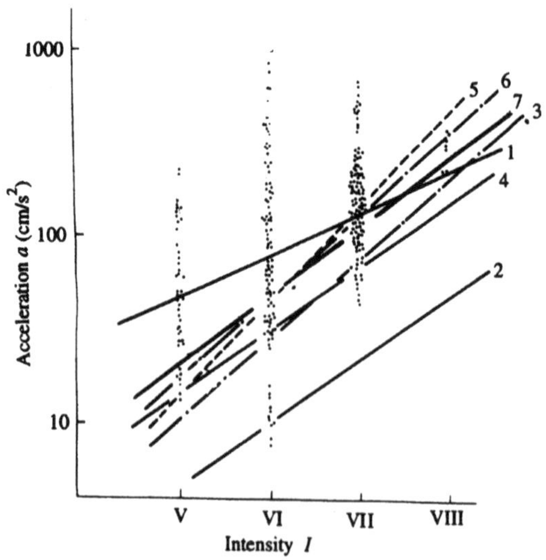

Figure 5.17 Relation of intensity and acceleration: 1 = data from this Figure 2 = Cacani, 1904 (Gutenberg and Richter, 1942, 1956); 3 = Ishimoto (1931a–c); 4 = Gutenberg and Richter (1956); 5 = Davison (1921); 6 = Hershberger (1956); 7 = China Intensity Scale 1980 (Liu, 1980).

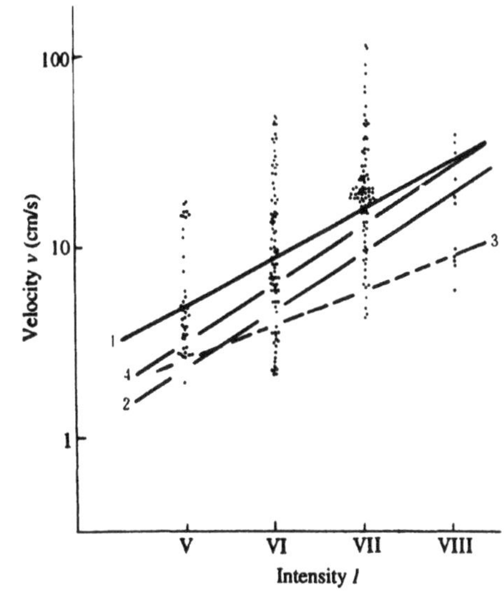

Figure 5.18 Relation of intensity and velocity: 1 = data from Figure 2 = Neumann (1960); 3 = Kanai and Suzuki (1968); 4 = China Intensity Scale (Liu, 1980).

a conversion relation between intensity and any single parameter of ground motion; intensity is too simple and rough to be used in engineering design.

5.4.2 Relation of intensity with several parameters of motion

Many studies (Liu *et al.*, 1982) have tried to correlate intensity directly with several parameters of ground motion, but with no better results.

A more satisfactory result was obtained by McGuire (1977). He obtained some relations between intensity, a ground motion parameter and a third or fourth parameter of either epicentral distance or magnitude from the US data by regression, as given in Table 5.6. The following conclusions may be emphasized.

1. If only one parameter of ground motion is used, either acceleration, velocity or displacement, the scatter is nearly the same.

Table 5.6 Relations between intensity and ground motion on soft sites ($\ln Y = c_1 + c_2 M + c_3 \ln(R + R_0) + c_4 I$)

Parameter Y	c_1	c_2	c_3	c_4	$\sigma_{\ln Y}$
a(cm/s^2)	2.71	–	–	0.601	0.781
	2.01	–	– 0.313	0.506	0.723
	1.81	0.904	– 0.901	–	0.696
v(cm/s)	– 1.51	–	–	0.543	0.770
	– 1.11	–	– 0.720	0.532	0.771
	– 1.58	0.997	– 0.710	–	0.715
d(cm)	– 1.47	–	–	0.415	0.790
	– 2.35	–	0.157	0.463	0.780
	– 2.67	0.863	– 0.398	–	0.740

2. If another parameter of magnitude M or distance R is added, scatter will be reduced by 6–12%.

Figure 5.19 shows a set of data from the 1971 San Fernando earthquake, which gives one a visual idea. If the distance is not considered, acceleration of the same intensity varies over a large range, for example, 0.021g–0.36g for intensity VI; if distance is considered, acceleration varies between 0.021–0.078 for $R=73$ km, to 0.11–0.33 for $R=27$. The ratio of the upper to lower boundaries of the range is 17 for the former and reduces to 3 for the latter.

When McGuire performed his study, he considered adding two parameters M and R to the relation of intensity and acceleration Y (or velocity or displacement), i.e.

$$I=c_1+c_2\ln Y+c_3M+c_4\ln(R+R_0) \tag{5.31}$$

and ground motions are considered as functions of M and R, i.e.

$$\ln Y=c_5+c_6M+c_7\ln(R+R_0) \tag{5.32}$$

where Y may be acceleration, velocity, duration or other parameter of ground motion, and R_0 is a constant. Therefore, the last two terms in equation 5.31 may be considered to represent velocity or parameter other than acceleration through equation 5.32.

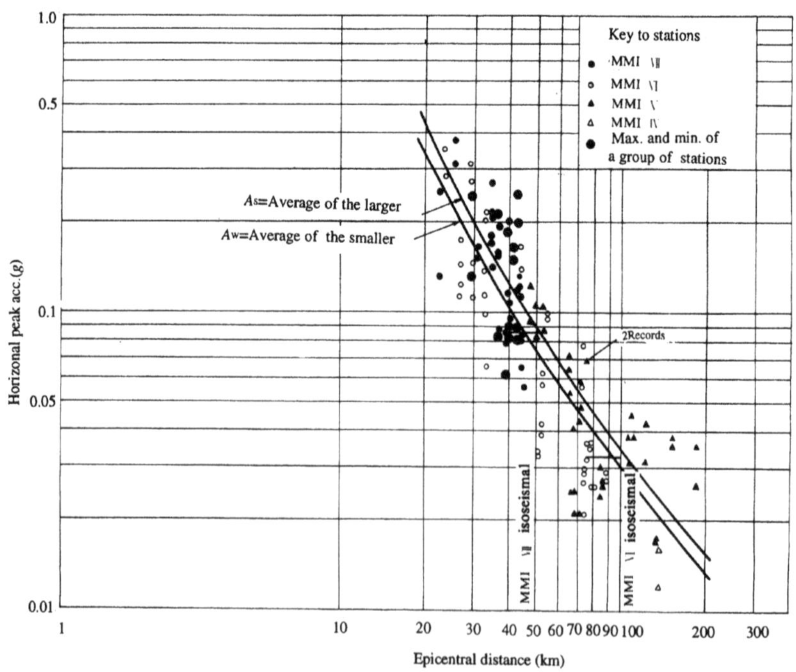

Figure 5.19 Attenuation of horizontal peak acceleration from 1971 San Fernando earthquake.

The first author (Hu and Zhang, 1983; Hu *et al.*, 1989) suggested adding the M term to the relation between acceleration and intensity. On the basis of the attenuation laws of acceleration and intensity of a region, such as equations 5.31 with $c_2=0$ and 5.32, if the R term is eliminated, the result will be a relation of I, a and M. The attenuation laws of acceleration and intensity for the west coast of the USA are used as an example here. Attenuation of intensity of the San Andreas region is taken from Chandra's result (1979)

$$I = I_0 + 2.065 - 0.00594\mathrm{R} - 2.065\lg(\Delta + 10) \tag{5.33}$$

and the relation between epicentral intensity I_0 and magnitude M

$$I_0 = 1.5(M - 1)$$

is used. The acceleration attenuation of Battis (1981)

$$\ln a = 5.832 + 1.212m_b - 2.076\ln(\Delta + 25) \tag{5.34}$$

is taken together with the relation

$$m_b = 1.276 + 0.749M$$

If the distance Δ term is canceled from equations 5.33 and 5.34, a relation among intensity I, acceleration a/g and magnitude M is obtained and the result is given in Table 5.7.

It is clear that for one intensity, the peak acceleration may vary by nearly ten times when the magnitude varies.

5.5 Ground motion evaluation

Ground motion evaluation is one of the most important links in earthquake engineering; it is based on the results of seismology and provides engineers with the earthquake load or action to design their structures, in the same way as wind pressure in wind-resistant design and flood level in dam design against flooding.

Table 5.7 Relation of acceleration to intensity and magnitude

	Acceleration at intensity				
M	VI	VII	VIII	IX	X
5.5	0.135	–	–	–	–
6.0	0.082	0.284	–	–	–
6.5	0.048	0.179	0.583	–	–
7.0	0.030	0.104	0.389	1.154	–
7.5	0.022	0.064	0.228	0.833	–
8.0	0.018	0.044	0.136	0.502	1.747
8.5	0.017	0.034	0.091	0.292	1.100

5.5.1 Three approaches of ground motion evaluation

There are three approaches to evaluating the ground motion of a site or a region. The first and the oldest one is to evaluate the earthquake intensity first and then to evaluate the ground motion through a conversion from intensity. The second one is to evaluate directly the ground motion from an attenuation relationship between the ground motion and the source (usually magnitude) and propagating medium (usually distance) obtained empirically from acceleration data. The third one is to compute directly the ground motion from energy released from earthquake source to the site by wave propagation theory.

The first approach has been used for a long time. The advantage is to use fully the intensity data, usually available in seismic regions, but the disadvantage is that ground motion evaluated this way is too inexact to satisfy the modern requirements of earthquake engineering.

The second approach has been widely used since the 1960s when acceleration data had been accumulated in some countries such as USA and Japan. The key to it is to have acceleration data for the region of interest. The United States has perhaps most data of this kind, coming directly from acceleration observation, and Canada can use them because they are close together and the tectonic and geological background may be similar. Japan also has many such data. The advantage of this approach is its simplicity in physical idea and reliability, and the disadvantage is the lack of data for many regions in the world. The deduced ground motion from movement and overturning of objects has been used in some cases, but it is only equivalent in nature and can give only a single parameter, such as effective acceleration, not the spectrum, and its accuracy is unknown. A method of evaluation in terms of attenuation of both intensity and ground motion of a foreign region with many acceleration data and the intensity attenuation data of the local region has been suggested (Hu and Zhang, 1983; Battis, 1981), which can give spectra under certain assumptions.

The third approach has developed very fast in recent years (Bolt, 1987) and may find its application in practice soon. In the following, attenuation relations will first be introduced and these approaches then explained.

5.5.2 Ground motion attenuation relations

(a) Regressional formulas and data available

Regressional formulas used may be written in a general form as follows (Campell, 1981, 1985):

$$\ln y = c_1 + c_2 M + c_3 \ln(R + R_0) + c_4 M^2 + c_5 R + c_6 j_s + c_7 j_v + c_8 p \qquad (5.35)$$

where ln is the natural logarithm, M the magnitude of the earthquake, R the distance of the site from the center of the earthquake, p the probability of exceedance of y; and $c_i (i = 1, 2 \ldots, 8)$ and R_0 are regressional constants, subscript 's' refers to site and 'v' to components of motion. For example, $j_s = 0$, 1 or 2 for rock, hard soil or soft site respectively; and $j_v = 0$ or 1 for horizontal or vertical component of ground motion. One way of representing the randomness is by adding a term, or by giving a standard error to the estimates $\ln y$; another way is by giving separate standard error to each coefficient c_i.

Some terms in this formula have physical meaning. The linear M term means the definition of earthquake magnitude, because M is defined as the logarithm of the ground motion amplitude y; the M^2 term represents the saturation of magnitude if the saturation of ground motion with increasing magnitude, especially the high frequency components, is considered. The logarithmic R term represents geometrical attenuation of motion, while the linear R term represents the material damping which becomes important at longer distances, say $R \geqslant 100$ km. Many engineers neglect the linear R term just for simplicity. The R_0 term represents the finite dimension of the earthquake source and may be constant or a function of magnitude M, varying from less than 1 km for $M \leqslant 4$ to 50 km or more for $M \geqslant 8$. At closer distances, say $R \leqslant R_0$, which are important to engineers, the definition of R is important. The following distances have been widely used:

- epicentral distance Δ from the instrumental or macroseismic center on the surface;
- fault distance D;
- focal distance R from the instrumental focus or released-energy center.

The form assumed in equation 5.35 for the site or components term includes an important assumption. Because the constants c_6 and c_7 are not functions of R or M, it assumes then that the attenuation curves or surfaces in Y, R and/or M coordinates are forced to be parallel. For example, several investigators have suggested attenuations for different site conditions as shown in Figure 5.20, which cannot be represented by equation 5.35. To obtain regressional curves shown in the figure, data for different site conditions have to be regressed separately.

There are also formulas different from equation 5.35, such as the following one suggested by Bolt and Abrahamson (1982):

$$\ln y = c_1 + c_3 \ln[(D + R_0)^2 + 1] + c_5(D + R_0) \tag{5.36}$$

where coefficients $c_i (i = 1, 3, 5)$ may be functions of magnitude M. The idea of equation 5.36 is to try to regress the data so that the regressed values will be defined mostly by the data near them; for example the near field regressed curve by the near field data. The disadvantage of the ordinarily used approach is that the near field (say, $R \leqslant R_0/2$) result is controlled by the middle field data because there are few data for the near field.

Figure 5.21 shows the general distribution of the available data in the United States. It is seen here that most data occur for the middle field (say, D = 10–100 km) of moderately strong earthquakes (say, M = 5–6.7), and that we are strongly in need of data for the near field of strong earthquakes. Bolt and Abrahamson (1982) (Figure 5.22) have shown that the regressional result in the near field will not change much if data in the range $D \leqslant 8$ km were deleted, i.e. the near field regression is not decided by the near field data.

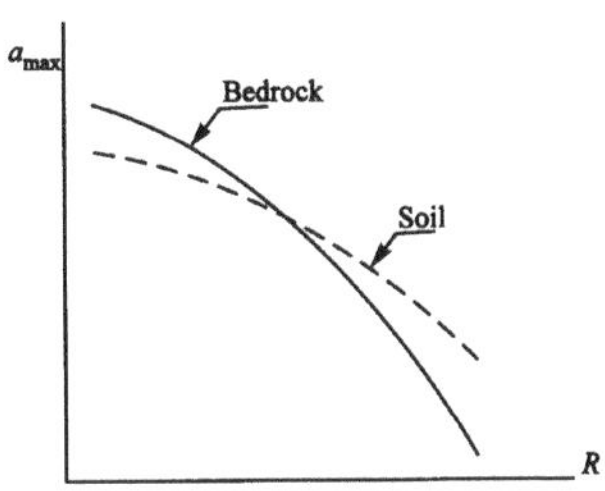

Figure 5.20 Effect of site conditions on acceleration attenuation.

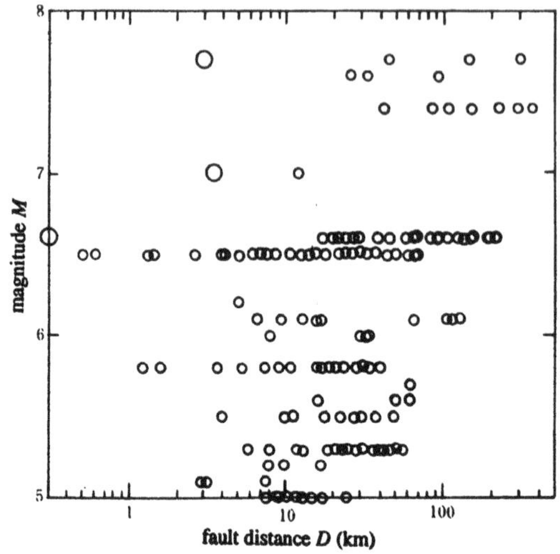

Figure 5.21 Distribution of available acceleration data.

In the near field, the selection of the distance R should be made with care. If the $\ln R$ term is used, R should not be epicentral distance, because $\ln 0$ is singular. With epicentral distance or fault distance, $\ln(\Delta^2+H^2)$ or $\ln(\Delta+\text{const.})$ have been used if near field values are needed, where H is the focal depth.

In regression, the observed values of Y are considered to be random or of random error and the regressional principle is to make the sum of the error, or the difference of the observed and the regressed values, a minimum, such as in the least squares principle. Let y_m be the regressed value of y; the least squares principle is

$$J=\sum(y-y_m)^2=\min. \tag{5.37a}$$

$$J=\sum(\ln y-\ln y_m)^2=\sum\ln(y/y_m)=\min. \tag{5.37b}$$

If other parameters M and R are also in error, then

$$J=\sum d^2=\min. \tag{5.38}$$

where d is the shortest distance from each observation data point to the regressional surface $y=f(x_1, x_2, \ldots, x_j)$, and y and $x_i (i=1, 2, \ldots, j)$ are variates of random error (Hu and Huo, 1988).

(b) Comparison of some attenuation relations

Most currently available data from strong motion acceleration records concentrate on the USA and Japan, from which empirical attenuation relations are obtained. When plotted as in Figure 5.22, the data always show scattering so large that data observed for the same distance, similar site conditions and same magnitude may differ from each other by one or

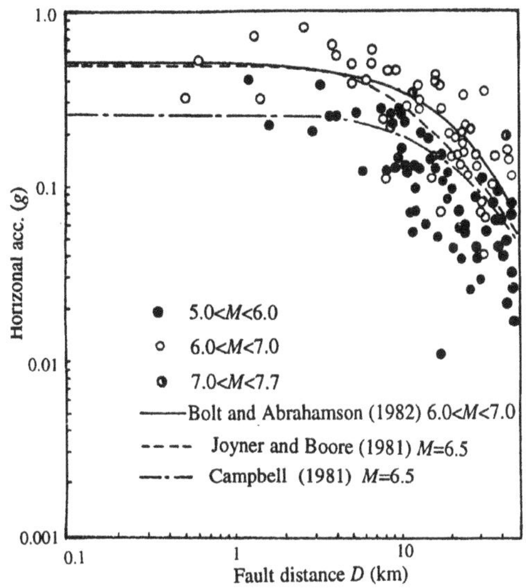

Figure 5.22 Comparison of attenuation for the near field.

two orders of magnitude. Attenuation relations derived from these data by different investigators are also quite different, as shown in Figure 5.23 and others later in this section. The reasons for the large scattering include the following.

1. There are many other factors in addition to M and R, such as source mechanism of each earthquake, geometrical and mechanical details of fault rupture, and variation of the wave propagation path in nonuniform media.
2. Describing site conditions in terms of site types like rock and soil is too simple as it does not represent the real variation of complicated layering and of mechanical properties of rock or soil or the difference in dynamic properties of different sites of the same site type.
3. Data selection may differ; for example, sometimes only the larger of the two peaks of the two horizontal components is used, sometimes both peaks, sometimes the vector of these two peaks and sometimes the peak of the vector accelerogram; sometimes all data are kept available and sometimes the so-called 'outliers' will be deleted.
4. The method of regression may differ. For the regression formula, sometimes equation 5.35 is used and sometimes some terms are neglected; sometimes the instrumental focal distance is used and sometimes the fault distance; the set of data used will differ. The regression principle used may also be different: mostly equations 5.37 are used, sometimes equation 5.38 and sometimes another.
5. Data may be concentrated in a limited range of magnitude (M = 5–6.7) and distance (D = 10–100 km), which, together with the large scattering, causes extrapolation to be sensitive to small differences in other details. The following comparisons of some commonly used attenuation relations will help the understanding of this.

(i) Horizontal peak acceleration (cm/s² or as indicated)

1. Aptikaev and Kopnichev (1980). Worldwide records; no site types:

$$\log a_m = 1.70 + 0.28M - 0.80\log R \quad (\text{for } a_m \geqslant 160\,\text{cm/s}^2) \qquad (5.39a)$$

$$= 0.80 + 0.80M - 2.3\log R \quad (\text{for } a_m < 160\,\text{cm/s}^2) \qquad (5.39b)$$

If type of faulting is considered, a factor given in Table 5.8 may be applied. This may be the only paper considering the effect of faulting type, but the sufficiency of data is not clear.

2. Joyner and Boore (1981) $(M = 5\text{–}7.7)$:

$$\log(a/g) = -1.02 + 0.249M - \log R - 0.00255R + 0.26p \qquad (5.40)$$

3. Schnabel and Seed (1973) for rock sites in graphs:
4. Hu *et al.* (C1982) for rock sites:

$$\log a_{max} = 1.712 + 0.657M - 2.183\log(R+30), \quad \sigma_{ln} = 0.47 \qquad (5.41)$$

5. McGuire (1978). $j_s = 0$ for rock sites and 1 for soil:

$$\ln a_{max} = 3.4 + 0.89M - 1.17\ln R - 0.2j_s, \quad \sigma_{ln} = 0.62 \qquad (5.42)$$

6. Battis (1981):

$$\ln a_{max} = 5.832 + 1.212m_b - 2.076\ln(\Delta+25), \quad \sigma_{ln} = 0.7 \qquad (5.43)$$

$$m_b = 1.276 + 0.749M_L$$

7. Donovan and Bornstein (1978):

$$a_{max} = c_1\exp(c_2M)\cdot(R+25)^{-c_3},$$

$$\sigma = 0.30g\,(a_{max} > 0.3g) - 0.48g\,(a_{max} = 0.05g) \qquad (5.44a)$$

$$c_1 = 2.154\times10^6R^{-2.1} \qquad (5.44b)$$

$$c_2 = 0.046 + 0.445\log R \qquad (5.44c)$$

$$c_3 = 2.515 - 0.486\log R \qquad (5.44d)$$

8. Japan Road Association (J1980):

(a) rock $\quad a_{max} = 46.0\times10^{0.208M}(\Delta+10)^{-0.686} \qquad (5.45a)$

Table 5.8 Effect of faulting type on peak acceleration

Faulting type	$\log(a_0/a_m)$	$\log(T_{d0}/T_d)$
Reverse	0.35 ± 0.13	0.19 ± 0.15
Reverse slip	0.11 ± 0.17	0.12 ± 0.14
Strike	0.22 ± 0.08	0.30 ± 0.08
Strike dip	0.06 ± 0.13	0.23 ± 0.32
Dip	− 0.06 ± 0.20	0.35 ± 0.10

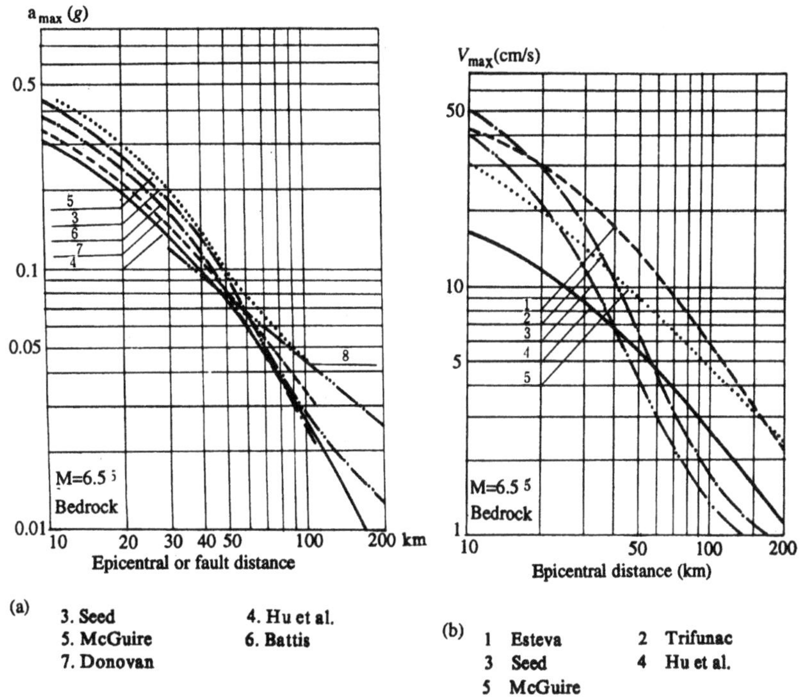

Figure 5.23 Comparison of attenuation: (a) acceleration; (b) velocity. (1 = Esteva and Villaverde, 1973; 2 = Trifunac and Brady, 1976; 3 = Schnabel and Seed, 1973; 4 = Hu *et al.*, 1972; 5 = McGuire, 1978; 6 = Battis, 1981; 7 = Donovan and Bornstein, 1978; 8 = Aptikaev and Kopnichev, 1980.)

(b) hard site $= 24.5 \times 10^{0.333M}(\Delta + 10)^{-0.924}$ (5.45b)

(c) Moderately hard site $= 59.0 \times 10^{0.261M}(\Delta + 10)^{-0.886}$ (5.45c)

(d) Soft site $= 12.8 \times 10^{0.432M}(\Delta + 10)^{-1.112}$ (5.45d)

Figure 5.23(a) compares some of these relations. It can be seen that, in the range M = 6–6.6 and R = 30–100 km, where data concentrate, all relations give similar results, but away from this region results may be quite different.

(ii) Near-field peak acceleration Here 'near field' refers roughly to a region bounded by line segments connecting the following points: $M = 5$, $D = 5$ km; $M = 6.5$, $D = 35$ km; $M = 7.5$, $D = 45$ km.

1. Bolt and Abrahamson (1982) ($M = 6.0$–7.7):

$$a_{max} = 1.6 \times 10^3[(D + 8.5)^2 + 1]^{-0.19}\exp[-0.026(D + 8.5)], \quad \sigma = 0.09g \quad (5.46)$$

2. Campell (1981, 1985):

$$a_{max} = 0.0185\exp(1.28M) \cdot (R + 0.147\exp(-0.732M))^{-1.75}g,$$

$$\sigma_{ln} = 0.384 \quad (5.47)$$

There is disagreement regarding acceleration attenuation at the near field. More investigators believe that near-field acceleration saturates, or that (Figure 5.22):

- a_{max} remains constant for $M \geqslant 7.5$ (roughly);
- a_{max} stays constant for $D \leqslant 10\,\text{km}$ when M is large.

(iii) Horizontal peak velocity (cm/s)

1. Kanai *et al.* (1966; Kanai, 1983), for rock sites:

$$\log v_{max} = 0.61M - [1.66 + (3.6/R)]\log R - [0.631 + (1.83/R)] \tag{5.48}$$

2. Esteva and Villaverde (1973):

$$v_{max} = 32\exp(M)\cdot(R+25)^{-1.7}, \quad \sigma = 0.74 \tag{5.49}$$

3. Trifunac and Brady (Trifunac, 1976):

$$\log v_{max} = 8.357 - 1.087p - 1.059M + 0.201M^2 + 0.134j_s + 0.344j_v - \log A_0\cdot(R) \tag{5.50}$$

   where $j_s = 2$ (rock), 1 (hard soil) and 0 (alluvium); $M = 3$–7.7; the last term takes values given in Table 5.9.
4. McGuire (1978) ($M = 4.5$–7.7):

$$\ln v_{max} = -1.00 + 1.07M - 0.96\ln R + 0.07J_s, \quad \sigma_{ln} = 0.64 \tag{5.51}$$

5. Hu *et al.* (C1982), for rock sites: values are given in Table 5.11.

Figure 5.23(b) compares these relations.

(iv) Response spectrum with 0.05 damping

1. Katayama (1980):

$$S_s(T) = f_M(T)\, f_\Delta(T)\, f_s(T) \qquad (\text{cm/s}^2) \tag{5.52}$$

   where $f_M(T)$, $f_\Delta(T)$ and $f_s(T)$ represent respectively the effect of magnitude, distance and site, given in numerical tables.
2. Boore and Joyner (1984; Joyner and Boore, 1982) ($\sigma = 0.26-0.37$):

Table 5.9 Relation between $\log A_0$ and epicentral distance R

R(km)	$\log A_0$
0	1.40
10	1.61
20	1.83
30	2.08
40	2.31
⩾50	2.52

Table 5.10 Coefficients of Boore–Joyner spectrum attenuation (cm/s)

T	c_1	c_2	$-c_3$	c_4	$-c_5$	c_6	h	σ
0.1	2.24	0.30	0.09	1.0	0.0067	− 0.06	10.6	0.27
0.15	2.46	0.34	0.10	1.0	0.0063	− 0.05	10.3	0.27
0.2	2.54	0.37	0.11	1.0	0.0061	− 0.03	9.3	0.27
0.3	2.56	0.43	0.12	1.0	0.0057	0.04	7.0	0.27
0.4	2.54	0.49	0.13	1.0	0.0055	0.09	5.7	0.30
0.5	2.53	0.53	0.14	1.0	0.0053	0.12	5.2	0.32
0.75	2.46	0.61	0.15	1.0	0.0049	0.19	4.7	0.35
1.0	2.41	0.66	0.16	1.0	0.0044	0.24	4.6	0.35
1.5	2.32	0.71	0.17	1.0	0.0034	0.30	4.6	0.35
2.0	2.26	0.75	0.18	1.0	0.0025	0.32	4.6	0.35
3.0	2.17	0.78	0.19	1.0	0	0.29	4.6	0.35
4.0	2.10	0.80	0.20	0.98	0	0.24	4.6	0.35
a/g	0.49	0.23	0	1.0	0.0027	0	8.0	0.28
V_{max}	2.17	0.49	0	1.0	0.0026	0.17	4.0	0.33

$$\log S_v(T) = c_1 + c_2(M-6) + c_3(M-6)^2 - c_4 \log R + c_5 R + c_6 j_s \quad \text{(cm/s)} \tag{5.53}$$

where $R=(D^2+h^2)^{1/2}$; the coefficients are given in Table 5.10.

3. Hu *et al.* (1982):

$$\log S_a(T) = c_1 + c_2 M - c_3 \log(R+30) \tag{5.54}$$

The values of the coefficients are given in Table 5.11.

Table 5.11 Coefficients of Hu spectrum attenuation $S_a(T)/g$

T	$-c_1$	c_2	$-c_3$
0.05	0.1589	0.4980	2.2068
0.10	0.0769	0.5568	2.3262
0.15	0.1202	0.5477	2.2329
0.20	0.2956	0.5421	2.1144
0.24	0.1724	0.5067	2.0701
0.30	0.5052	0.5041	1.9139
0.40	0.9060	0.5360	1.8678
0.50	1.7960	0.5355	1.4391
0.60	2.2812	0.6528	1.6281
0.80	2.9491	0.7641	1.7001
1.00	3.3992	0.8303	1.7260
1.40	3.9294	0.8224	1.5119
2.00	4.2943	0.7829	1.3107
3.00	4.5818	0.7702	1.1947
5.00	5.3647	0.8948	1.2980
8.00	4.3741	0.5631	0.8798
PD[a]	1.3500	0.6180	1.1787
PV[a]	0.2690	0.6041	1.5320
PA/g[a]	1.2801	0.6574	2.1826
SD[a]	1.4816	− 0.2070	0.3474
SV[a]	1.1664	− 2.2782	1.5558
SA[a]	0.6536	0.6036	− 0.6284

[a] PD = Peak displacement; PV = peak velocity; PA/g = peak acceleration; SD = spectral displacement; SV = spectral velocity; SA = spectral acceleration.

(v) Duration of strong motion $T_d(s)$

1. McGuire (1978), 90% energy duration:

$$\ln T_d = 0.19 + 0.15M + 0.35\ln R + 0.73j_s + 0.23j_v, \quad \sigma_{\ln} = 0.47 \qquad (5.55)$$

2. Aptikaev and Kopnichev (1980), $a_{max}/2$ duration:

$$\log T_d = -1.30 + 0.20M + 0.50\log R \qquad (5.56)$$

The effect of faulting type is given in Table 5.8, where it may be seen that actual duration may be 1.3 to 2.2 times the value given by equation 5.56. Figure 5.24 shows some of the results.

Hu and Tao (1984) suggested a method to estimate the correlated amplitudes and duration of ground motion for low-cycle fatigue analysis.

5.5.3 Ground motion assessment in regions of few acceleration data

From intensity attenuation relations in different regions, it is quite possible that ground motion attenuation is dependent on the geological background of the region. Since accumulation of strong acceleration data is a slow process, it is an important question then for regions with few acceleration data to obtain some applicable attenuation relations.

Two approaches have been used for a long time. One is simply to use the attenuation relation obtained in the United States or Japan for another region, with or without subjective modification; the other is to estimate the intensity first and then convert it to peak acceleration through a relation obtained between intensity and peak acceleration (section 5.4). A simple and better method has been discussed in section 5.4.2 (Hu and Zhang, 1983), which makes full use of attenuation relations of acceleration and intensity for a reference region and attenuation of intensity for the region of interest.

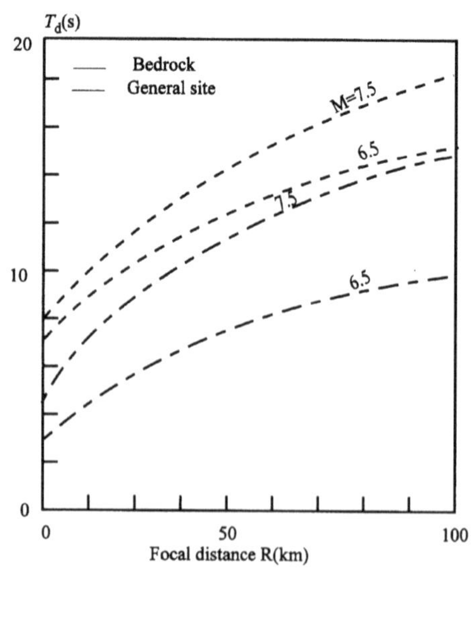

Figure 5.24 Comparison of duration.

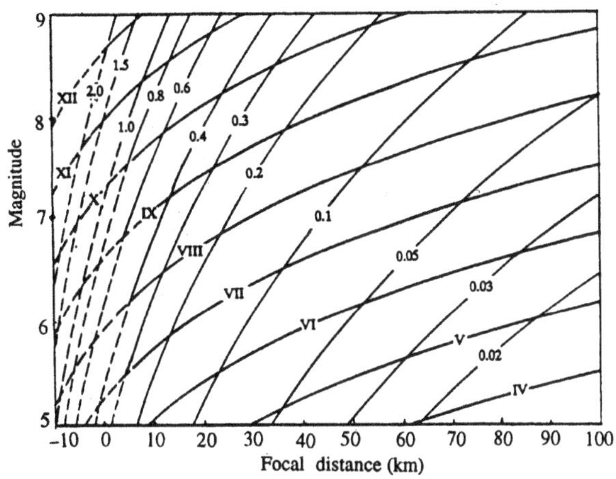

Figure 5.25 A plot of relations among I, a, M and R for western USA: $\ln(a/g) = 3.363 + 0.530M - 2.216\ln(R + 25)$; $I = 5.867 + 1.500M - 2.100\ln(R + 25)$.

Figure 5.25 plots both intensity and acceleration attenuations. Any point in the figure gives a set of I, a, M and R and three of them are independent, which is the conversion relation suggested. It is seen here that, for the same value of intensity I, peak acceleration may vary by more than one order of magnitude, if M or R varies. It can also be seen that, for smaller values of M or R, the peak acceleration for a given I is higher than those for a larger value of M or R.

5.6 Simulated earthquake motion

Earthquake ground motion at a site may be expressed in the frequency domain as follows:

$$A(\omega) = O(\omega)P(\omega)G(\omega)I(\omega) \tag{5.57}$$

where $A(\omega)$, $O(\omega)$, $P(\omega)$, $G(\omega)$, and $I(\omega)$ are respectively the spectra of ground motion, source, path, local site effect and instrument. All the spectra are complex functions so as to maintain both amplitude and phase information; or in the time domain

$$a(t) = o(t) * p(t) * g(t) * i(t) \tag{5.58}$$

where $a(t)$, $o(t)$, $p(t)$, $g(t)$ and $i(t)$ are respectively the Fourier transformation, and $*$ represents convolution.

Engineering methods start from the response spectra obtained from past acceleration records by regression; seismological methods start from the foregoing formulas, by physical and mathematical analyses based on fault rupture mechanisms and wave propagation. It is hoped that a hybrid method may be possible to combine both of them.

5.6.1 Scaling method (Seed et al., 1969; Hadjian, 1972)

This method starts from a selected accelerogram recorded under conditions similar to the case studied and then modifies this accelerogram by multiplying the time and acceleration ordinates separately by adequate factors to satisfy the prescribed spectrum. This method was commonly used in the 1960s.

5.6.2 Engineering method

This method is perhaps the one most widely used in practice, especially for nuclear power plant design. Although random impulses have been used earlier and the ARMA (autoregression-moving average) method has been studied recently, the trigonometrical series is certainly the most popular.

(a) Methodology

On the basis of given parameters of the response spectrum $S_a(T)$, the target spectrum and the strength envelope $f(t)$, select the number (M) of ordinates of $S_a(T)$ to be controlled within given allowable error ε. Usually, M is around 50–100 with the corresponding periods approximately located with constant frequency intervals, and $\varepsilon = 5$–10%. The number N of terms of the trigonometrical series is usually 2048 or 4096. The second step is to select an initial time history $a_0(t)$, preferably a real record of adequate characteristics, to start the iteration. The initial time history may be generated as follows:

$$a_0(t) = f(t) \cdot \sum A_k \exp[i(\omega_k t + \phi_k)] \tag{5.59a}$$

$$A_k = A(\omega_k) = [4S(\omega_k)\Delta\omega]^{1/2}, \quad \omega_k = k\Delta\omega \tag{5.59b}$$

$$S(\omega_k) = \frac{[\varsigma/(\pi\omega_k)]\, S_a^2(\omega_k)}{\ln[-(\pi \ln p)/(\omega_k T_d)]} \tag{5.59c}$$

where $\Delta\omega = 2\pi/T_d$ and p may be taken as 0.85 or greater. Usually the phase angle ϕ_k is taken as uniformly but randomly distributed in the range 0–2π, but it is better to use those from a real record under similar conditions (Ohsaki, 1979). The summation is taken from N_1 to N_2; $N_1 \geqslant 1$ and $N_2 < (N/2) + 1$. Let the periods of the controlling ordinates of the response spectrum be $T_i (i = 1, \ldots, M)$, then $N_1 \Delta\omega < 2\pi/T_M$ and $N_2 \Delta\omega > 2\pi/T_1$. The values of N and M are selected so that there are at least three components of Fourier series between two consecutive controlling periods to facilitate the convergence of the iteration. The time increment $\Delta t = T_d/N$ is to be much less than the smallest period T_1, say $T_1/8$. The initial $a_0(t)$ is obtained from equation 5.59. The next step is to calculate the response spectrum $S_a^c(T)$ of $a_0(T)$, the calculated spectrum. A modification of the Fourier amplitudes of those components closer to the kth controlling period than other controlling periods, but with the phase spectrum unchanged, is then made for those controlling ordinates where the difference of the ordinates of the target and the calculated spectra is greater than the error allowed, and a new time history $a(t)$ is obtained. This step is repeated until all controlling ordinates of the response spectrum are close to the target within the allowable error (Unruh and Kana, 1981; Hu and He, 1986).

(b) Some improvements

In the modification of the Fourier amplitudes, the ordinary method is to modify all components close to the kth controlling ordinate in the same way. An improvement has been suggested (Hu and He, 1986) to modify these amplitudes differently by considering the contribution of each component to the response spectrum. Because the phase angles are usually random, consecutive components may have opposite contributions to the response spectrum and the same modification is certainly unwise and theoretically unreasonable. The new method suggests modifying those with a negative contribution by multiplying by a factor $1/c$ if those with a positive contribution are multiplied by c. Numerical examples prove the suggested modification is effective, good even for obstinate points where the calculated spectral ordinate refuses to approach the target by the ordinary method.

For the selection of phase angles (Ohsaki, 1979), it is better to use those from recorded accelerograms of similar earthquake magnitude, distance and site conditions.

5.6.3 Seismological method

The seismological method may be described by the following formula:

$$a(t) = \sum\sum S_{ij}(t) \cdot a_{ij}(t + \tau_{ij}, R_{ij})\, H(t - t_{ij}) \tag{5.60}$$

It solves the problem in two steps: one is to find the ground motion $a(t, R)$, or Green's function for the problem, at a given site due to an element of earthquake source or fault rupture; the other is to sum up the contributions of motions due to all elements in the time domain. The planar source of an earthquake is divided into $N = n_1 n_2$ elements and each element is considered an element earthquake; the summations run over $i = 1, 2, \ldots, n_1$ and $j = 1, 2, \ldots, n_2$; $S_{ij}(t)$ is the source function or energy of the element ij; τ_{ij} is the total time lag of the arrivals at the site of the wave from the element ij with respect to that from the first element 11; R_{ij} is the distance between the element source ij and the site; and $a_{ij}(t, R)$ is the wave arrived at the site, where R is the distance away from the source ij but with a unit strength ($S_{ij} = 1$); $H(t)$ is the step function.

If the wave $a_{ij}(t, R)$ from a unit source is computed theoretically, it will be referred to as the theoretical method; if it is obtained from earthquake observation, it is referred to as the semi-empirical method. The latter method has attracted the interest of many seismologists and engineers, because it is not easy to calculate $a_{ij}(t, R)$ accurately for complicated and unknown nonuniform media along the transmission path as well as fine details of the element source, but they are included in the measured ground motion. Small earthquakes occur more frequently than strong earthquakes; if records are obtained from some of them, it is then possible to use equation (5.60) to simulate the ground motion time history at the same site for a stronger earthquake occurring at the same source.

Hartzell (1978) was first to suggest using the ground motion record in place of the theoretical Green's function; equation (5.60) thus becomes

$$a(t) = \sum [a_i(t) * Q_i(t)] \, H(t - \tau_i) \tag{5.61a}$$

and the small earthquake record $a_i(t)$ is

$$a_i(t) = O_i(t) * Q_i(t) * P_i(t) * I_i(t) \tag{5.62b}$$

where $O(t)$ is the source effect, $Q(t)$ a scaling factor, $P(t)$ the response of the transmitting media of the crust and $I(t)$ the instrument function. The scaling factor takes care of the difference in mechanism and size of the small earthquake and the large earthquake, which was first expressed as a constant, but more recently as a spectrum by Irikura (1988), Aki (1984, 1988), Luo and Hu (1989) and many others.

6 Earthquakes in China

6.1 Seismic Belts in China

China lies at the southeast end of the Eurasian plate with the Pacific plate pushing underneath to the west with a velocity of 9 cm/year and the Indian plate to the north with a velocity of 6 cm/year. Under these general tectonic forces, the major seismic belts shown in Figure 6.1 were generated; the zones in the eastern part of China are roughly in the NNE direction, nearly parallel to the plate boundary with the Pacific plate and roughly in the NW direction to the west, nearly parallel to the plate boundary with the Indian plate.

6.2 Earthquake history of China

In the long history of China, written records have existed for four thousand years, and earthquake records can naturally be traced far back in history, perhaps being the earliest in the world. The earliest dependable one was in 1831 BC; 'Taishan shocked in the seventh year of the King Fa of Xia' was recorded in the Bamboo Book Annals, and many other earthquake records were found in the following dynasties, because earthquakes were believed to represent the will of gods and very important events closely related to the rise and fall of a kingdom.

The historical records of earthquake occurrence and damage to large cities were indispensable treasures of China, which enable us to study and understand better the past seismicity of this part of the world. In earlier days, earthquake records were scattered among the palace records and throne memorials, or notes and books on special events and disasters as well as poems, fiction, tablets and tombstones; but they gave only rather simple information on time and general location of the earthquake. During the Ming dynasty (1368–1644), local chronicles started to become popular and were produced regularly, recording local important events, including earthquakes, in detail. The local chronicle was a printed collection of events important to an area, produced by its local government, which listed the governmental events and recorded disasters. Earthquakes were considered a disaster and there was usually a section assigned to them. Each district, province or county had its own local chronicle which was then revised, added to and printed every year. In the Ming and Qing dynasties, there were more than 2000 countries and a very strong earthquake would be felt and recorded in several provinces and many counties; this formed the basic historical earthquake database in China.

For example, of the Guanzhong Earthquake of the Shanxi Province on 23 January 1556, 'True Records of Jiajing of Ming' recorded the following:

> On Ren-Yin day of the 12th month of Jiajing 34th year, Shānxi, Shǎnxi and Henan Provinces felt an earthquake at the same time with thunderous sounds,

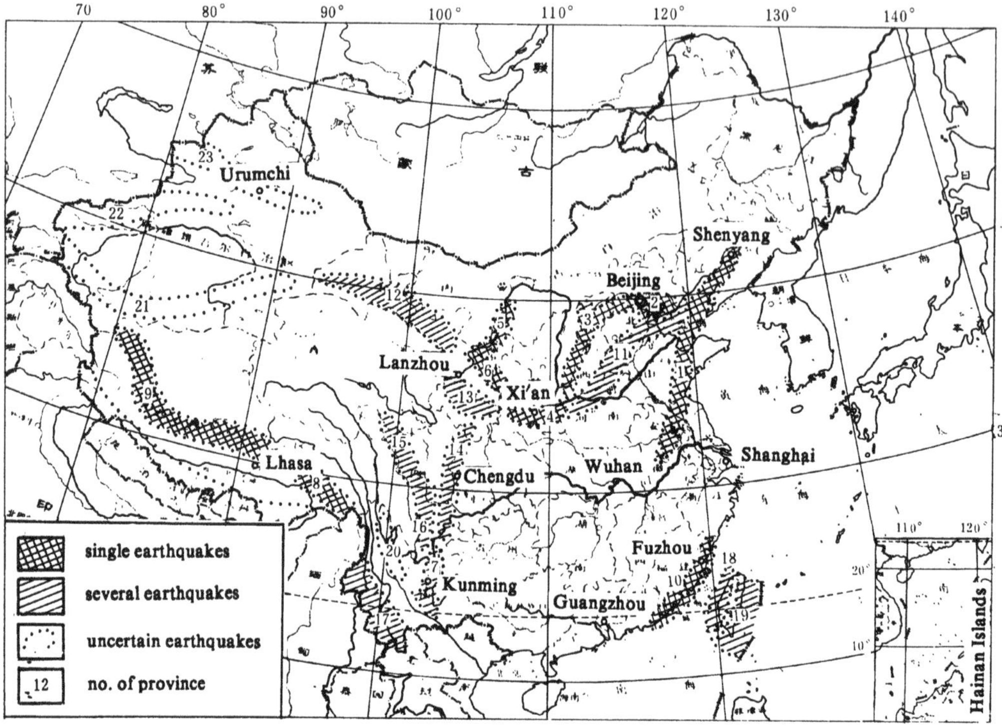

Figure 6.1 Seismic belts in China.

> and fowls cried and dogs barked. Counties of Weinan, Huazhou, Zhaoyi and Sanyuan of Shānxi Province and Puzhou County of Shǎnxi Province were damaged the most: either springs with fishes newly induced by the earthquake; or city walls and buildings sunk underground, or flat ground turned to a hill; or several shocks in a day or shock all day. Rivers rose or overflowed, Mountains of Huayue and Zhongnan shouted, and rivers were dammed for several days. Earthquake-killed officers, army and civilians with names listed on the reports to the government reached more than 820 000 and those without names listed were countless.

In addition to this record, there were other county chronicles of 211 counties which recorded this earthquake. The 1977 edition of the *China Earthquake Catalog* collected all historical records available of this earthquake and others.

In the 1960s, the Chinese government organized a group of historians and earthquake specialists to search through all the dependable literature and collect and collate historical earthquake records, in order to compile a 'Chronological Table of China Earthquakes (occurred before 1955)'. This included more than 8000 earthquakes, with roughly 1000 causing damage over 24 provinces, of which 165 earthquakes were in Yunnan, 125 in Gansu, 98 in Shanxi, 87 in Hebei, 62 in Shanxi, 58 in Sichuan, 46 in Taiwan and 46 in Henan.

6.3 Seismicity in China

In Figure 6.1 the geographical characteristics of seismicity in China are outlined; this section will concentrate on the temporal characteristics of seismicity of seismic belts or provinces. Two indices are used to describe the temporal characteristics of seismicity: the strength or magnitude and the frequency, and the temporal distribution of various magnitudes of earthquakes in a belt. The frequency is sometimes referred to as the return period of a certain magnitude. The prediction methodology currently used is to carry out statistical analysis of the temporal distribution of past earthquakes and then to predict the future trend of seismicity over a short time for that belt.

6.3.1 Periodicity of seismicity

The long history of earthquake records of nearly one thousand years reflects a general trend that, within a region, seismicity usually shows an intermittent process of relatively quiescent and active episodes. The time from the beginning of one episode, quiescent or active, to the end of the following episode is one complete period of seismicity of the region. This pattern holds for a small region as well as a large province or long belt.

Figure 6.2 shows this pattern for some seismic provinces of China, from which the following results may be drawn for China.

1. For one seismic belt or province, seismicity periods are roughly the same, but the seismicity periods of different provinces may be different,

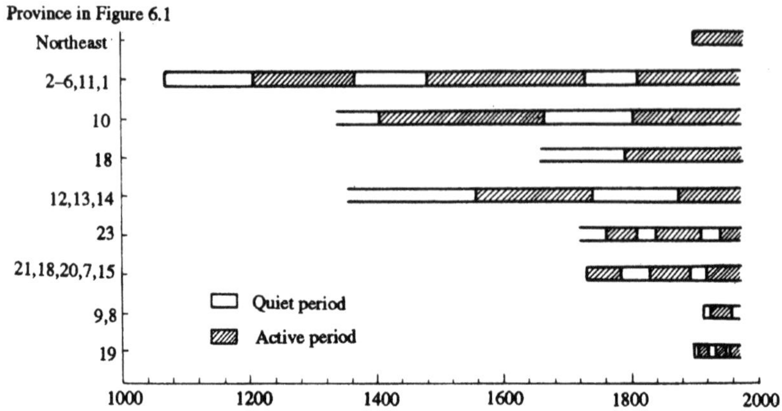

Figure 6.2 Seismicity periods of several provinces in China.

which seems to indicate that seismicity period is a characteristic of the seismic belt. Figure 6.2 shows the longest period to be about 300 years, such as in north China, south China and the northern part of the Qinghai-Tibet plateau; medium periods about 100 years, such as in the central part of Xinjiang and the central part of Qinghai-Tibet; and the shortest a few tens of years, such as in east Taiwan and the southern part of Qinghai-Tibet.

2. Within a quiescent episode in any seismic province, generally no or only a few earthquakes occur with a magnitude more than 7, and those with magnitudes between 6.0 and 6.9 are also infrequent; but many earthquakes occur with magnitudes greater than 6.0 in one active episode of a province. With only a few exceptions, an earthquake with a magnitude ≥ 8 has occurred in each episode in each province.
3. The active episodes of several provinces with a long seismicity period in the eastern part of China occurred roughly at the same time, and the same is true for the quiescent and active episodes of different provinces. This is perhaps because these provinces are all located near the same boundary of the Eurasian plate and are thus subject to the same tectonic forces.

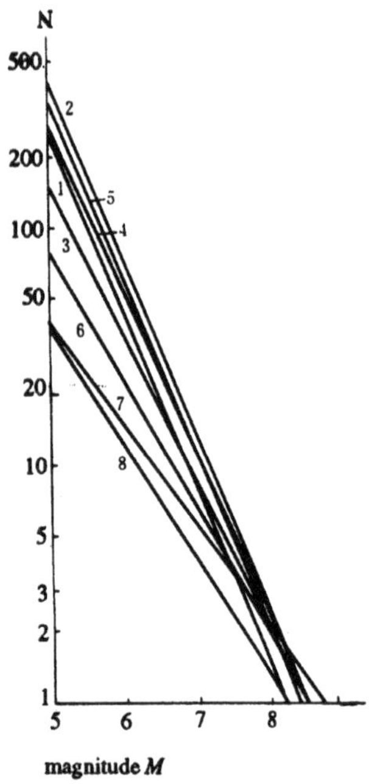

Figure 6.3 Magnitude–frequency relations: 1=Taiwan (July 1952–November 1972); 2=Taiwan (Jan-uary 1923–June 1952); 3=central Xinjiang (1915–1983); 4=southern Qinghai–Tibet (1917–1959); 5=central Qinghai–Tibet (1896–1983); 6=northern Qinghai–Tibet (1740–1983); 7=northern China (1369–1739); 8=northern China (1740–1983).

6.3.2 Magnitude–frequency or earthquake recurrence relation

The magnitude–frequency relation

$$\log N = a - bM$$

suggested by Gutenberg (section 2.4) has been applied to investigate this relation of various provinces of China and some results are given in Figure 6.3, and the following features are obtained for parameters a and b for those provinces with relatively complete data.

1. For the same province, the values of parameters a and b are roughly stable, such as the straight lines 1 and 2, and 7 and 8.
2. Those provinces with shorter seismicity periods, say 100 years or shorter, have higher values of a and b, 5.0–6.5 for a and 0.60–0.75 for

b (e.g. east Taiwan, the southern and central parts of Qinghai-Tibet, central Xinjiang etc.); on the other hand, those provinces with long seismicity periods of 300–400 years have lower values of *a* and *b*, 3.5–5.0 for *a* and 0.40–0.60 for *b* (e.g. north china, the north of Qinghai-Tibet, west Taiwan etc.).

3. The straight lines in Figure 6.3 all intersect the magnitude axis at $N=1$ (the average number of earthquakes with magnitude in the range $(M, M+\Delta M)$) at $M>8$, which means that many seismic provinces in China are very likely to experience one earthquake of magnitude 8 or more.

With a known magnitude–frequency relation for a province, together with a trend analysis of current status in the seismicity period, it is possible to evaluate the expected average number of earthquakes of given magnitude for a current episode and for the near future in that province, but with large uncertainty.

From the results shown in Figure 6.2, two possibilities may be noted.

1. For those provinces with a fairly short history of earthquake records such as the Qinghai-Tibet Plateau and northeast China, the reliability of the magnitude–frequency relation and of the seismicity period and episodes thus obtained is low.
2. Even those provinces with relatively long-standing records, such as north China, still experienced less than three seismicity periods, which is not really sufficient from the statistical point of view.

Moreover, it can be seen from Figure 6.2 that the seismicity periods and episodes for one province are rather different; and the empirical magnitude–frequency relation is subject to error at large magnitudes, because magnitude M saturates at roughly 9, as discussed in section 2.6.2, but the straight line magnitude–frequency relation allows no upper limit for M. The results given by this relation are only averages, subject to error and are random in nature.

6.4 Damage caused by earthquakes in China

Generally speaking, in China, Taiwan has the highest seismicity, in both strength and frequency; arranging in decreasing order, the southwest comes next; the northwest has high strength but with frequency similar to the west; and the lowest are northeast, mid-south and the lower reaches of the Yangzi River. The damage caused by a few large earthquakes will be described in the following.

6.4.1 Haiyuan Earthquake of 1920 (Lanzhou Earthquake Research Institute et al., C1980)

A very large earthquake of magnitude $M_S=8.5$ occurred on 16 December 1920 at the border of Gansu and Ningxia provinces, with surface rupture 220 km long and a focal depth of 29 km. The following three features deserve special mention.

1. Enormous number of casualties and very wide influence. The number of deaths was more than 230 000. The area of intensity IX and above reached more than 20 000 km^2, and the radius of the area over which the earthquake was felt was about 1500 km, more than half the length of China.

2. Large and numerous areas of abnormal intensity. The largest was at Tongwei, where a highly abnormal maximum intensity X occurred over an area 130 km long by 20 km wide, in a region of normal intensity VIII. Although it might have been caused by an earthquake at Tongwei triggered by the main shock, most believe that it was due to either local geological conditions or superposition of direct waves and waves reflected from the Moho discontinuity.

3. Damage to buildings. Old-style timber-framed buildings have the best earthquake resistance and adobe dome dwellings the worst. However, because of bad building practices with regard to foundations, the timber-framed buildings almost all collapsed in regions of intensity X or XI, though not many collapsed in intensity IX region. Buildings of the combined style completely collapsed in regions of intensity X. Bearing-wall buildings almost completely collapsed in regions of intensity IX. Adobe dome buildings could hardly survive in regions of intensity VIII, and partly collapsed in regions of intensity VII.

6.4.2 Tonghai Earthquake of 1970

A strong earthquake of magnitude 7.7 occurred on 5 January 1970, in the mountain area at the border of Tonghai, Jianshui and Eshan counties. Focal depth was 13 km and Figure 6.4 shows the fault rupture and the

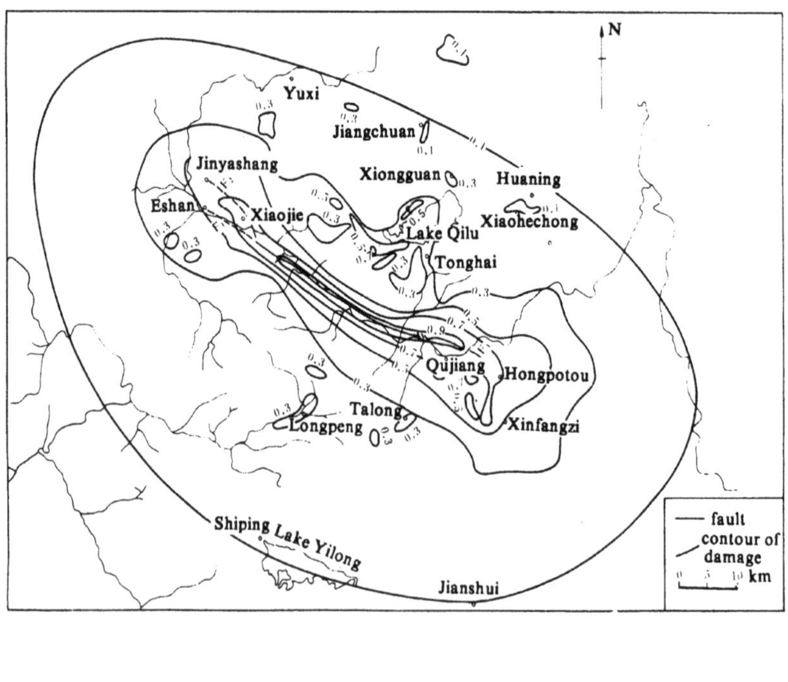

Figure 6.4 Actual contours of damage index of Tonghai earthquake.

isoseismals in terms of the damage index. Three special features are worth mentioning:

1. clear surface fault rupture and narrow inner isoseismals;
2. clear effect of site conditions;
3. thorough field investigation.

Both the number of specialists and scope of profession, as well as the length of time in the field, were perhaps a record in China. To study seriously the site effects in the field, more than 40 specialists – geologists, seismologists and engineers – were well organized and spent nearly 12 months collecting first-hand information on damage and site conditions, including soil and local geography, over 1390 villages and towns.

6.4.3 Zhaotong or Yongshan-Daguan earthquake of 1974

An earthquake of magnitude 7.1 occurred on 11 May 1974 in hilly country at the border of Yunnan and Sichuan provinces, with focal depth 14 km. The intensity distribution is shown in Figure 6.5. On this map, figures by the villages are intensity ratings and those in parentheses are of low reliability either because there were not enough buildings or there was

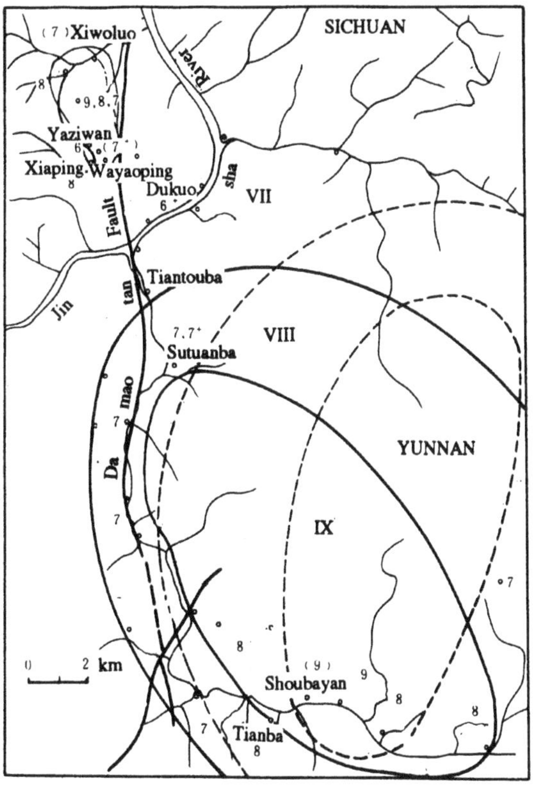

Figure 6.5 Isoseismal map of 1974 Zhaotong earthquake.

insufficient investigation. Two features of this earthquake are presented below.

1. Disorderly pattern of intensity. The disturbed area is a hilly country with villages, often located on the tops or slopes of hills or on terraces; local site conditions overshadow attenuation. There were no new surface fault ruptures found and it was not clear which major fault nearby was responsible for this earthquake. This, together with the disorderly intensity distribution, made the plotting of the isoseismals difficult.

2. Great effect of topography and soil on intensity. In strongly disturbed areas, many villages were founded on rocks. For example, for those villages by the Mugan River near the west edge of the meizoseismal area, even when close to Damaotan fault, the intensities were lower than the average; on the contrary, the abnormally high intensity area located on the north of the meizoseismal area was due to a group of hill tops or a ridge crest, or loose soil.

6.4.4 Haicheng earthquake of 1975

A strong earthquake of magnitude 7.3 occurred on 14 February 1975 in the district of Haicheng and Yingkou of the Liaoning province, with focal

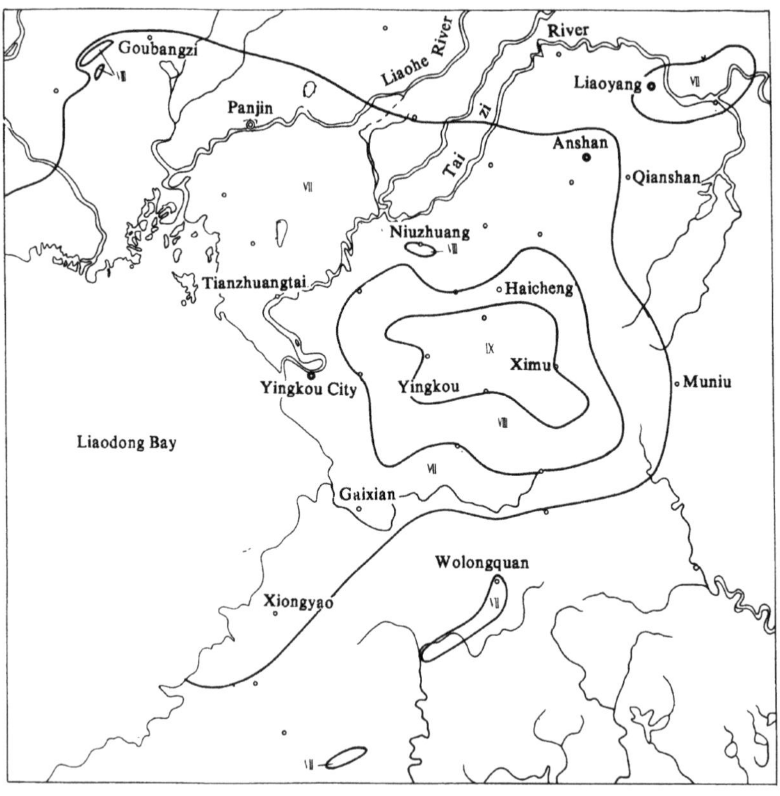

Figure 6.6 Isoseismal map of Haicheng earthquake.

depth 12–15 km; the isoseismal map is shown in Figure 6.6. The features are as follows.

1. Disorderly pattern of intensity in meizoseismal area. The meizoseismal area was located in hilly country to the east of the railway of Anshan-Haicheng-Yingkou county. Faults in the disturbed area consist of two sets, with the major one in the NE direction and the other in the NNW. Although there were many surface ground fissures induced by this earthquake, most of them were secondary fissures in soft soils and could not be identified as being related to crust structures. The set of en echelon fissures, extending in the EW direction and 5.5 km long, was found along an old fault and cut through the saddle part of the mountain ridge or bedrock, not affected by local topography and looked like earthquake faulting. However, both its length and scale were short for such a large earthquake and located at the far corner of the meizoseismal area, where some ordinary brick buildings were almost intact, as shown in Figure 6.7. It was not considered the causative fault for this earthquake. The aftershock distribution shown in Figure 6.8 seems to indicate that the major energy-releasing faults were directed NW or NNW.

The starfish shape of the meizoseismal area was considered to be the influence of a conjugated faulting formation in this district, and the disorderly pattern of intensity distribution was the effect of the local topography and site conditions.

2. Wide experience of engineering works. The strongly disturbed area was densely populated and industrially developed, with the large Anshan Steel Factories within the zone of intensity VII and some branch factories in higher intensity areas. Tables 6.1–6.3 give some examples of damage

Figure 6.7 Intact building near surface fault rupture.

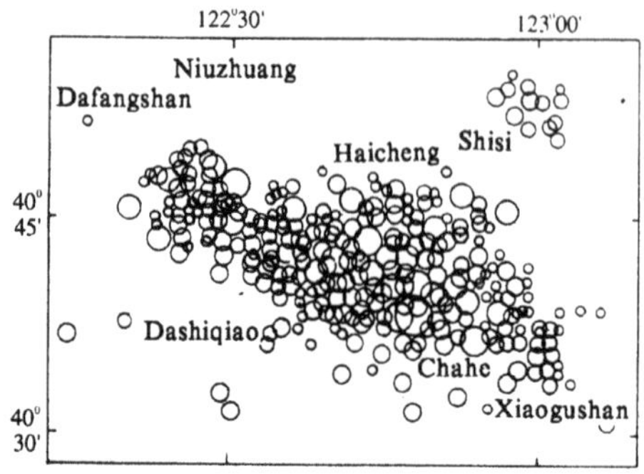

Figure 6.8 Fore-and aftershocks of the Haicheng earthquake.

matrices of multistory brick buildings, beam-type bridges and earth dams. The following points are worthwhile mentioning.

- The most serious structural damage was on the west side of the disturbed area, while the fault rupture was located on the east side.

Table 6.1 Damage to multistory brick buildings (%)

Intensity zone	Collapse	Serious damage	Moderate damage	Slight damage	Intact	Total no.
IX	7.7	56.7	16.4	12.5	6.7	104
VIII	0	21.0	29.0	21.0	29.0	90
VII	1.0	44.7	13.8	16.0	24.5	94
VI	0	0	10.0	45.0	45.0	20

Table 6.2 Damage to beam-type bridges (%)

Intensity zone	Part	Type	Collapse	Serious damage	Moderate damage	Slight damage	Intact
IX	West	railway	–	–	–	–	–
		highway	–	–	–	–	–
	Rest	railway	0	1	5	21	73
		highway	0	1	7	27	65
VIII	West	railway	0	0	22	78	0
		highway	4	17	21	7	41
	Rest	railway	0	0	0	0	100
		highway	0	0	8	27	65
VII	West	railway	0	3	21	55	21
		highway	4	3	15	21	57
	Rest	railway	0	0	0	0	100
		highway	0	0	3	6	91

Table 6.3 Damage to earth dams

Intensity zone	Cracks			Other damage	Intact	Total no.
	Long. and trans.	Trans.	Long.			
IX	9	0	0	2	0	11
VIII	4	1	3	1	1	10
VII	9	0	3	3	18	33

- To the west of the railway line, nearby villages with fewer sandboils and waterspouts showed heavier structural damage than a group of nearby villages with heavy sandboils and waterspouts but with no ground failure.
- Sand liquefaction had a strong effect on damage to engineering works and the heavy damage on soft ground was partly due to ground failure.
- Damage to brick chimneys showed different patterns, such as several horizontal cracks, top portion twisted or shifted, or even overturned, as shown in Figures 6.9–6.10, most of the damage being concentrated in the top half of the chimney.
- Some industrial buildings were heavily damaged because of insufficient lateral bracing systems, such as that shown in Figure 6.11.

3. Different patterns of damage in plain and hilly areas. The west part of the disturbed area was alluvial or bay-shore plain where soils were very soft, with a high water table, and was full of ground failures. On the contrary, the eastern part of the disturbed area, including the mesoseismal area, was hilly and hard ground; the damage pattern was primarily due to vibration of the structure, such as shear or crush failures of bridge piers (Figure 13.6), quite different from those due to ground failure on soft plains.

4. Different damage attenuation in one-story brick buildings and tall smokestacks. Figures 6.12 (a) and (b) show respectively the attenuation of damage of tall brick smokestacks and one-story brick buildings under

Figure 6.9 Twisted top portion of a brick chimney.

Figure 6.10 Badly cracked brick chimney.

Figure 6.11 Damage to a mechanical repair workshop.

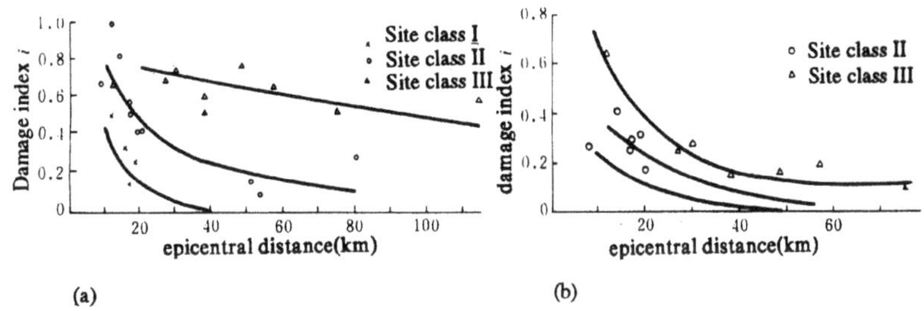

Figure 6.12 Attenuation of damage: (a) brick smokestacks; (b) brick buildings.

different site conditions. It is obvious that damage to flexible smokestacks attenuates much slower on soft sites.

5. Successful earthquake prediction and low casualty figure. This earthquake had been successfully predicted with regard to the location and time, and government action was taken before the earthquake. For example, some communities showed movies in open fields and attracted people out of their houses, in addition to issuing announcements of a future earthquake.

6.4.5. Tangshan Earthquake of 1976

On 28 July 1976, a very strong earthquake of magnitude $M_s = 7.8$ occurred right under Tangshan City in Hebei province, with focal depth 16 km, epicenter intensity XI, but surface ground rupture less than 6 km. No prediction was made for this earthquake. Isoseismals are shown in Figure 6.13. This earthquake had many features in common with the 1975 Haicheng earthquake, but the number of casualties was much greater.

1. Extensive casualties. Although the engineering damage was quite comparable with the Haicheng earthquake the previous year, for which a

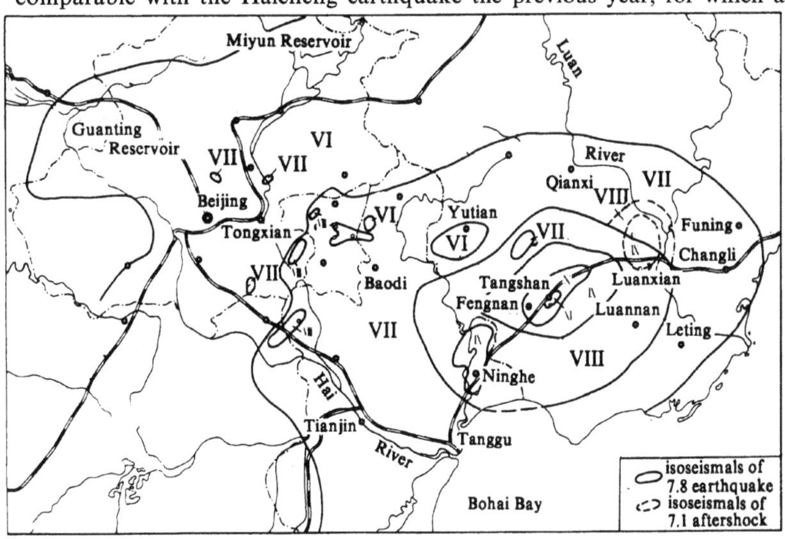

Figure 6.13 Intensity distribution of 1976 Tangshan earthquake.

successful prediction had been made, the number of casualties in the Tangshan earthquake reached 240 000, about 1000 times that in the Haicheng earthquake. In comparison with the Haicheng earthquake, the contrast is very clear in this terrible result from a strong earthquake which was unpredicted.

2. Almost total destruction of the city. Since the magnitude was nearly 8, and buildings and other engineering works in Tangshan area were not designed to withstand earthquakes, the whole city was almost destroyed. More than 98% of multistoried buildings collapsed or were seriously damaged beyond repair in areas of intensity X and XI. In Tangshan City, all brick smokestacks and brick cylinder-supported towers collapsed; all bridges were damaged. Figure 6.14 shows a bird's-eye view of part of the meizoseismal area. The ground was deformed and rails were bent like waves. River slopes slid and bridges were badly damaged. An earth dam across the Dou River 20 km to the north of Tangshan City was seriously damaged due to liquefaction of the foundation sand; the dam top settled by more than 1 m (Figure 10.3).

3. Obvious effect of accumulated damage. About 15 hours after the main Tangshan earthquake, a strong aftershock of magnitude 7.1 occurred near the Luan county, which intensified damage; its isoseismals are shown by the

Figure 6.14 Bird's-eye view of part of Tangshan City.

Figure 6.15 Damage of a highway bridge over Luan River in Tangshan earthquake.

dashed lines in Figure 6.13. For example, the Luanhe highway bridge of 35 simple spans of reinforced concrete beams, 50 km east of Tangshan, which was damaged but opened for transportation after the main shock, collapsed (Figure 6.15) in an aftershock. No serious ground deformation was found near the bridge and there was no liquefaction near piers or abutments; the damage was thus considered to be mainly vibration-induced.

4. Many abnormal intensity areas. It can be seen from Figure 6.13, that the largest high abnormality area was Ninghe IX inside intensity VIII, and the largest low abnormality area was Yutian VI inside intensity VII. The latter was also a low abnormality area for other historical earthquakes. In the Ninghe area, ground failure was visible, but in Yutian, the bedrock surface was shallow and no ground failure was noticed.

5. Damage far away from epicenter. Miyun reservoir was about 150 km west of Tangshan and of intensity VI. During the main shock, the protection layer of the earth dam below water level on the reservoir side slid along the core wall to the heel of the dam due to liquefaction (Figure 10.21). This dam was equipped with several accelerographs before the earthquake; the records from the main shock showed that the peak acceleration was $0.16g$ at the top and $0.05g$ at the bottom of the dam, and the strong part of the duration was about 20–30 s.

Tianjin City was about 100 km away and on very deep and soft soil, a high abnormality area of intensity VII inside VI. The damage to low buildings was normal, but buildings of longer natural periods and tall smokestacks were heavily damaged by an intensity of VIII. The damage is described in detail in Liu (C1985–86).

6.4.6 North Taiwan Earthquake of 1986 (CEER, C1986)

An earthquake of magnitude 6.8 occurred about 40 km northeast of Hualien with focal depth 10 km, on 15 November 1986. The intensities were given on the 7-grade scale and the highest was VI (roughly IX on the 12-grade scale) at Ilan, and V (roughly VIII on the 12-grade) at Taipei. This earthquake was followed two hours later by an earthquake of magnitude 6.3, 18 km northeast of Hailien with a focal depth 27 km. Figure 6.16 shows the intensity and peak acceleration distribution.

Preliminary statistics showed 12 dead, 36 buildings collapsed and 26 partially collapsed. Damage to highways, railways, power, communications, harbour, water and gas pipelines were reported. The lessons learned here were to strengthen the requirements of licensing and review of the design of buildings taller than seven stories and of important buildings, and to

Figure 6.16 Intensity and acceleration distribution.

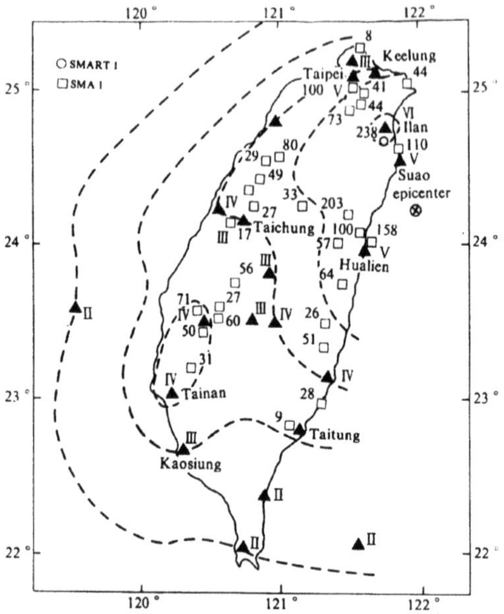

increase the level of design in Taipei City. A maximum peak acceleration of more than 0.2*g* was recorded at the SMART-1 array, and the maximum at Taipei was 0.1*g*.

7 Earthquakes in the USA

7.1 Seismicity in the USA

The USA is located on the North American Plate which is adjacent to the Pacific Plate on the West Coast, and the Eurasian and African Plates at the Mid-Atlantic Ridge (Figure 7.1). Most earthquakes have occurred in the vicinity of the boundary between the Pacific and North American Plates. The well-known San Andreas fault zone is a portion of this boundary where the Pacific Plate is sliding northwest past the North American Plate at a rate of about 5–8 cm per year. Earthquakes on the San Andreas fault typically occur at relatively shallow depths of 5–40 km below the surface, with varying rupture lengths. In the 1906 San Francisco and 1857 Fort Tejon earthquakes, the rupture lengths were over 400 km. The Aleutian Subduction Zone is another portion of this boundary where the Pacific Plate is being subducted under the North American Plate, creating earthquakes along the zone of contact (called the Benioff Zone). Earthquakes at the subduction zone have usually occurred at great depths. Earthquakes have also occurred in the interior of the plates, most likely caused by the build-up of strains from pressures developed at the plate boundaries. The most notable ones are the New Madrid earthquakes of 1811–1812 in the Mississippi River Valley and the Charleston, South Carolina, earthquake of 1886. Figure 7.2 shows the locations of the notable historic earthquakes in the USA.

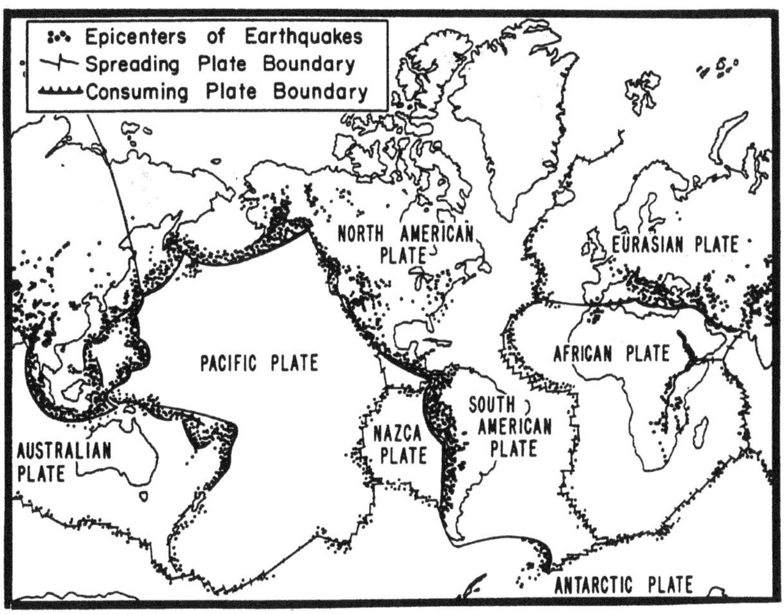

Figure 7.1 World map of seismic zones and major crustal plates. (Source: US Geological Survey.)

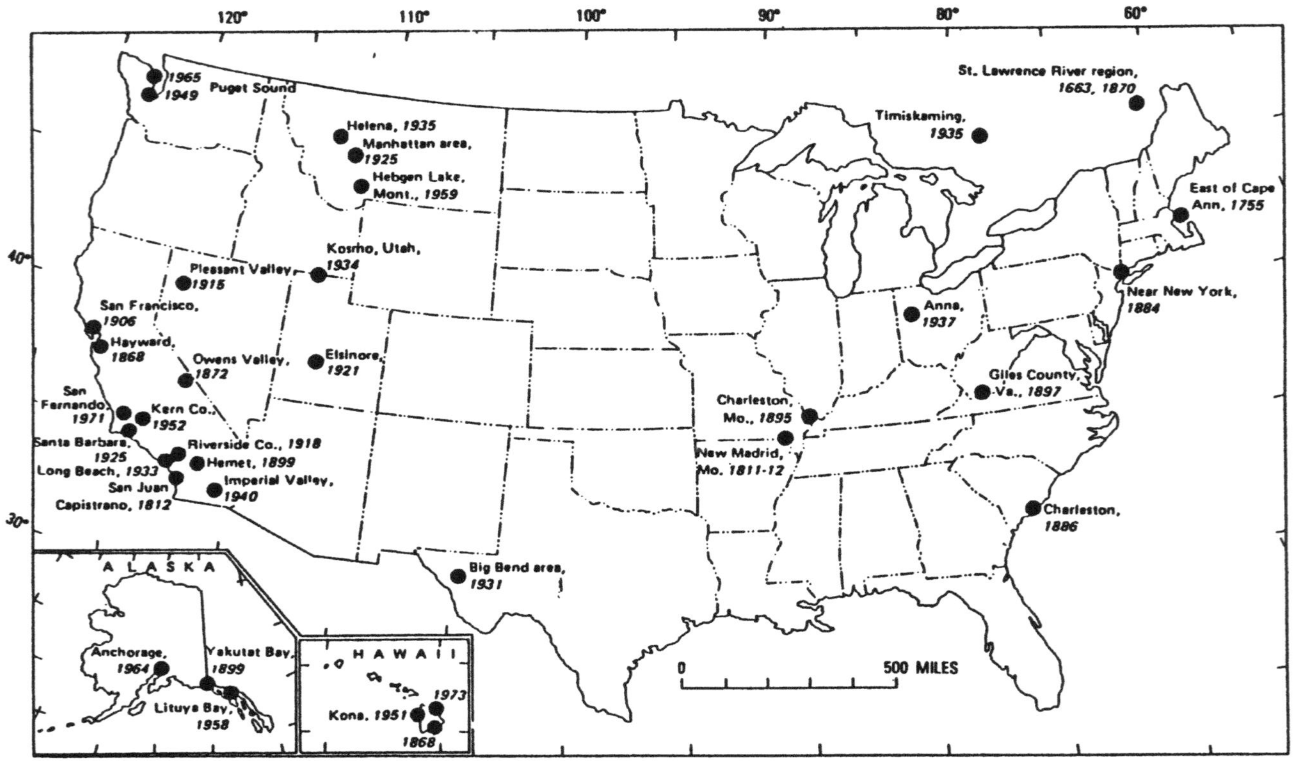

Figure 7.2 Location of notable historic earthquakes in the United States.

7.2 Damage caused by major earthquakes

7.2.1 The New Madrid Earthquakes of 1811–1812

During the winter of 1811–1812 four great earthquakes occurred in the Mississippi Valley, each having a body-wave magnitude m_b above 7.0 (Table 7.1 and Figure 7.3). There had been no other earthquakes larger than these within the conterminous USA prior to that time. Their magnitudes are comparable to those of the largest California earthquakes, and, because of low attenuation of seismic intensity in the eastern and central USA, they were felt by much larger areas than shocks of similar magnitudes in California. They were felt with intensity equal to or greater than MMV

Table 7.1 Relative sizes of the 1811–12 earthquakes (Hopper, 1985)[a]

Date	Local time (a.m.)	I_0	M_S	m_b	Epicenter	
					Lat. °N.	Long. °W.
16 December 1811	2:15	XI	8.6	7.2	35.8	90.30
16 December 1811	8:15	–	–	7.0	36.0	90.0
23 January 1812	9:00	X–XI	8.4	7.1	36.2	89.8
7 February 1812	3:45	XI–XII	8.7	7.3	36.5	89.6

[a] Intensity estimates from Nuttli (1981); magnitude estimates from Street and Nuttli (1984); and epicenter estimates from Russ (personal communication, 1983) and Street (1982). – indicates no data.

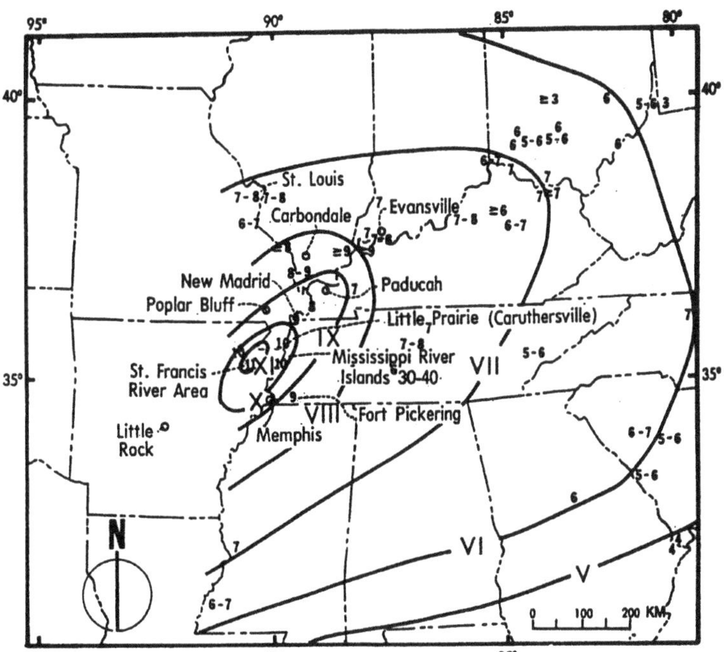

Figure 7.3 Isoseismal map of the 16 December 1811 earthquake in northeast Arkansas; first major shock of the New Madrid series. (After Nuttli, 1981.)

(enough to cause alarm),* over $2.5 \times 10^6\,km^2$ which includes the entire eastern USA (Nuttli, 1973). The area of intensity VII (mainly architectural damage) and greater covers parts of Illinois, Indiana, Ohio, Kentucky, Tennessee, Alabama, Mississippi, Louisiana, Arkansas and Missouri. Because of the low population density in 1811, the effects of these earthquakes west of the Mississippi River are not known, but they can be estimated. It is clear that large parts of Kansas, Oklahoma, and Texas were shaken at the MMV level.

The maximum intensities of the four 1811–1812 earthquakes range from X to XII (Table 7.1). In the epicentral area of the first shock (16 December 1811, 2:15 a.m.), the St. Francis River area of northeastern Arkansas, a lake was uplifted and drained, while other places subsided as much as 12 ft (3.7 m, Figure 7.4). Sand and other materials were thrown from fissures or cracks in the swampland. The greatest disturbance occurred along the Mississippi River between Islands 30 and 40 along the Tennessee–Arkansas border north of Memphis (Nuttli, 1973). According to Fuller (1912, p.10), 'Great waves were created, which overwhelmed many boats and washed others high upon the shore, the return current breaking off thousands of trees and carrying them out into the river. High banks caved and were precipitated into the river, sand bars and points of land gave way, and whole islands disappeared'. Uplifted areas caused pounding or waterfalls along the Mississippi. Landslides were extensive along the river banks as far up the Ohio River as Indiana, but particularly severe along the Chickasaw Bluffs on the Mississippi river north of Memphis. The road between New Madrid and Arkansas was made impassable by the earthquake. The area of marked earth disturbances extended from Cairo to Memphis and from Crowley's Ridge to Chickasaw Bluffs.

The people in the epicentral area in 1811–1812 were able to survive as well as they did only because of their lifestyle. There were only about 5000 people living in the area of intensity X and greater, and they occupied light wood-framed structures, the kind least susceptible to earthquake damage. Transportation was by horse, boat and foot, and most escaped on foot; the small population was able to feed itself after the earthquake by hunting wild geese (Nuttli, 1981).

7.2.2 Charleston earthquake of 1886

On 31 August 1886, Charleston, South Carolina, and the surrounding area was struck by an earthquake of MMI X. The loss of life, including wounded who afterwards died, was about 60 (Dutton, 1889). Property damage was estimated at $5 million (in 1886 dollars). Two epicenter points were mapped by Dutton near Ladson and Rantowles as shown in Figure 7.5 reproduced from Dutton (1889). The Charleston earthquake of 1886 was unique because it was the most damaging earthquake on record to strike the eastern US (Figure 7.6). Although three larger eastern earthquakes occurred in 1811–1812 in the Mississippi River Valley centered near New Madrid, Missouri, only one death and only slight property damage were reported

* See section 4.1.2(a) for the entire modified Mercalli intensity (MMI) scale.

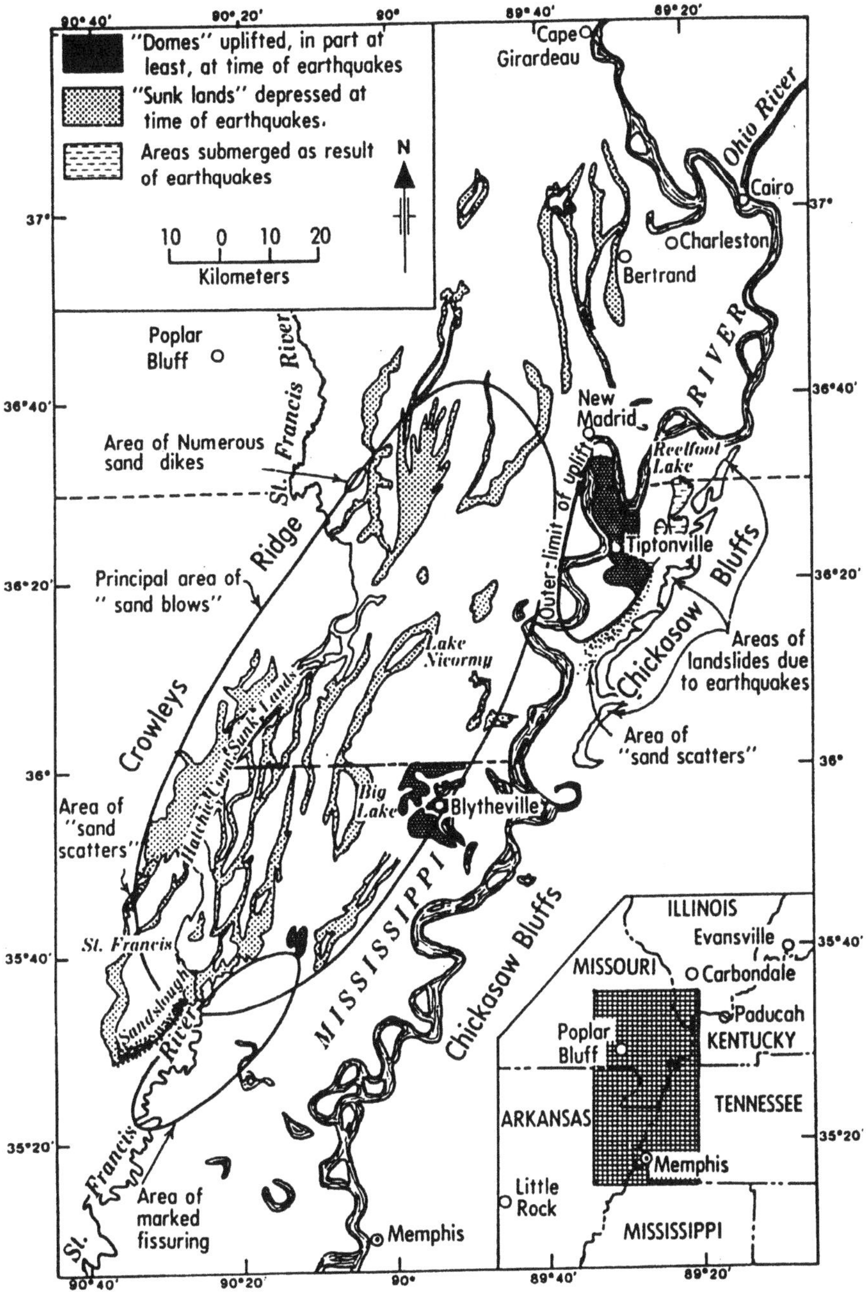

Figure 7.4 Map of the ground effects of the New Madrid district. (After Fuller, 1912.)

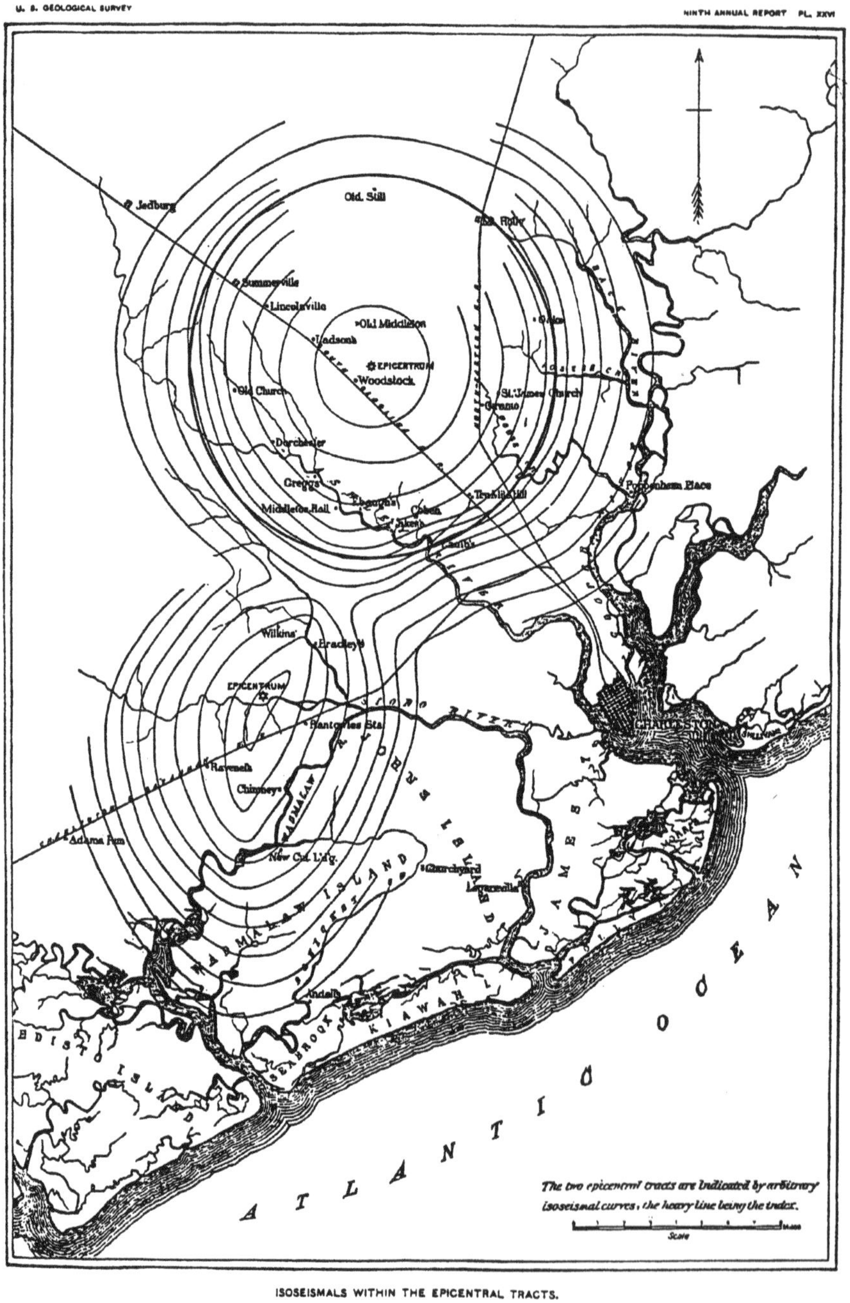

Figure 7.5 Charleston 1886 earthquake epicenters. (From Dutton, 1889.)

Figure 7.6 Charleston earthquake of 1886 which caused serious damage in Charleston, SC. (Source: US Geological Survey.)

because the epicentral area was lightly inhabited at that time. Much of the damage occurred in the peninsular part of Charleston where most of the population of 50 000 lived and worked. Within the epicentral tract of the Charleston earthquake, extending to a radius of approximately 50 miles (80.5 km) from the Summerville area, notable effects included ground fissures and craterlets; water, sand and mud fountains; railroad rails bent and tracks displaced; earth and water waves; sulfur gas released; and generally loud earthquake sounds. Within the city, hardly any building totally escaped injury and few had escaped serious damage. Although the number of buildings completely demolished was not great, several hundred buildings had lost a large portion of their walls. Damage to chimneys was pervasive. Less than 100 out of 14 000 chimneys escaped injury. Ninety-five percent of the 14 000 were broken off at the roof and went to the ground (Parker and Stewart, 1886).

Another unique aspect of the 1886 Charleston earthquake was the extremely large area over which it was felt, as shown in Figure 7.7 (from Bollinger, 1977). The Charleston earthquake was felt as far away as Boston; Milwaukee and Green Bay, Wisconsin; and La Crosse, Wisconsin, 960 miles (1546 km) from the epicentre, the most distant locality within the US. The tremors were felt faintly in Bermuda, almost exactly a thousand miles from the epicentre. Thus the motion was sufficient to attract the attention of the unexpecting observer throughout a circular area with a radius of a thousand miles (1600 km), an area of approximately three million square miles (7.8×10^6 km^2). Within this area, distinct, well-defined tremors were felt without the aid of instruments. Damage was caused as far away as Chicago.

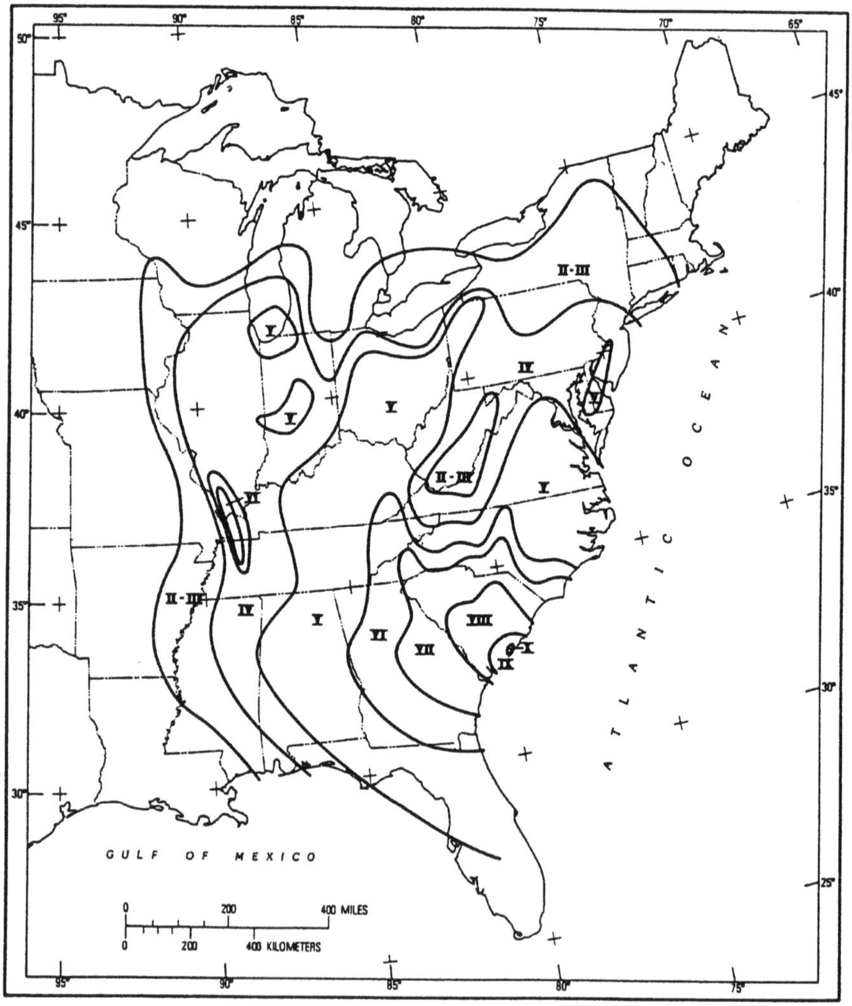

Figure 7.7 Isoseismal map of the eastern United States, contoured to show the broad regional patterns of the reported intensities for the 1886 Charleston earthquakes. Contoured intensity levels are shown in Roman numerals. (From Bollinger, 1977.)

7.2.3 San Francisco earthquake of 18 April 1906

This was one of the greatest known shocks in California. The destructive intensity of this earthquake extended 400 miles (640 km), and was felt over 375 000 square miles (9.7×10^5 km^2) including water area.

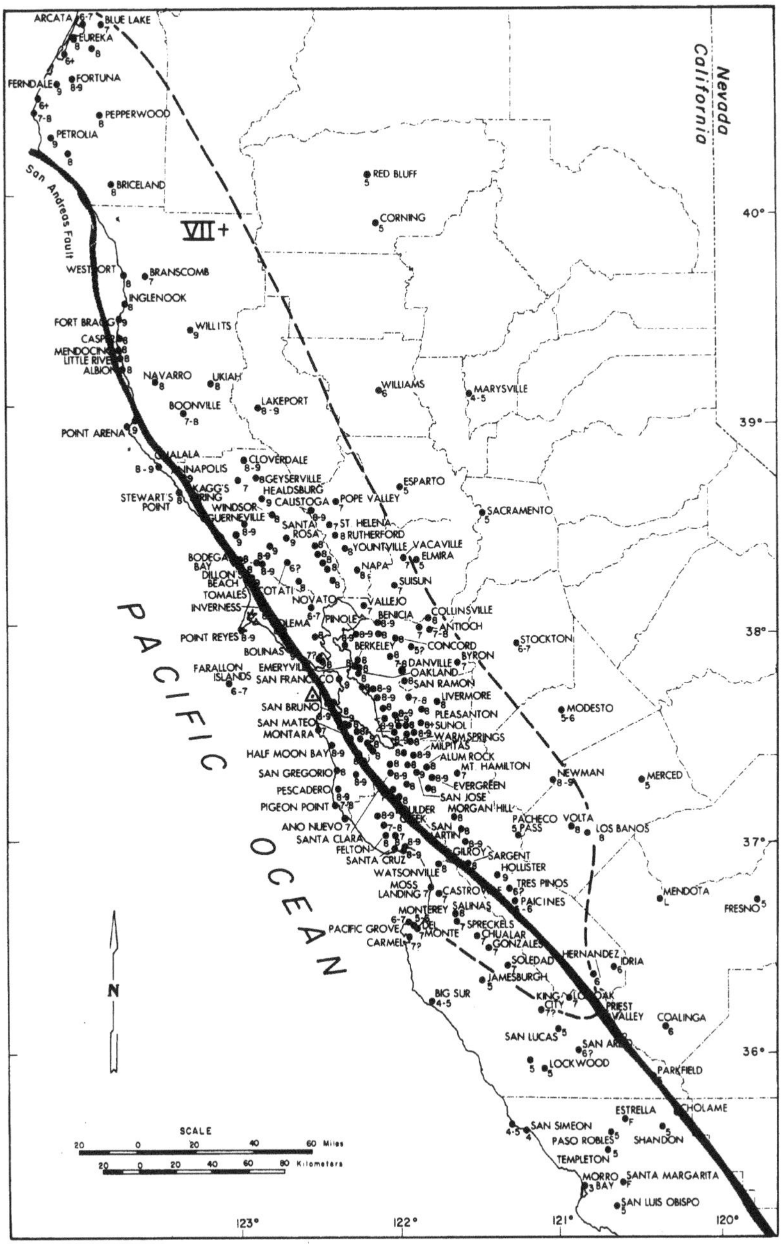

Figure 7.8 Intensity distribution of the 18 April 1906 San Francisco earthquake.

The earthquake was associated with the largest known length of slip along a fault plane in the contiguous USA. The observed extent of displacement on the San Andreas Fault was about 180 miles (290 km) in a N35°W direction. Assuming that the slip in the Cape Mendocino region is an extension of this, the total distance was 270 miles (435 km) from San Juan, San Benito County, to Telegraph Hill, Humboldt County (Figure 7.8). The greatest slip, 21 ft (6.3 m), was in Marin County. In many places, dislocation of fences and roads indicated the amount of movement (Figure 7.9). Motions of 10–15 ft (3–4.5 m) were not uncommon. In the region north of the Golden Gate some vertical slipping occurred, nowhere more than 3 ft (0.9 m) and, in general, not more than 1 ft (0.3 m). Features of the movement included cracks and landslides.

The region of destructive intensity extended 400 miles (644 km). The total area over which the shock was felt included most of California and parts of Nevada and Oregon. On or near the rift, buildings were destroyed and trees

Figure 7.9 Fence in Marin County offset by 8.5 ft (2.6 m) during the 1906 earthquake. The San Andreas fault runs left to right across the field at the spot where the fence is broken. (Source: US Geological Survey.)

were uprooted or broken off. One house was torn apart. Roads crossing the fault line were rendered impassable and pipelines were broken. This had an important effect in San Francisco, as lack of water supply let the fire get out of control, although it probably would have done so in any case. The amount of damage was unevenly distributed owing to subsurface conditions.

Buildings and structures in all parts of the city and county of San Francisco were damaged (Figure 7.10). Much of it was moderate in amount and character. Most chimneys toppled, plaster on walls and ceilings cracked, and there were many forms of permanent damage. In small districts on made land, where the intensity was augmented, pavements were buckled, arched and fissured; brick and frame houses of ordinary construction were severely damaged or destroyed; sewers and water mains were broken; and heavy streetcar tracks were bent into wave-like forms.

At Santa Rosa, although 19 miles (31 km) from the rift, destruction was great, and the intensity was higher than at most other points of comparable distance. The district lies directly inland from the region of greatest motion on the San Andreas Fault.

Another region of violent shock detached from the main San Andreas Rift Zone was in the Los Banos region of the western San Joaquin Valley where, more than 30 miles (48 km) from the rift, the intensity was fully IX. A long belt of high intensity extending northwest and southeast suggested activity of some other fault.

There was considerable cracking of alluvial surfaces. In addition to the great rift fissures, there were branch fissures. Avalanches and earth slumps occurred, and earth flows were observed where the ground contained excess water.

Trees swayed violently and were thrown down or snapped off in some cases. Springs and artesian wells either increased or decreased in flow. Where jets of water spurted through holes or fissures, a few mud or sand craterlets were formed. Although no tsunami (seismic seawave) was reported, water in the harbor was disturbed. Vibrations from the shock were felt on offshore vessels.

The great shock and resulting fires caused at least 700 deaths, about 50 of which were in Santa Rosa. According to figures compiled many years afterward by John R. Freeman, earthquake damage in San Francisco was $20 million and $4 million outside the city, but these estimates appear to be far below the actual amount.

7.2.4 Alaska earthquake of Good Friday 1964

This earthquake, one of the most violent ever recorded, was accompanied by vertical displacement over an area of 170 000–200 000 square miles ($(4.4–5.2)\times 10^5\,km^2$). The magnitude was about 8.3 and epicenter intensity IX–X. Vertical displacements ranged from 38 ft (11.4 m) of uplift to 7.5 ft (2.25 m) of subsidence relative to sea level, with seaward horizontal shifts of roughly 64 ft (19.2 m) and subsidiary reverse faulting with dip-slip offsets to 26 ft (7.8 m). The area affected included approximately 700 000 square miles ($1.8\times 10^6\,km^2$) of Alaska, and portions of western Yukon Territory and British Columbia, Canada.

Figure 7.10 San Francisco's California Street after the earthquake and fire of 18 April 1906.

Anchorage, about 80 miles northwest of the epicenter, sustained the greatest earthquake damage, although it was greater than the apparent intensity because of geological features. Violent ground motion triggered numerous rock slides, snow avalanches, and landslides throughout south-central Alaska. Fractures or cracks developed, chiefly in the unconsolidated deposits; mud spouts, slumping and subsidence resulted from compaction; and sandboils were produced. Cracks and pressure ridges were observed in lake and river ice.

About 30 blocks of dwellings and commercial buildings were severely damaged or destroyed in downtown Anchorage. The J.C. Penney Company store, a new five-story building, was damaged beyond repair (Figure 7.11).

The school system was hit hard. Denali Grade school was damaged by ground cracks. Anchorage High School was severely damaged. Government Hill Grade School, astride a landslide, was nearly a total loss. Water mains and gas, sewer, telephone and electricity systems were disrupted.

Landslides caused greatest devastation in Anchorage. Great slides occurred in the downtown business section, at Government Hill and at Turnagain Heights. Less destructive slides occurred in many other locations.

Figure 7.11 J.C. Penney Co. store in Anchorage after the earthquake of 27 March 1964; note that the building across the street appears undamaged.

Figure 7.12 Part of land in Turnagain Heights area of Anchorage, devastated by 1964 earthquake.

The Turnagain Heights slide was the largest, most complex, and most devastating landslide. It extended west to east along the bluff line about 8600 ft (2.58 km). The slide destroyed about 75 homes. A total area of about 130 acres ($5.25 \times 10^5 m^2$) was completely devastated by displacements that broke the ground into numerous deranged blocks, which were collapsed and tilted at odd angles (Figure 7.12).

This earthquake generated a tsunami that devastated towns along the Gulf of Alaska and left serious damage at Alberni and Port Alberni, Canada, along the west coast of the United States and in Hawaii.

7.2.5 San Fernando earthquake of 1971

This magnitude 6.4 earthquake was located in southern California, about 14 km north of San Fernando. Intensity XI was assigned to the destroyed Olive View Hospital, where three people were killed. In all, 58 people died, 49 of them at the San Fernando Veterans Administration Hospital; over 2000 injuries were reported.

Damage to public and private property was estimated at over $500 million. Hundreds of businesses and private houses had to be vacated; thousands of them sustained appreciable damage. The collapse of newly built, earthquake-resistant buildings at the Olive View Hospital in the northern Sylmar area reduced this $23.5 million complex to a total loss. Four of its five-story wings pulled away from the main building; three toppled. The second story of a two-story building in the complex dropped to ground level (Figure 7.13). This damage resulted from severe ground

Figure 7.13 Damaged parking shelter at Olive View Hospital after the 9 February 1971 San Fernando earthquake.

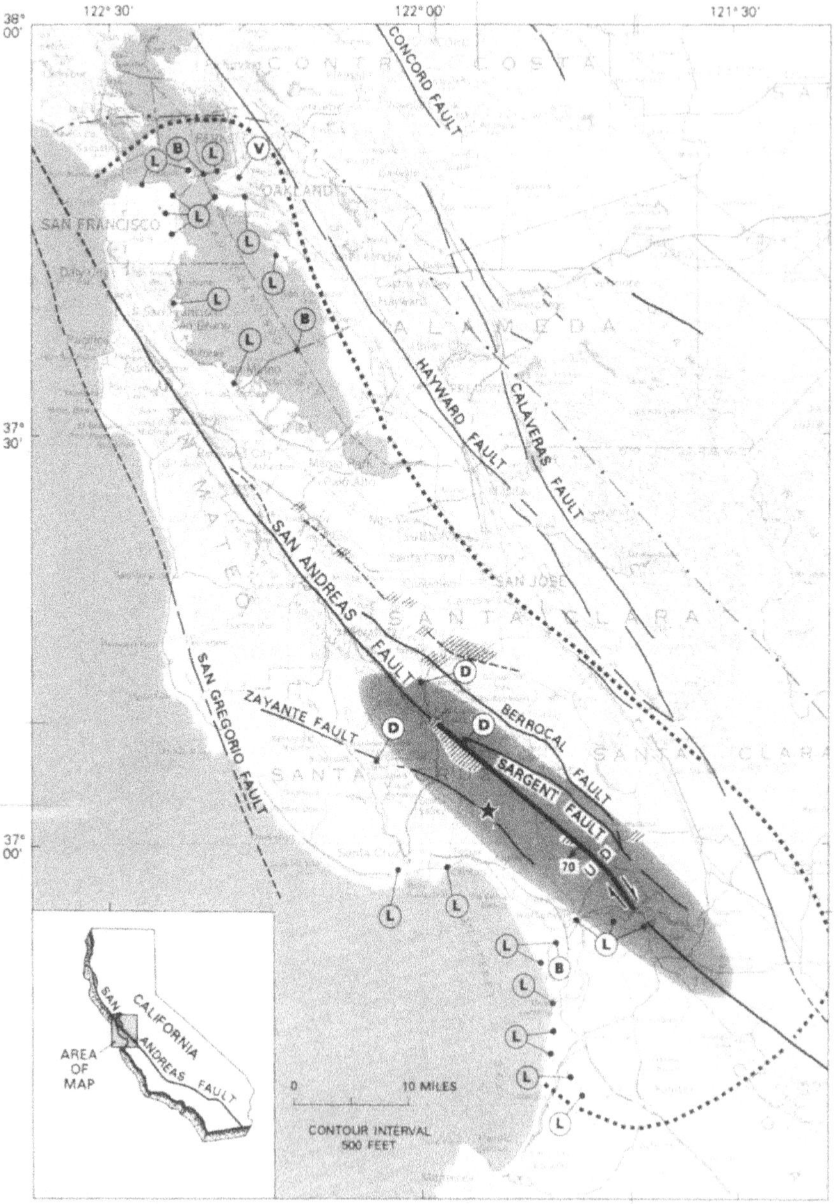

Figure 7.14 Main shock epicenter and inferred fault rupture of Loma Prieta earthquake, relative to areas of larger after shocks, abundant ground cracks and landslides, and to limits of structural damage.

shaking, not from ground faulting. Older, unreinforced masonry buildings collapsed at the Veterans Administration Hospital; freeway overpasses collapsed. The near failure of the Lower Van Norman Dam led authorities to evacuate thousands of residents in the area. Public utilities and facilities of all kinds were damaged, both above and below ground.

The earthquake created a zone of discontinuous surface faulting, named the San Fernando fault zone, that extends from the Bee Canyon area about 20 km eastward across the Sylmar–San Fernando area to the Big Tujunga Canyon area (north of Sunland). Severe ground fracturing and landslides were responsible for extensive damage in areas not associated with faulting. The most damaging slide occurred in the Upper Lake area of the Van Norman Lakes, where highway overpasses, railroads, pipelines and practically all structures and facilities in the slide were damaged severely. Many roads were blocked by numerous and widespread landslides and rockfalls.

No foreshocks were recorded, but 174 aftershocks of magnitude 3.0 or greater were recorded up to 07:36 on 10 February. Two of these shocks were of magnitude 5.8. The record from the main shock as recorded on the Pacoima Dam accelerograph revealed the highest acceleration ever measured – 1.25*g* horizontally, 0.72*g* vertically. This dam sustained $1.5 million damage to its abutments.

7.2.6 Loma Prieta earthquake of 17 October 1989

The magnitude 7.1 Loma Prieta earthquake is the largest earthquake to strike the San Francisco Bay region and environs, home to more than 5.9 million people, since the great San Francisco earthquake of 1906 (Figure 7.14). It was felt over an area of approximately 400 000 square miles (10^6 km^2) from Los Angeles (to the south) to the Oregon–California State

Figure 7.15 Collapsed cypress structure on Interstate 880, showing close-up of failed column that supported second deck; only four 5 cm diameter steel reinforcing bars were used in this joint (foreground).

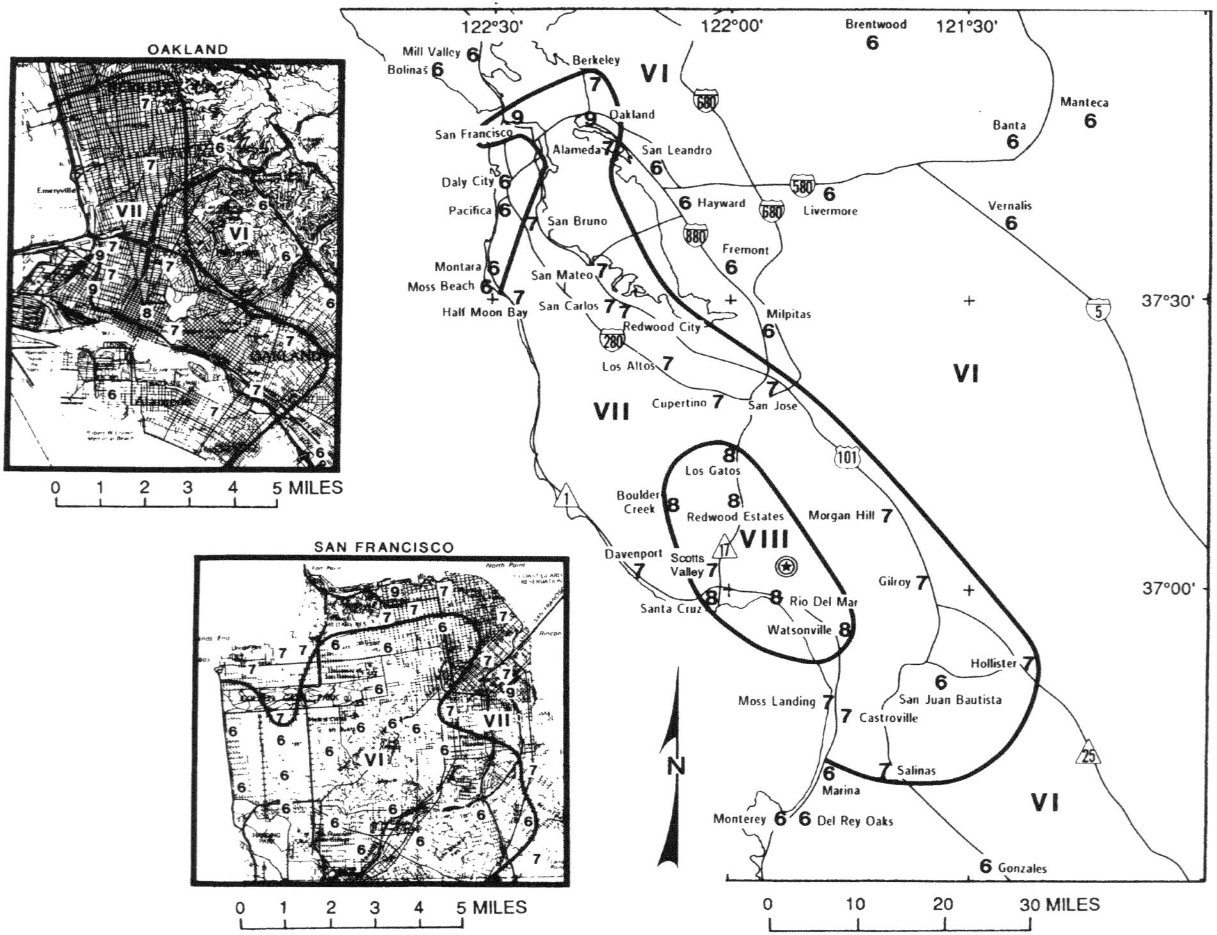

Figure 7.16 Intensity distribution of Loma Prieta earthquake; location of main shock epicenter is shown by a circled star. Insets show more detailed assessments in cities of San Francisco and Oakland.

line (to the north), and to western Nevada (to the east). Within about 15 seconds of seismic shaking of the region extending from Monterey Bay to northern San Francisco Bay, the Loma Prieta earthquake resulted in 62 known deaths, 3757 injuries and more than 12000 people homeless; over $6 billion property damage; and disruption of transportation, utilities and communications.

An unusual aspect of the earthquake is the absence of recognizable primary surface faulting; instead, a zone as much as 3 miles (4.8 km) wide of numerous ground cracks along and near the surface trace of the San Andreas fault suggests that the strain is distributed over a broad area. Displacement across many of these cracks is large enough to damage houses and roads; many of them also appear to have been the locus for land sliding. Similar features were observed in this region in 1906. This type of distributed cracking constitutes a previously unappreciated category of earthquake hazard, one that can extend well beyond the well-defined and narrow fault trace that typifies most of the San Andreas fault.

As observed in the 1906 earthquake, and predicted in maps of future earthquake effects, seismic shaking was locally amplified. Significant amplification occurred in areas of unconsolidated deposits and artificial fill over unconsolidated deposits that are widely distributed around the margins of Monterey Bay and San Francisco Bay and along lakes, lagoons and rivers. Amplified ground motion, which can affect structures on such deposits at great distances from the earthquake source, is suspected to be an important factor in damage to structures such as the Cypress section of Interstate 880 (Figure 7.15).

A preliminary assessment of the regional distribution of modified Mercalli intensity resulting from the Loma Prieta earthquake is shown in Figure 7.16. The intensity scale subjectively groups observations on earthquake effects into similar qualitative levels of shaking and then ranks the shaking levels into ascending order I–XII. Ground failure phenomena such as landslides, sandblows and liquefaction are generally relegated to MMI IX and higher. However, research and observations since the scale was developed shows that such effects can occur at lower shaking levels depending on a number of physical properties of the surficial materials, such as water content, permeability and degree of consolidation, and on the slope angles of hillsides and bluffs.

8 Earthquakes in Japan

8.1 Seismicity in Japan

Japan lies on the east side of the Eurasian plate where it collides with the Pacific plate and the Philippine plate, with the Pacific and the Philippine plates pushing westbound and underneath, as shown in Figure 8.1. In the same figure is shown also a small-scale sea-ridge in the Japan Sea. This plate-tectonic background controls the seismicity of Japan.

As a nation, Japan is most famous for frequent and strong earthquakes; strong earthquakes may occur almost anywhere in Japan. On the basis of tectonic background and the frequency and focal depth of the earthquake, the major earthquake belts in Japan may be grouped into three belts as follows (Figure 8.2).

8.1.1 The Outerside Belt

The Outerside Belt runs offshore of the Pacific Ocean side of Japan, with the northeastern part following roughly the subduction zone of the Pacific

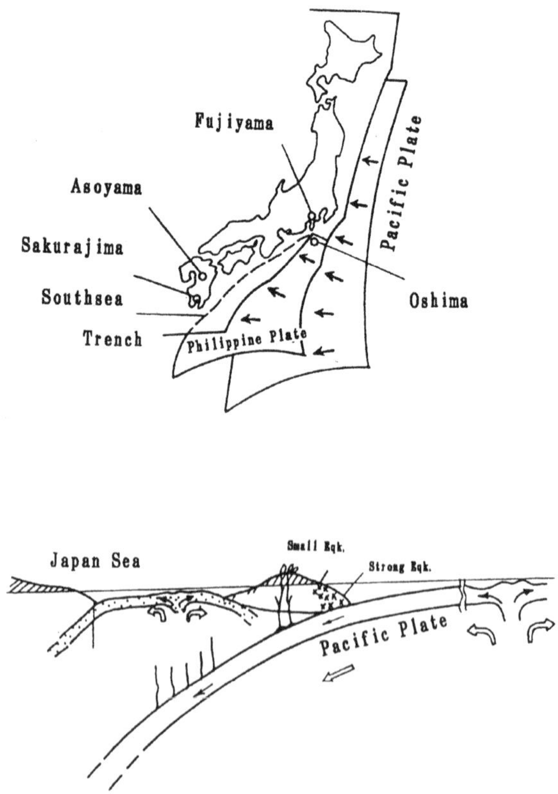

Figure 8.1 Plates and type of earthquakes around Japan.

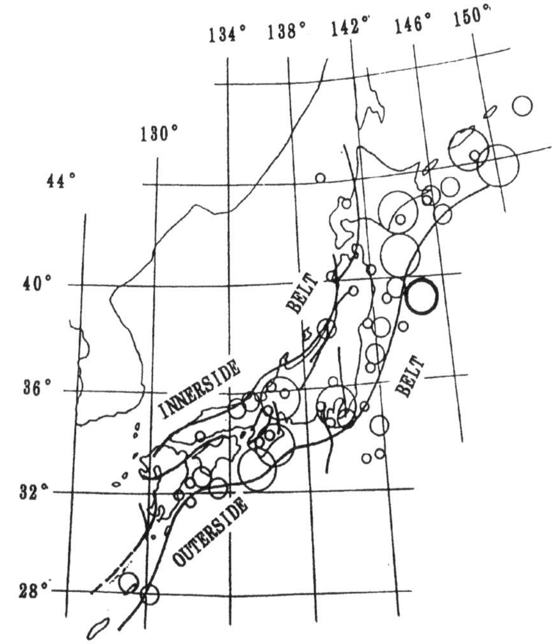

Figure 8.2 Earthquake belts and epicenters of major earthquakes (1888–1980) around Japan.

plate and the southwestern part the South Sea trench of the Philippine plate, and the junction of these two parts is thought to be in the Kanto district.

In the northeastern part, most of the earthquakes have had a focal depth in the range 20–70 km, but in a region near the coast of the Kanto district they have been of shallow depth, 10 km or less. The largest earthquake in this region was of magnitude 8. The average rate of large earthquakes is about once in 10–20 years. Direct damage has not usually been serious because large earthquakes occurring here have been mostly of deep focus and far away from the coast in the Pacific Ocean; indirect damage was primarily from tsunami due to the earthquake. Typical examples are the 1952 (M = 8.1) and the 1968 (M = 7.9) Tokachi-oki earthquakes offshore of Hokaito.

In the southwestern part, most of the earthquakes have had a focal depth less than 30 km and with a return period of 30–50 years for the strong earthquakes, somewhat longer than those occurring on the northeastern part. The maximum magnitude is also 8. Because of their shallow focus and closeness to the shoreline, the earthquakes on this part may cause considerable loss; tsunami is also a serious problem. A typical example is the 1944 Tonankai earthquake (M = 8.0).

8.1.2 The Innerside Belt

The Innerside Belt runs roughly along the Japan Sea shoreline of Japan, which is nearly the subduction zone of the second-order small Japan Sea plate thrusting underneath the Japan islands, with the southwestern part showing a narrow belt and the northeastern part showing a much wider zone. The focal depth of earthquakes on the southwestern part is within 30 km, mostly 10–20 km, and they are less frequent than those on the Outerside Belt, but their magnitude may be as high as 7.5. These earthquakes may cause heavy loss to the nearby inland cities because of their shallow focal depth, occasional inland occurrence and tsunami, such as in the 1948 Fukui earthquake.

8.1.3 The Inland Belts

There are several belts between the Innerside and the Outerside Belts as shown in Figure 8.2, with three belts running roughly northwesterly from the Kanto district to the Niigata area, three belts northeasterly from the south end of Honshu Island, and one belt through the southern part of Kyushu Island, running north.

Strong earthquakes occurring inland between 135–137°E are mostly related to these belts. The focal depth of most strong earthquakes has been less than 50 km and a majority of them occurred within 30 km. Because they occurred inland and with shallow focal depth and because of the dense population and developed industries, damage was usually very high. The 1891 Nobi earthquake of magnitude 7.9 is a typical example which was accompanied by several faults of roughly 100 km long extending in a NNE–SSE direction and with offsets of 2–6 m vertically and 2–8 m horizontally. Its aftershocks lasted for more than 10 years and its impact helped the founding of the Imperial Earthquake Investigation Committee in Japan.

Earthquakes in other inland regions are of medium magnitude, but considerable damage may be induced in their epicentral areas.

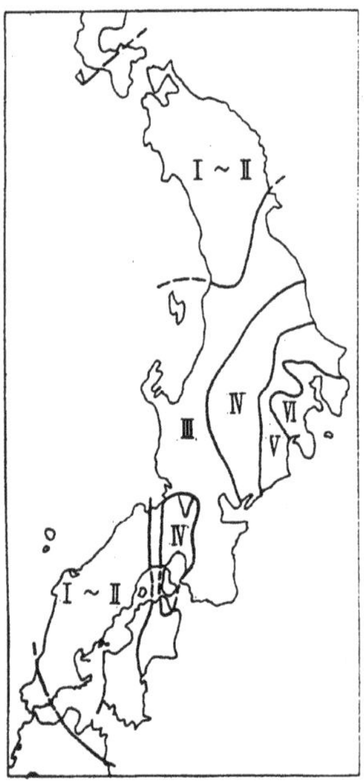

Figure 8.3 Isoseismals for the 1923 Kanto earthquake.

8.2 Damage caused by some specific earthquakes

8.2.1 Kanto earthquake of 1 September 1923

The magnitude was estimated more than 8 and the focal depth 0–10 km. Land on the northwestern part of the Kanto district was lowered by 1.4 m maximum and the coast on the southeastern part of Kanto was raised by 1.8 m maximum, with no fault rupture on land. The isoseismals for the earthquake are shown in Figure 8.3.

The total number of casualties reached 143 000, of which 90% was due to the induced fires, and the number of totally destroyed houses reached 128 000; but the fires burned another 447 000 houses, which shows clearly that the indirect loss in very large cities may be much heavier than the direct loss from the earthquake.

This earthquake is famous for the following features. First is the induced fire disaster; there were about 120 fires after the earthquake, starting from the early morning next day. Such heavy loss was due partly to the wooden houses and very strong wind at the time.

Second was the knowledge gained from the earthquake of the important effect of site conditions on earthquake damage. According to detailed investigations of damage and site conditions, it was perhaps for the first time realized that damage of wooden houses on hard sites uptown was less than that on soft sites downtown; but the opposite was true for adobe houses, i.e. damage of adobe houses on soft sites was less than that on hard sites, as shown in Figure 5.7, where the intensity rating is the JMA intensity scale and its intensity V is roughly intensity VIII on the MM intensity scale.

Third was that this earthquake helped to show the adequacy of earthquake design of buildings by considering the earthquake action as a horizontally applied equivalent static load with the magnitude of kW, where W is the total weight of the building and k is the so-called earthquake coefficient, usually taking roughly the value of one-tenth.

The fourth feature is of seismological engineering significance. The record of the main shock by the typical seismological instrument used showed a motion which might be taken for simplicity to be a simple harmonic motion of period 1.35 s and displacement amplitude 4.43 cm, and a rough estimation of the peak acceleration of $0.1g$ was obtained from simple harmonic waves. The design peak acceleration of $0.05–0.1g$ or $k = 0.05–0.1$, used for several decades afterwards, seemed quite reasonable. It took quite some time to realize that this is not the right way to reach a correct design value and that earthquake ground motion is much more complicated than simple harmonic motion. Also the actual maximum ground acceleration may be much higher than $0.1g$, although it may be acceptable to use $k = 0.1$ for the design of some structures.

Two years after the Kanto earthquake, in 1925, the Earthquake Research Institute was founded.

8.2.2 Niigata earthquake of 16 June 1964

This earthquake of magnitude 7.5 occurred on the Innerside Belt in Japan Sea, offshore north of Niigata and with a focal depth of 40 km; its isoseismals are shown in Figure 8.4.

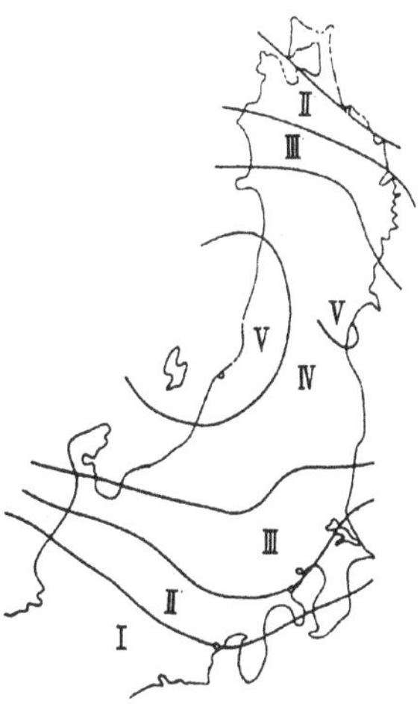

Figure 8.4 Isoseismals for 1964 Niigata earthquake.

Soil liquefaction data from this earthquake are widely known. The sketch shown in Figure 10.2 of the apartment buildings inclined or even leveled to the ground and the collapse of five out of 12 simple spans of the Showa Bridge over the Shinnano River are the often-quoted phenomena from this earthquake which show the effect of soil liquefaction. Our understanding of soil liquefaction and its effect on buildings, bridges, lifelines, roads and other engineering works has advanced much since this earthquake. Failure due to nonequal settlements from liquefaction is of a static nature, and the building may be inclined as a whole, so that it is not wise to strengthen the superstructure at all in the case of liquefaction. Acceleration records for the basement of an apartment building on a liquefied area were obtained before and after the liquefaction process of the foundation soil, which showed maximum peak acceleration of $0.16g$ horizontally, and lengthening of the motion (roughly with a period of 6 s) after the liquefaction. It is interesting to note that the leveled apartment building was otherwise nearly intact; even the windows and walls were not damaged.

A second feature of this earthquake is the danger of earthquake-induced fire and explosions from oil storage tanks, which can be highly dangerous in big modern cities.

Tsunami with a maximum height of 4 m occurred near the shore of the west coast.

8.2.3 Tokaichi-oki earthquake of 16 May 1968

Strong earthquakes occurred on the northern part of the Outerside Belt, offshore to the southeast of Hokaito, and the isoseismals for one of magnitude 7.9 with very shallow focal depth, which occurred on 16 May 1968, are shown in Figure 8.5.

The experience gained of modern structures is perhaps the main feature of this earthquake. A specialty committee from the Japan Society of Architecture summarized the lessons learned from it as follows.

1. The average strength of concrete was much lower than the design requirement.
2. The casting and reinforcement of the concrete were inadequate.
3. The foundation soil was not uniform.
4. The shearing strength and the ductility of columns and piles were insufficient.
5. The stress conditions from loading in two horizontal directions was not adequately considered.
6. Torsional effects were not sufficiently considered.
7. The earthquake loading considered for small parts, such as penthouses, on top of buildings was too small.
8. The design and construction of overlapping at joints and constructional joints were not good.
9. The complicated layering of the soils under buildings made the ground motion at the foundation very complicated.

Figure 8.5 Isoseismals for 1968 Tokaichi-oki earthquake.

Ground motion records show that components with a long period of $T = 2.7$ s were predominant at Hachinohe, 165 km from the epicenter, which is of importance to long-period structures.

Another important impact of this earthquake was its effect on the Japanese policy of earthquake disaster reduction, with more emphasis on engineering measures than on prediction and on the application of prediction than on the prediction research.

8.2.4 Miyakiken-oki earthquake of 12 June 1978

This earthquake of magnitude 7.4 occurred on the northern part of the Outerside Belt, 100 km from the coast of Miyaki and at a depth of 40 km, with Sendai City the most seriously damaged. Maximum intensity on land was V on the JMA scale (roughly VIII on the MM scale).

Because Sendai City is the largest modern city to be damaged, after Tokyo City in the 1923 Kanto earthquake, the experience gained from the earthquake disaster to this city of modern facilities is particularly important in studying the formation and modeling of both direct and indirect disasters, especially those from lifeline systems, 'high-tech' facilities and socio-economic damage.

The old Sendai City is located on hard sites and was thus damaged less than the new city, which is located on alluvial sites. The damage showed a feature of modern cities, i.e. that the damage to modern buildings and lifeline systems seriously affects functioning of the city or society, although the casualty numbered only 28. For example, the failure of some transformers forced the interruption of the electricity supply, which not only caused inconvenience to the residents of about 680 000 houses, but also stopped automatic transportation signals in many cities. Damage to the highway, railway and bridges, the collapse of large gas storage tanks, and damage to various water, electricity, gas, transportation, communication and other utility systems brought heavy losses to modern cities.

Another feature of modern city damage is the loss of memory and data stored in computers for many banks, industries and organizations, which stopped the banks running their services to the market, which in turn affected the normal societal and economic functioning. This took quite some time to re-establish. The Japanese declared it a third type of earthquake disaster.

8.2.5 Urakawa earthquake of 21 March, 1982

An earthquake of magnitude 7.1 struck the South coastline of Hokaito and the estimated intensity in Urakawa City was IV (VI–VII on the MM intensity scale). No casualty and no induced fires or other disasters were reported.

This was the third strong earthquake to strike this district after the second world war; the residents felt 299 earthquakes in 1981 and most residents experienced damage from earthquake and tsunami. The earthquake resistance of buildings was raised, and furniture fixed to floors. The Japanese considered that the mitigation of disasters from this earthquake was the result of the disaster culture or, in other words, the earthquake preparedness. A steady effort has armed the city against the attack of a strong earthquake, not only in engineering works but also in society organization. For example, during the sudden attack of the 1982 Urakawa earthquake, a new movie was showing in a theater with 120 people, mostly students from primary or high schools, watching. Some of them were screaming, but all remained seated.

8.2.6 Nihonkai-Chubu earthquake of 26 May 1983

This earthquake of magnitude of 7.7 occurred on the northern part of the Innerside Belt in the Japan Sea, with focal depth 14 km. The source zone extended roughly 100 km in the northeast direction, with its northeast end only 50 km from the western coasts of Akita and Aomori Prefectures in northern Honshu. It was a typical ocean earthquake. The damage was mostly caused by ground failure and tsunami; casualties amounted to 104, of which 100 were from tsunami.

Part Three

Theory of Earthquake Response of Structures

9 Earthquake response analysis of structures

9.1 Evolution of the theory of earthquake response analysis

Progress in the theory of structural analysis is closely related to progress in the understanding of earthquake ground motion and dynamic properties of structures, such as the consideration of the spectral content of earthquake ground motion, the nonlinear behavior of structures, the uncertainty of ground motion evaluation and the multiple levels of structural behavior in design, as well as the development of computer technology. The corresponding stages in earthquake response analysis are those of static, spectrum and dynamic theory (section 13.1).

9.1.1 Static theory

As stated in section 13.1, in the static theory, the earthquake load is specified as a static horizontal force P equal to a fraction ($k=a_{max}/g$) of the weight W of the structure and a_{max} is the acceleration:

$$P=kW=\frac{W}{g}a_{max} \tag{9.1}$$

and the fundamentals and format of equation 9.1 served as the foundation of static seismic design from the 1920s until the spectral theory came in the 1950s and caused modification to it.

9.1.2 Response spectrum theory

The design principle can be modified slightly by introducing a factor $\beta(T)$ to equation 9.1 as follows:

$$V_0=k\beta(T)\,W \tag{9.2}$$

where V_0 is the base shear of the structure and $\beta(T)=S_a(T)/a_{max}$.

9.1.3 Random vibration

Following the accumulation of accelerograms of real earthquake ground motion, the complex and random properties of earthquake motion were gradually understood and random vibration theory, widely used already in automatic control, was applied to describe both the ground motion and the structural response.

The random vibration theory treated earthquake motion as a random process and only statistical properties could be found and used to describe it. For example, no maximum could be given without a probability meaning. Through this theory, the modal combination necessary in response spectrum analysis was solved both theoretically and practically; a more important application of this theory was the development of seismic design theory from a factor of safety to reliability.

9.1.4 Numerical analysis of seismic response of structures

Following the rapid development and wide application of computers and computing technology in the 1960s and 1970s, together with the fast accumulation of accelerograms, it was realized that an equivalent static method of response spectrum analysis was not enough to control the safety of structures and that it was necessary to have a real dynamic method of structural analysis, namely time history analysis. At the first anniversary conference of the 1971 San Fernando earthquake, a majority of participants agreed upon the statement that the response spectrum theory stated only half of the problem and that there were other important factors to be considered in seismic design. Housner and Jennings (1972) pointed out: 'We conclude that the building code should be revised to reflect more realistically the actual accelerations, deformations and strengths of structures subjected to strong ground shaking. . . . first an increase to realistic levels of the simplified loadings . . . it is to the economic advantage of building to have realistic dynamic analyses made in the design of most multistory structures. Such dynamic analyses should be required for major structures'.

9.1.5 Vibration of nonlinear structures

Since the 1960s, it has been realized that nonlinear properties of ordinary structures, soils and foundations have to be considered or even used to resist the earthquake attack with reasonably economical construction cost. Efforts have been made to extend the response spectrum method to nonlinear response analysis (Newmark and Rosenblueth, 1971; Riddel and Newmark, 1979; Nau and Hall, 1982). Capability of inelastic or plastic deformation is usually expressed in terms of the ductility factor $\mu = e_{max}/e_y$, where e_{max} and e_y are respectively the maximum allowable and the elastic limit deformation of the structure or part of the structure. Emphasis on the importance and usefulness of ductility and the necessity of consideration of deformation response on structural safety and its application in seismic design codes were great advances in earthquake engineering in the 1960s and 1970s. Perhaps firstly in Japan (Section 14.2.1) and later in other design codes for structures other than nuclear power plants (such as the API code, Section 13.4.9), a two-level design method was specified: one to satisfy the functional purpose of the structure and to remain in the elastic stage, for smaller ground motion, and another to guarantee the safety of the structure against much larger ground motion by allowing the structure to go beyond the elastic stage without collapse. Only when working in the inelastic stage of the structural response, can one realize the importance of the dynamic process of earthquake behavior. This is the real dynamic stage of response analysis.

9.2 Earthquake response of SDF systems and response spectrum

9.2.1 Numerical analysis

As shown in Figure 9.1, a linear SDF system of one mass m is supported on a horizontal spring. On the mass, there is a horizontal dynamic force

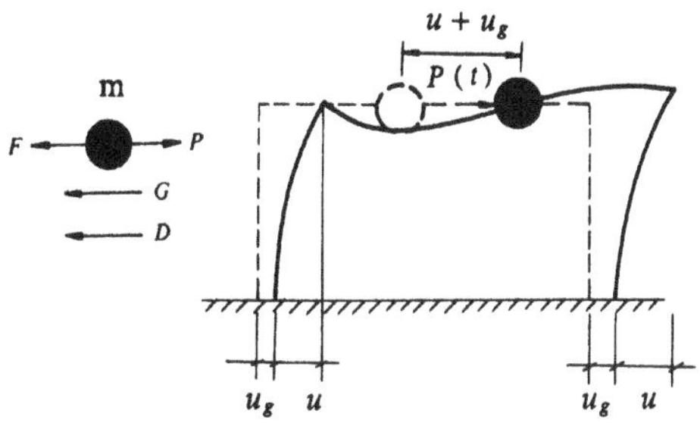

Figure 9.1 Response of an SDF System.

$P(t)$, and at the foundation a horizontal ground motion $\ddot{u}_g(t)$. The controlling equation of motion is

$$m\ddot{u}+C\dot{u}+ku=P-m\ddot{u}_g=P(t)$$

If only earthquake motion is considered, then $P(t)\equiv 0$, and the foregoing equation may be written in its common form

$$\ddot{u}+2\zeta\omega\dot{u}+\omega^2u=\frac{P}{m}-\ddot{u}_g=-\ddot{u}_g \tag{9.3}$$

where $\zeta\omega=C/(2m)$, $\omega^2=k/m$, ζ is the damping ratio, and ω the natural frequency of the system. Only small damping $\zeta\ll 1$ is considered.

Because earthquake motion $\ddot{u}_g(t)$ is irregular, equation 9.1 has to be solved numerically. For linear systems, the integration of the differential equation can be solved by a Duhamal integral as follows. On the basis of the superposition principle,

$$u(t)=\int_0^t \ddot{u}_g(\tau)h(t-\tau)\,d\tau \tag{9.4}$$

where $h(\tau)$ is the unit impulse response function or the response $u(t)$ of the system at time t when subjected to a unit impulse $\ddot{u}_g(0)=\delta(0)$ at $t=0$, and $\delta(t)$ is the Dirac function. For an SDF system,

$$h(t)=-\left[\left(\frac{1}{\omega_D}\right)\exp(-\zeta\omega t)\right]\sin\omega_D t \tag{9.5}$$

where $\omega_D=\omega(1-\zeta^2)^{1/2}$ is the damped natural frequency. The responses of relative displacement $u(t)$, relative velocity $\dot{u}(t)$ and the absolute acceleration $\ddot{u}(t)+\ddot{u}_g(t)$ of the SDF system with small damping under the action of ground motion $\ddot{u}_g(t)$ are

$$u(t)=-\frac{1}{\omega_D}\int_0^t \ddot{u}_g(\tau)\exp[-\zeta\omega(t-\tau)]\sin\omega_D(t-\tau)\,d\tau$$

$$\dot{u}(t)=-\frac{\omega}{\omega_D}\int_0^t \ddot{u}_g(\tau)\exp[-\zeta\omega(t-\tau)]\cos[\omega_D(t-\tau)+\alpha]\,d\tau$$

$$\ddot{u}(t)+\ddot{u}_g(t)=\frac{\omega^2}{\omega_D}\int_0^t \ddot{u}_g(\tau)\exp[-\zeta\omega(t-\tau)]\sin[\omega_D(t-\tau)+2\alpha]\,d\tau$$

where $\tan\alpha=\zeta/(1-\zeta^2)^{1/2}$. When damping is small, these equations are reduced to the following:

$$u(t)=-\frac{1}{\omega}\int_0^t \ddot{u}_g(\tau)\exp[-\zeta\omega(t-\tau)]\sin\omega_D(t-\tau)\,d\tau \quad (9.6a)$$

$$\dot{u}(t)=-\int_0^t \ddot{u}_g(\tau)\exp[-\zeta\omega(t-\tau)]\cos\omega_D(t-\tau)\,d\tau \quad (9.6b)$$

$$\ddot{u}(t)+\ddot{u}_g(t)=\omega\int_0^t \ddot{u}_g(\tau)\exp[-\zeta\omega(t-\tau)]\sin\omega_D(t-\tau)d\tau=-\omega^2u(t) \quad (9.6c)$$

There are several methods applicable to integrate the foregoing equations, all using a computer. But in the early days analog and graphic phase plane methods were used.

All numerical integration methods work step by step with a small time interval τ, in order to find the structural responses at the end of the present interval from those known at the end of the previous interval. The equation of motion now used is in the finite difference form as follows:

$$\ddot{u}_{t+\tau}+2\zeta\omega\dot{u}_{t+\tau}+\omega^2u_{t+\tau}=-\ddot{u}_g(t+\tau) \quad (9.7)$$

Newmark's method (Clough and Penzien, 1975, p. 263; Newmark and Rosenblueth; 1971 p. 15) and Wilson's θ method (Clough and Penzien, 1975, p. 268) are now perhaps the most commonly used.

Newmark's method assumes a linear change of acceleration within one time interval, and his result is

$$\dot{u}_t+\Delta t\dot{u}_t+(\ddot{u}_t+\ddot{u}_t+\Delta t)\frac{\Delta t}{2} \quad (9.8a)$$

$$u_{t+\Delta t}=u_t+\dot{u}_t\Delta t+\ddot{u}_t\frac{(\Delta t)^2}{2}+(\ddot{u}_{t+\Delta t}-\ddot{u}_t)\frac{(\Delta t)^2}{6} \quad (9.8b)$$

These equations, together with equation 9.7, can easily be solved for three unknown quantities at the end of the present interval $t+\Delta t$. The stability requirement is $\Delta t/T\leqslant 0.551$ and the convergence requirement is $\Delta t/T\leqslant 0.389$; but the routine is to take $\Delta t<T/8$.

Wilson's method assumes that the acceleration is linear in an interval $\tau=\theta\Delta t$ and $\theta>1$ to avoid divergence from oscillation. His result is

$$\dot{u}_{t+\tau}=\dot{u}_t+\ddot{u}_t\tau+(\ddot{u}_{t+\Delta t}-\ddot{u}_t)\frac{\tau^2}{2\Delta t} \quad (9.9a)$$

$$u_{t+\tau}=u_t+\dot{u}_t\tau+\ddot{u}_t\frac{\tau^2}{2}+(\ddot{u}_{t+\Delta t}-\ddot{u}_t)\frac{\tau^3}{6\Delta t} \quad (9.9b)$$

Similarly, these equations are solved simultaneously with equation 9.7 to find the acceleration after $t+\theta\Delta t$ and the acceleration at $t+\Delta t$ can be obtained by interpolation, and finally the results at the end of interval $t+\Delta t$ are computed from equation 9.9.

It is proved that, when $\theta > 1.37$, this method is stable. Usually θ is taken to be 2 or 1.4.

When using Newmark's method with a large value of $\Delta t/T$, the response oscillates about the correct solution. Wilson's method is designed to avoid this oscillation by using first a large step, say $\theta = 2$ or $\tau = 2\Delta t$, and finally the response at the middle point of the step, $\tau/2 = \Delta t$, will be closer to the correct solution. Another feature of Wilson's method is the effect of artificial damping introduced in the algorithm; and responses of higher modes will be suppressed for a multi-degree-of-freedom (MDF) system. It is an advantage if interest is concentrated only on responses of lower modes; it is a disadvantage, however, if responses of higher modes are needed. Newmark's method does not introduce artificial damping.

9.2.2 Response spectrum

Suppose there are a series of N oscillators, or SDF systems, with natural periods $T_i (i = 1, 2, \ldots, N)$ and the same damping ratio ζ. Under a given ground motion $\ddot{u}_g(t)$, the maximum absolute acceleration of each oscillator or SDF system $S_a(T, \zeta)$, where

$$S_a(T, \zeta) = |\ddot{u}(t) + \ddot{u}_g(t)|_{max}$$

can be found numerically through equation 9.6, which is the so-called acceleration spectrum. Similarly, the velocity spectrum S_v and displacement spectrum S_d can also be obtained. It is clear from equation 9.6 that response spectra are functions of period T and damping ratio ζ.

The routine way of computing the response spectra is to define a pseudo-velocity spectrum from equation 9.6:

$$\mathrm{PSV}(T, \zeta) = \left| \int_0^t \ddot{u}_g(\tau) \exp[-\zeta\omega(t-\tau)] \sin[\omega_D(t-\tau)]\, d\tau \right|_{max} \tag{9.10}$$

For small damping, the relative displacement and absolute acceleration spectra are then respectively

$$S_d(T, \zeta) = \frac{1}{\omega} \mathrm{PSV}(T, \zeta) \tag{9.11a}$$

$$S_a(T, \zeta) = \omega \mathrm{PSV}(T, \zeta) \tag{9.11b}$$

The difference between the pseudo-velocity spectrum $\mathrm{PSV}(T, \zeta)$, equation 9.10, and the velocity spectrum from equation 9.6

$$S_v(T, \zeta) = \left| \int_0^t \ddot{u}_g(\tau) \exp[-\zeta\omega(t-\tau)] \cos[\omega_D(t-\tau)]\, d\tau \right|_{max}$$

is not large if the duration of random input $\ddot{u}_g(t)$ is long enough.

It is also common, particularly in China, to use the amplification or normalized spectrum $\beta(T, \zeta) = S_a(T, \zeta)/\ddot{u}_{gmax}$ in place of the spectrum S_a, though it is still referred to as the acceleration spectrum.

9.2.3 Response spectrum theory and its fundamental assumptions

The response spectrum theory may be stated as follows: a structure may be considered as an MDF system; the maximum response of each mode can be obtained from the response spectrum; and a rough approximate maximum response of the structure may be obtained by a combination of the spectral responses of the first few modes.

There are at least three fundamental assumptions in this theory. Firstly, the structure is linear and of constant damping, and the superposition law holds. Secondly, the structural supports are rigid. Thirdly, the response spectrum can be given or specified before the earthquake occurs. Although none of these assumptions is likely to be true, it is still a simple and acceptable way to design a structure.

9.2.4 Properties of response spectrum

The response spectrum has the following major properties.

1. The relative values of displacement, velocity and acceleration of the SDF system with respect to the ground are zero for absolutely rigid systems, and the absolute acceleration of the system is equal to that of the ground.
2. The relative values of displacement, velocity and acceleration of the SDF system for absolutely flexible systems are equal to the corresponding values of the ground, and the absolute acceleration of the system is equal to zero.
3. The peak acceleration of the ground motion is controlled by the higher frequency components, peak velocity by the moderate frequency components and peak displacement by the lower frequency components. As shown in Figure 9.9 (section 9.7), Newmark and others suggested a way of estimating the response spectrum from the peak values of ground acceleration a, velocity v and displacement d.
4. All three response spectra can be plotted in one graph of log-scales as shown in Figure 9.8 (section 9.7) because of the approximate relations between them as given in Equation 9.11.
5. The relations among response spectra of different damping values may be given as follows:

$$\beta(T, \zeta) = \beta(T, \zeta = 0.05) \cdot \lambda(T, \zeta) \tag{9.12a}$$

$$\lambda(T, \zeta) = (16.6\zeta + 0.16)^{1/3}\left(\frac{0.8}{T}\right)^{\alpha} \tag{9.12b}$$

$$\alpha(\zeta) = \frac{0.05 - \zeta}{0.156 + 3.38\zeta} \tag{9.12c}$$

The Seismic Guidelines JEAG 4601 of Japan (Japan Association of Electricity, 1987) adopted

$$\lambda(T, \zeta) = \left[1 + 17(\zeta - 0.05)\exp\left(-\frac{2.5T}{T_{\mathrm{d}}}\right)\right]^{-1/2} \tag{9.13}$$

9.3 Earthquake response of MDF systems

Real structures are mostly MDF systems, such as buildings of several stories, bridges, dams, smokestacks etc. For the earthquake response analysis of real structures, there are two kinds of method available. One is to simplify the real structure into a continuous simple model, such as cantilever beam, for analysis. The other method is to reduce the real structure into a discrete model, which can be solved using a computer. The latter method is used increasingly because of the fast development of computers.

9.3.1 Equation of motion

As an example of the approach for MDF systems, Figure 9.2 shows a simplified sketch of a two-story building. Assume that all masses are concentrated at floor levels, that floors are rigid in the plane and that only ground motion in one horizontal direction is considered. Similarly to the case of the SDF system, the equation of motion in matrix form is

$$\boldsymbol{M}\ddot{\boldsymbol{u}} + \boldsymbol{C}\dot{\boldsymbol{u}} + \boldsymbol{K}\boldsymbol{u} = \boldsymbol{P} - \boldsymbol{M}\boldsymbol{I}\ddot{u}_{\mathrm{g}} \tag{9.14}$$

where $\boldsymbol{M}$, $\boldsymbol{C}$ and $\boldsymbol{K}$ are $n \times n$ matrices, $\ddot{\boldsymbol{u}}$, $\dot{\boldsymbol{u}}$, $\boldsymbol{u}$ and $\boldsymbol{P}$ vectors of rank n and $\boldsymbol{I}$ is a unit matrix, but $\ddot{u}_{\mathrm{g}}$ is a scalar process. For example,

$$\boldsymbol{M} = \begin{pmatrix} M_1 & 0 & 0 & \dots & 0 \\ 0 & M_2 & 0 & \dots & 0 \\ \vdots & & & & \vdots \\ 0 & . & . & \dots & M_n \end{pmatrix} = \begin{pmatrix} M_1 & & & & \\ & M_2 & & & 0 \\ & & \cdot & & \\ & 0 & & \cdot & \\ & & & & M_n \end{pmatrix}$$

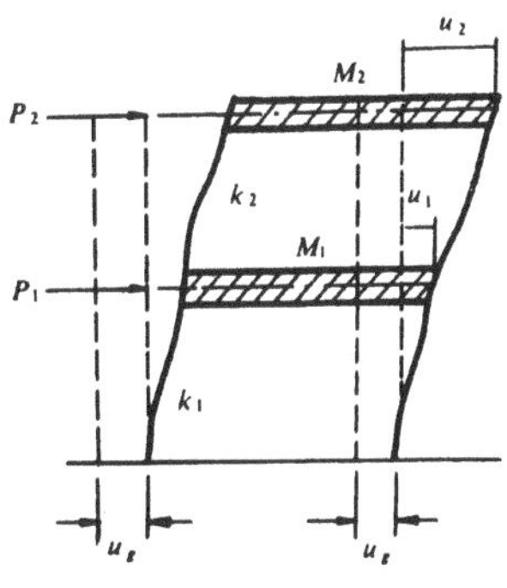

Figure 9.2 Example of an MDF System.

9.3.2 Natural frequencies and modes

The dynamic response of the structure is determined by forces $\boldsymbol{P}$ and/or excitations $\ddot{u}_{\mathrm{g}}$ applied to the structure as well as the natural properties of the structure. Dynamic properties of a structure include natural frequencies, modes and damping, which are related to the structure only. Let $\boldsymbol{P} \equiv \boldsymbol{0}$, $\ddot{u}_{\mathrm{g}} \equiv 0$ and neglect damping, i.e. $\boldsymbol{C} \equiv \boldsymbol{0}$, equation 9.14 becomes

$$\boldsymbol{M}\ddot{\boldsymbol{u}} + \boldsymbol{K}\boldsymbol{u} = \boldsymbol{0} \tag{9.15}$$

The natural frequencies are the eigenvalues and the modes are the eigenvectors of the foregoing equation.

Let the solution be

$$\boldsymbol{u} = \boldsymbol{U}\sin\omega t$$

Equation 9.15 becomes

$$[-\omega^2 \boldsymbol{M} + \boldsymbol{K}]\boldsymbol{U} = \boldsymbol{0} \tag{9.16}$$

Only when $\boldsymbol{U}$ is not identically zero, can the system be in a state of vibration and equation 9.15 have a nontrivial solution. If so, we then have the eigen or characteristic equation

$$|\omega^2 \boldsymbol{M} + \boldsymbol{K}| = 0 \tag{9.17}$$

Eigenvalues or natural frequencies ω_i $(i = 1, 2, \ldots, n)$ can be obtained from this equation, and, by substituting them in equation 9.15 the eigenvectors or the vibrational modes can be obtained.

Take the structure shown in Figure 9.12 as an example, together with the following numerical data: $k_1 = 30$ t/cm, $k_2 = 20$, $M_1 = M_2 = 0.1\,\mathrm{t \cdot s^2/cm}$. The characteristic equation is

$$-\omega^2 = \begin{vmatrix} 0.1 & 0 \\ 0 & 0.1 \end{vmatrix} + \begin{vmatrix} 50 & -20 \\ -20 & 20 \end{vmatrix} = \begin{vmatrix} 0.1\omega^2 - 50 & 20 \\ 20 & 0.1\omega^2 - 20 \end{vmatrix} = 0$$

Two roots of the equation are $\omega_1^2 = 100$ and $\omega_2^2 = 600$ or $T_1 = 0.628$ s and $T_2 = 0.256$ s; and the modes are (taking arbitrarily the value of u_1)

$$\boldsymbol{\Phi} = [\boldsymbol{\Phi}_1 \boldsymbol{\Phi}_2] = \begin{vmatrix} u_1(1) & u_2(1) \\ u_1(2) & u_2(2) \end{vmatrix} = \begin{vmatrix} 1 & 2 \\ 2 & -1 \end{vmatrix}$$

where ω_1 and $\boldsymbol{\Phi}_1$ are respectively the natural frequency and mode for the first, and ω_2 and $\boldsymbol{\Phi}_2$ for the second. The modes are shown in Figure 9.3. It is well known that the modes are mutually orthogonal:

$$\boldsymbol{\Phi}_i^T \boldsymbol{M} \boldsymbol{\Phi}_j = 0 \qquad (i \neq j) \tag{9.18a}$$

$$\boldsymbol{\Phi}_i^T \boldsymbol{K} \boldsymbol{\Phi}_j = 0 \qquad (i \neq j) \tag{9.18b}$$

9.3.3 Modal analysis

Because the natural modes of a system are complete orthogonals, any responses $\boldsymbol{u}(t) = [u_1(t)\, u_2(t) \ldots u_n(t)]^T$ of a linear system can be expanded in a series of natural modes $\boldsymbol{\Phi}_1, \boldsymbol{\Phi}_2, \ldots, \boldsymbol{\Phi}_n$:

$$\boldsymbol{u}(t) = \sum \boldsymbol{\Phi}_i q_i(t) = \boldsymbol{\Phi} \boldsymbol{q}(t)$$

where $\boldsymbol{q}$ is a vector of amplitude of the generalized coordinates of modes. The equations of motion in terms of the natural modes are

$$\ddot{q}_i + 2\zeta_i \omega_i \dot{q}_i + \omega_i^2 q_i = P_i^*(t)/M_i^* \quad (i = 1, 2, \ldots, n) \tag{9.19}$$

where

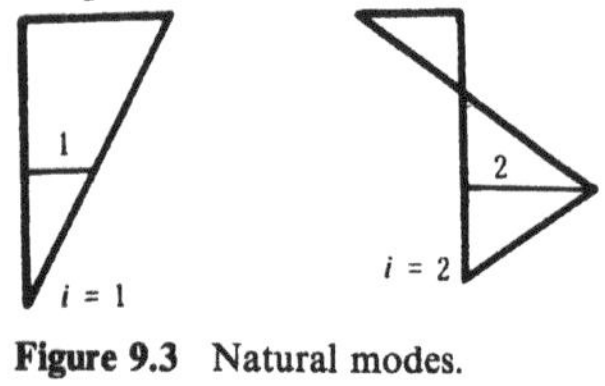

Figure 9.3 Natural modes.

$$\left.\begin{array}{ll} \text{generalized mass} & M_i^* = \boldsymbol{\Phi}_i^T \boldsymbol{M} \boldsymbol{\Phi}_i \\ \text{generalized damping} & C_i^* = \boldsymbol{\Phi}_i^T \boldsymbol{C} \boldsymbol{\Phi}_i \\ \text{generalized stiffness} & k_i^* = \boldsymbol{\Phi}_i^T \boldsymbol{K} \boldsymbol{\Phi}_i \\ \text{generalized load} & P_i^* = \boldsymbol{\Phi}_i^T (\boldsymbol{P} - \boldsymbol{M I} \ddot{u}_g) \\ \text{modal damping} & \zeta_i = C_i^* / (2\omega_i M_i^*) \\ \text{natural frequency} & \omega_i = (k_i^2 / M_i^*)^{1/2} \end{array}\right\} \tag{9.20}$$

In the case of earthquake motion only, $\boldsymbol{P}(t) = \boldsymbol{0}$, and

$$\frac{P_i^*(t)}{M_i} = -\left(\frac{\boldsymbol{\Phi}_i \boldsymbol{M I}}{\boldsymbol{\Phi}_i^T \boldsymbol{M} \boldsymbol{\Phi}_i}\right) \ddot{u}_g = -\gamma_i \ddot{u}_g \tag{9.21}$$

where

$$\gamma_i = \frac{\boldsymbol{\Phi}_i \boldsymbol{M I}}{\boldsymbol{\Phi}_i^T \boldsymbol{M} \boldsymbol{\Phi}_i} = \frac{\sum M_j u_i(j)}{\sum M_j u_i^2(j)} \tag{9.22}$$

is the participation factor of the ith mode; the summation is over $j = 1, 2, \ldots$. If the nondimensional form of mode function

$$Y_i(j) = \frac{u_i(j) \sum M_j}{\sum M_j u_i(j)} \tag{9.23}$$

is used, the equation of motion is

$$\ddot{q}_i + 2\zeta_i \omega_i \dot{q}_i + \omega_i^2 q_i = -\gamma_j \ddot{u}_g \qquad (i = 1, 2, \ldots, n) \tag{9.24}$$

The final result is then that, if the response is written as

$$\boldsymbol{u}(t) = \sum \boldsymbol{\Phi}_i \gamma_i q_i(t) \tag{9.25}$$

instead of equation 9.19, the equation of motion for each mode is

$$\ddot{q}_i + 2\zeta_i \omega_i \dot{q}_i + \omega_i^2 q_i = -\ddot{u}_g \qquad (i = 1, 2, \ldots, n) \tag{9.26}$$

which is identical to that for the SDF system, equation 9.3.

If $\gamma_i \boldsymbol{\Phi}_i = \gamma_i [Y_i(1)\ Y_i(2) \ldots Y_i(n)]^T$ is taken as the mode and is used instead of $\boldsymbol{\Phi}_i = [u_i(1)\ u_i(2) \ldots u_i(n)]^T$ in equation 9.20, the generalized mass will be

$$M_i^* = \gamma_i^2 \sum M_j Y_i^2(j) \tag{9.27}$$

and the mode participation factor and generalized mass are then

$$\gamma_i = \frac{\sum M_j Y_i(j)}{\sum M_j Y_i^2(j)} = \frac{\left[\sum M_j u_i(j)\right]^2}{\left[\sum M_j\right] \sum M_j u_i^2(j)} \tag{9.28}$$

$$M_i^* = \frac{\left[\sum M_j Y_i(j)\right]^2}{\sum M_j Y_j^2(j)} = \frac{\left[\sum M_j u_i(j)\right]^2}{\sum M_j u_i^2(j)} \tag{9.29}$$

From equation 9.28, it is then obtained that

$$\gamma_i = \frac{M_i^*}{\sum M_j} \quad \text{and} \quad \sum \gamma_i = 1 \tag{9.30}$$

The advantage of using nondimensional modes is to have positive and gradually decreasing participation factors $\gamma_i (i = 1, 2, \ldots)$.

It is seen that the original coupled equation 9.14 can be decoupled into the n equations 9.26.

9.3.4 *Methods of earthquake response analysis*

There are three methods to find the earthquake response for a linear MDF system. The first is the step-by-step integration method as explained in section 9.2.1; the second is the response spectrum method to find a safe maximum of the response from the maxima of several modal responses; and the third is to solve the equation of motion in the frequency domain as explained in section 9.4.

(a) Step-by-step integration

When this method is applied to an MDF system, the procedure is the same as for an SDF system. If equation 9.14 is used, the only change is to use matrices in place of the scalar quantities, and to solve the simultaneous equations instead of equations 9.26. If the decoupled equations 9.26 are used to solve the modal responses first, the final solution has to be combined according to equation 9.25.

Of course, in equation 9.25, the responses $q_i(t)$ can be superposed only at the same time t, not the maxima of modes. If maxima or the processes are found for some quantities, such as inertia forces at each floor, the maximum story shear or deformation cannot be found from the maxima of inertia forces. It is free from mistakes if the modal responses of the quantity required are obtained first for each mode and then combined either according to equation 9.25 or by another method of modal superposition to obtain the total response.

Another important difference of integration between MDF and SDF systems is in the selection of the time interval of integration. When the decoupled equations 9.26 are used, the requirement is to have a sufficiently small value of $\Delta t/T_i$, and Δt is different for different modal response. When the coupled equation 9.14 is used, the selection of Δt is controlled by the highest natural frequency of the system desired; the value of Δt becomes very small and integration is difficult by Newmark's method. Because of the artificial damping introduced, Wilson's method is helpful. In practice, integration through the decoupled equations is preferred.

(b) Response spectrum method

Instead of finding the structural response for a given ground motion $\ddot{u}_g(t)$ in the step-by-step integration method, the response spectrum method finds the maximum structural response from the given response spectrum $S_a(\omega)$ of the ground motion.

The response spectrum method, when applied to an MDF system, gives only an approximate result, but in a very easy way. Given an MDF system, the characteristic equation 9.15 is solved first for natural frequencies, modes and their participation factors. The second step is to select the damping ratios for each mode (section 9.6). The third step is to read the maximum response for each mode from the given spectrum $S_a(T, \zeta)$ of the given ground motion. The last step is to combine the maximum responses of all modes considered to have the maximum response of the system. The approximation comes from the last step.

9.4 Response analysis in the frequency domain

For linear systems, analyses in time domain and in the frequency domain are entirely equivalent, and frequency domain analysis has been widely used since the development of the FFT algorithm. Table 9.1 shows their advantages and disadvantages. An example of a frequency-related system is the soil–structure interaction, where the dynamic property of the soil foundation is frequency-related. The nonlinear system is popular since structures under very strong earthquake motion usually go beyond the elastic limit. There is no exact method of analysis for a system of both nonlinear and frequency-dependent properties.

9.4.1 Frequency-domain analysis of an SDF system

(a) Solution in Fourier Series

In the equation of motion of an SDF system (equation 9.3)

$$m\ddot{u} + C\dot{u} + ku = P - m\ddot{u}_g = p(t) \tag{9.31}$$

In the case of earthquake problems, the loading term $p(t)$ is usually highly irregular (Figure 9.4). In mathematics, it can be expanded into simple harmonic components as

$$p(t) = a_0 + \sum[a_j\cos\omega_j t + b_j\sin\omega_j t] \quad (j = 1, \infty)$$

$$= \sum C_j\exp(i\omega_j t) \qquad (j = -\infty, \infty)$$

Table 9.1 Comparison of time and frequency domain analyses

System	Can be used in	
	Time domain	Frequency domain
Linear, with constant parameters	Yes	Yes
Nonlinear	Yes	No
Linear, with frequency-dependent parameters	No	Yes
Nonlinear, with frequency-dependent parameters	No Yes, with frequency-averaged parameters	No Yes, withleariztion

where coefficient $C_j=(a_j-ib_j)/2$ for $j>0$ or $C_j=(a_i+ib_j)/2$ for $j<0$ is complex, and $C_0=a_0$; $i=(-1)^{1/2}$; $\omega_{-j}=\omega_j$; $C_j=C^*_{-j}$. The Fourier development gives the following result:

$$p(t)=\sum C_j\exp(\mathrm{i}\omega_j t) \tag{9.32}$$

$$C_j=\frac{1}{T_\mathrm{p}}\int_0^{T_\mathrm{p}} p(t)\exp(-\mathrm{i}\omega_j t)\,\mathrm{d}t$$

where T_p is the total duration of motion, as shown in Figure 9.4, which includes a dynamic part and a quiescent part, and the latter is added for the convenience of Fast Fourier Transform (FFT) computation. In FFT, the total duration T_p is divided into N intervals of Δt, and $N=2^M$, where M and N are integers. The most commonly used N value is 1024, 2048 or 4096. The selection of $\Delta t=T_\mathrm{p}/N\leqslant T_{\min}/2$ is controlled by the accuracy requirement of high-frequency (period $T_{\min}$) components; and T_p should be several times the longest period considered.

(b) Transfer function

The transfer function plays an important role in frequency domain analysis. It is the ratio of the stationary response to the simple harmonic input. When $p(t)=C\exp(\mathrm{i}\omega t)$, the stationary response of an SDF system is

$$u(t)=H(\mathrm{i}\omega)\,C\exp(\mathrm{i}\omega t)$$

and the transfer function is then

$$H(\mathrm{i}\omega)=\frac{u(t)}{p(t)}=\frac{1}{-m\omega^2+\mathrm{i}C\omega+k}$$

$$=\frac{(1/k)}{1-\beta^2+2\mathrm{i}\zeta\beta} \tag{9.33a}$$

where $\beta=\omega/\omega_0$ is the ratio of the input motion frequency ω and the natural frequency $\omega_0=(k/m)^{1/2}$ of the system. The transfer function is complex; its modulus $|H(i\omega)|$ is the ratio of the amplitudes of the response and the input, and the phase angle of $H(i\omega)$ is the difference between the phase angles of the response and the input. There are many transfer functions. For earthquake motion $\ddot{u}_\mathrm{g}(t)$,

$$\ddot{u}+2\zeta\omega_0\dot{u}+\omega^2u_0=-\ddot{u}_\mathrm{g}$$

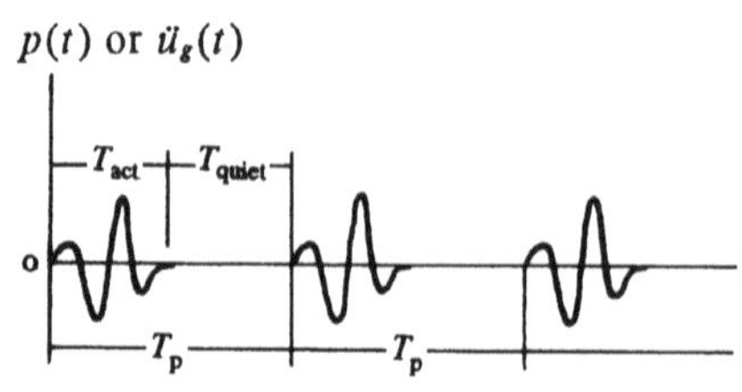

Figure 9.4 Irregular process $p(t)$ or $\ddot{u}_g(t)$.

and the transfer function of the relative displacement $u(t)$ of the system to the input $\ddot{u}_g(t)$ is, from equation 9.33a,

$$H_u(i\omega) = (\omega_0^2 - \omega^2 + 2i\zeta\omega_0\omega)^{-1} = [(\omega_0^2 - \omega^2)^2 + 4\zeta^2\omega_0^2\omega^2]^{-1/2}\exp(i\phi) \quad (9.33b)$$

$$\phi = \tan^{-1}\left(\frac{2\zeta\omega_0\omega}{\omega_0^2 - \omega^2}\right)$$

From the relations between the spectra of a function and its derivative, equation 3.20, it is easy to find the transfer functions of relative velocity $\dot{u}(t)$ and the absolute acceleration $\ddot{u}(t) + \ddot{u}_g(t)$ to $\ddot{u}_g(t)$:

$$H_{\dot{u}}(i\omega) = \omega H(i\omega) = \frac{\omega}{\omega_0^2 - \omega^2 + 2i\zeta\omega_0\omega} \quad (9.33c)$$

$$H_{\ddot{u}+\ddot{u}_g}(i\omega) = 1 + \omega^2 H(i\omega) = \frac{\omega_0^2 + 2i\zeta\omega}{\omega_0^2 - \omega^2 + 2i\zeta\omega_0\omega} \quad (9.33d)$$

With the transfer function found, the corresponding response in the time domain can be obtained from equation 9.32:

$$u(t) = \sum H_u(i\omega_j)\, C_j(\omega) \exp(i\omega_j t) \quad (9.34)$$

9.4.2 Frequency domain analysis of an MDF system

The procedure is exactly the same as for the SDF system. Take earthquake motion input as an example. The equation of motion is

$$\boldsymbol{M}\ddot{u} + \boldsymbol{C}\dot{u} + \boldsymbol{K}\boldsymbol{u} = -\boldsymbol{M}\boldsymbol{I}\ddot{u}_g(t) \quad (9.35)$$

The Fourier transformation of this equation is

$$(-\omega^2\boldsymbol{M} + i\omega\boldsymbol{C} + \boldsymbol{K})\, \boldsymbol{U}(i\omega) = -\boldsymbol{M}\boldsymbol{I}\ddot{U}_g(i\omega) \quad (9.36a)$$

or, in terms of the transfer function matrix,

$$\boldsymbol{H}(i\omega) = \boldsymbol{U}^0(i\omega)$$

which is the ratio of the stationary system response matrix $u(t) = U^0(i\omega)\exp(i\omega t)$ to input in the frequency domain, equation 9.35a is then

$$\boldsymbol{U}(i\omega) = \boldsymbol{H}(i\omega) * \boldsymbol{I}\ddot{U}_g(i\omega) \quad (9.36b)$$

The computation of transfer function matrix $H(i\omega)$ is time-consuming. The response matrix $\boldsymbol{u(t)}$ can then be found from $\boldsymbol{U}(i\omega)$ by a Fourier transformation. To avoid the computation of $H(i\omega)$, the modal response in frequency domain can be found from equation 9.26:

$$Q_j(i\omega) = H_j(i\omega)\, \ddot{U}_g(i\omega) \quad (9.37)$$

After finding the modal response $q_j(t)$ in the time domain by Fourier transformation, the total response can be found from equation 9.25.

As mentioned before, any other response can be found in the same way if the corresponding transfer function is used.

9.5 Vertical earthquake response and P–Δ effect

The vertical earthquake action is important in at least the following cases:

1. for those structures where the earthquake load is mainly vertical, such as cantilever structures;
2. for those structures sensitive to vertical earthquake motion, such as gravity dams, whose stability is maintained by their own weight;
3. for highrise structures where the P–Δ effect may be very important;
4. for those structures located in the epicentral area, where vertical earthquake motion may be much larger than expected.

9.5.1 Upthrowing phenomena

There have been many records of objects thrown up in the air in historical earthquakes. The earliest dependable record is perhaps the one for the Assam, India, earthquake, reported by R.D. Oldham. He wrote that stones originally lying on the ground jumped in the air like beans on a drum. The upthrow seems a very quick action so that the edge of the original shallow depression is still undisturbed. Other records mentioned that columns were thrown out of the original foundation with no sign of horizontal vibration. Upthrowing has been reported also in many other countries.

Upthrowing means that the objects have been accelerated over $1.0g$. But care should be taken when calculating acceleration from upthrowing. Two factors are important: the interaction of the object and the support or ground, and the frequency characteristics of the peak acceleration.

9.5.2 Vertical response of structures

The methodology for vertical and horizontal response analysis is the same in principle. However, the difference in vibrational properties of the structure has strong influences in response analysis. There are at least the following three influences.

Firstly, the amplitude and spectral content of earthquake motion vary with the magnitude and distance of the earthquake, but differently in the horizontal and vertical directions; the relative amplitude of the vertical motion is higher for sites in the epicentral region.

Secondly, the possible variation of both amplitude and frequency content is to be considered, when analyzing the response in the vertical direction. On the basis of observational and theoretical results the following can be stated.

1. The predominant period of horizontal earthquake motion is close to that of the second or higher mode of the multistory buildings (say, over ten stories) in the horizontal direction, but the predominant period of the vertical earthquake motion is close to that of the fundamental period in the vertical direction.
2. For multistory buildings, the main response comes from the first mode for vertical responses, but possibly the higher modes for the horizontal responses.

3. The peak acceleration of the vertical motion is roughly half that of the horizontal, but may be around 1.0 for strong motion in the epicentral area (Figure 5.6).
4. For maximum acceleration response at top, the amplification in the vertical direction is larger than that in the horizontal direction (Hu and Hu, C1983).

Thirdly, the vertical vibration response of vertical members is mainly from axial forces while the horizontal response is from shear and bending. The hysteretic property for different loads may be different, especially in the nonlinear range; the P–Δ effect will produce a descending branch in the force–deformation curve.

9.5.3 P–Δ effect

In an elastic system, the P–Δ effect may also introduce nonlinearity. This effect is very important for highrise structures and inelastic structures.

For a single mass system as shown in Figure 9.5, the P–Δ effect is the secondary moment, produced by the axial force $F(t)$ and the vertical inertial force $M(\ddot{v}+\ddot{v}_g)$ on the deflection u, which is equivalent to an additional horizontal force

$$[F(t)+M(\ddot{v}+\ddot{v}_g)]\frac{u}{H}$$

where $\ddot{v}_g$ is the vertical ground acceleration and $\ddot{v}$ is the relative vertical acceleration of the mass. Because vertical vibration is also considered, there are now two degrees of freedom and two equations of motion as follows:

$$M\ddot{u}+C_u\dot{u}+k_u u=P(t)-M\ddot{u}_g-[F(t)+M(\ddot{v}+\ddot{v}_g)]\frac{u}{H} \quad (9.38a)$$

$$M\ddot{v}+C_v\dot{v}+k_v v=F(t)-M\ddot{v}_g \quad (9.38b)$$

where C_u, C_v and k_u, k_v are respectively the damping coefficient and stiffness of the system in the horizontal and vertical direction. It is seen here that the last term in equation 9.38a makes the problem non-linear.

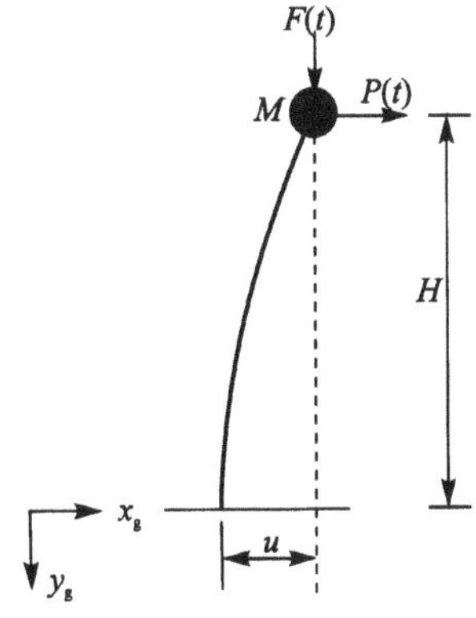

Figure 9.5 Effect of vertical vibration.

The nonlinear equations of motion may be solved by the methods described in section 9.7. In many practical cases, vertical dynamic amplification is neglected when considering horizontal vibration, i.e. $\ddot{v}$ is taken as zero, and the equations of motion are independent.

When considering simultaneously the horizontal and vertical vibrations and also the *P*–Δ effect, because of nonlinearity, the response has the following features.

1. Vertical action will increase the effect of the horizontal vibration; the greater the horizontal response without vertical action, the greater is the combined effect of the total response.
2. For bending of cantilevered structures, consideration of vertical ground motion increases the bending at the upper part of the structure.
3. The *P*–Δ effect is large for highrise structures and structures of long natural period.
4. The *P*–Δ effect is even larger for nonlinear structural response, which increases the nonlinear deformation and permanent deformation; this is consistent with the introduction of a descending branch in the hysteretic curves by the *P*–Δ effect.

9.6 Theory of damping

Damping is a term used to describe the energy-absorbing capacity of the material or structure at the elastic stage. In earthquake engineering problems, in addition to radiation or geometric damping, in principle, damping may be classified into two categories: internal and external. Internal damping is also called material damping, which represents the energy dissipated as heat by the internal friction of the material. For a structure, this will be in the form of friction at joints between members. External damping means energy dissipation from the interaction between the system studied and the external liquid, gas, or electric and magnetic field. Damping may also be classified into dry frictional and viscous according to the mechanical mechanism. The former is used for friction between the contact surfaces of two solids. This section discusses only viscous damping.

The damping ratio of ordinary structures in the linear range is usually about 5%, which is considered small. For small damping, the difference between different damping theories is not large and viscous damping is popularly used because of its simplicity.

9.6.1 Viscous and complex damping theories

Viscous damping theory assumes that the damping force is directly proportional to velocity; the stress strain relation is then

$$\sigma = E\varepsilon + C'\dot{\varepsilon} \tag{9.39}$$

where C' is the damping coefficient, E the elastic modulus and ε and $\dot{\varepsilon}$ are respectively strain and strain rate. For an SDF system subject to harmonic excitation, the equation of motion is

$$m\ddot{u}+C\dot{u}+ku=p_0\sin\omega t \tag{9.40}$$

Complex damping theory assumes that the damping force is proportional to the elastic force ku but in phase with velocity. The total stress is then

$$\sigma=E\varepsilon_0\exp(i\omega t)+\nu E\varepsilon_0\exp\left[i\left(\omega t+\frac{\pi}{2}\right)\right]=(1+i\nu)\,E\varepsilon \tag{9.41}$$

where ν is a constant related to damping. For an SDF system subject to simple harmonic excitation, the equation of motion is

$$m\ddot{u}+(1+i\nu)\,ku=p_0\exp(i\omega t) \tag{9.42}$$

which is a differential equation with complex coefficients, and its solution will be complex. The response of the system under an input of the real part of the right-hand term is given by the real part of the solution.

The stationary response to excitation $p_0\sin\omega t$ according to viscous damping theory is

$$u(t)=\frac{p_0}{k}(D_1^2+D_2^2)\frac{1}{2\sin(\omega t+\theta)} \tag{9.43}$$

$$D_1=1-\beta^2,\quad D_2=\frac{C\beta}{m\omega_0},\quad \theta=\tan^{-1}\left(\frac{D_2}{D_1}\right)$$

$$\Delta W=\frac{\pi C\omega p_0^2}{k^2}(D_1^2+D_2^2)^{-1}$$

where $\beta=\omega/\omega_0$, and according to complex damping theory

$$u(t)=\frac{p_0}{k}(D_1^2+\nu^2)^{-1/2}\sin(\omega t+\theta) \tag{9.44a}$$

$$\theta=\tan^{-1}\left(\frac{\nu}{D_1}\right) \tag{9.44b}$$

$$\Delta W=\pi\nu k\frac{p_0^2}{k^2}(D_1^2+\nu^2)^{-1} \tag{9.44c}$$

where ΔW is the energy dissipated in one cycle, and $C/(m\omega_0)=2\zeta\beta$.

The logarithmic decrement is, for viscous damping,

$$\lambda=\frac{2\pi}{[(2m\omega_0/C)^2-1]^{1/2}}=\frac{2\zeta\pi}{(1-\zeta^2)^{1/2}} \tag{9.45}$$

and for complex damping,

$$\lambda=\pi\nu \tag{9.46}$$

A comparison of these two sets of results shows that, if

$$\nu=D_2=\frac{C\omega}{m\omega_0^2}$$

the results for stationary vibration of these two sets will be indentical. For free vibration, this relation becomes $\nu = C/(m\omega_0)$ which also makes the results for free vibration identical.

In the earthquake engineering field, the routine is to use $C/m = 2\zeta\omega_0$, where ζ is the ratio to critical damping. Therefore,

$$\nu = \frac{2\zeta\omega}{\omega_0} \tag{9.47}$$

This is the condition that the viscous and complex damping theories will give the same results.

As indicated by Equation 9.47, ζ and ν cannot be both constant for varying frequency ω. Because earthquake motion is complicated and its spectrum is usually wide, in the case of small damping, say $\zeta < 0.05$ or 0.1, structural response is usually controlled by those components of frequency near ω_0 and other components do not contribute much. Therefore, the damping coefficient is that for $\omega \approx \omega_0$

$$\nu \approx 2\zeta \tag{9.48}$$

9.6.2 Equivalent damping (Iwan and Gates, 1979)

When damping is not as assumed above, an equivalent damping is used as a simple substitute. The criterion of equivalence is the approximately equal energy dissipation ratio, i.e.

$$\zeta = \frac{\Delta W}{4\pi W} \tag{9.49}$$

which can be used, in principle, only for small damping. However, for large elasto-plastic responses, this kind of equivalent damping is often used just for simplicity and ease of analysis.

9.6.3 Rayleigh damping

The damping matrix of an MDF system can be written in terms of the mass and stiffness matrices,

$$\boldsymbol{C} = a\boldsymbol{M} + b\boldsymbol{K} \tag{9.50}$$

where a and b are constants. $\boldsymbol{C}$ is orthogonal because both $\boldsymbol{M}$ and $\boldsymbol{K}$ are. The equation of motion 9.35 may now be decoupled in terms of generalized coordinates of natural modes $\boldsymbol{\Phi}$,

$$\ddot{q}_i + 2\zeta_i\omega_i\dot{q}_i + \omega_i^2 q_i = \frac{p_i^*}{M_i^*}$$

where ζ_i and ω_i are respectively the damping ratio and frequency of the ith mode and

$$\zeta_i = \frac{a}{2\omega_i} + \frac{b\omega_i}{2} \quad (i = 1, 2, \ldots) \tag{9.51a}$$

If the ith and $(i+1)$th modal damping ratios are known, the foregoing equation gives the constants a and b in terms of the known damping ratios:

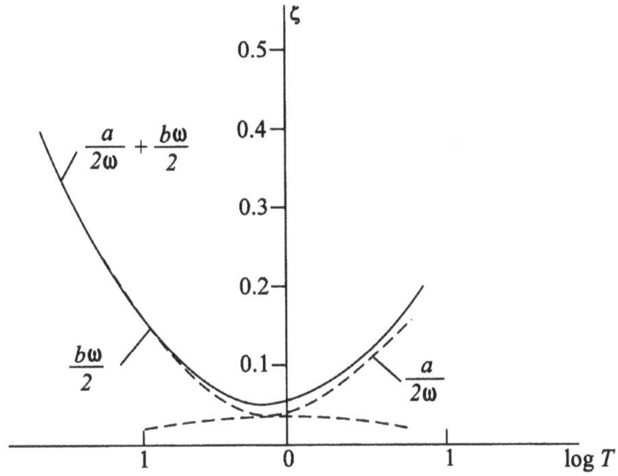

Figure 9.6 Variation of modal damping.

$$a=2\frac{(\zeta_i/\omega_i)-(\zeta_{i+1}/\omega_{i+1})}{(1/\omega_i^2)-(1/\omega_{i+1}^2)} \tag{9.51b}$$

$$b=2\frac{\zeta_{i+1}\omega_{i+1}-\zeta_i\omega_i}{\omega_{i+1}^2-\omega_i^2} \tag{9.51c}$$

This result shows that there are two modal damping ratios that one can assign because there are two constants a and b to be determined; other modal damping ratios will be determined automatically from equation 9.51a.

Figure 9.6 shows the curves determined by the first, second and both terms of equation 9.51a for $\zeta_i=\zeta_{i+1}=0.05$, $\omega_i=2\pi/1.0$, and $\omega_{i+1}=2\pi/0.3\,\text{Hz}$. It can be seen that the damping ratio for period $T=2\pi/\omega>1.0$ or $<0.3\,\text{s}$ increases so fast that the responses of those modes will be greatly suppressed. It is important to realize then, when using Rayleigh damping, that the responses of those modes between the modes with assigned damping are greatly enlarged because the damping ratios are too small and for those modes outside the responses are greatly suppressed.

9.6.4 Constant damping

A few field measurement data do not indicate any definite change of damping with modes; a constant damping ratio model is then sometimes used.

The constant damping model may be derived from the following assumption. As pointed out in section 9.6.1, complex damping assumes that the damping force is proportional to the elastic restoring force, but in phase with velocity. For an SDF system, the damping force is

$$C\dot{u}=cK|u|\frac{\dot{u}}{|\dot{u}|}=cK\left|\frac{u}{\dot{u}}\right|\dot{u} \tag{9.52}$$

where both C and c are constants related to damping. For stationary vibration,

$$\left|\frac{u}{\dot{u}}\right| = \frac{1}{\omega} \tag{9.53}$$

and the damping force vector, damping vector and modal damping are respectively

$$\boldsymbol{C}\dot{\boldsymbol{u}} = \frac{c\boldsymbol{K}\dot{\boldsymbol{u}}}{\omega} \tag{9.54}$$

$$\boldsymbol{C} = \frac{c\boldsymbol{K}}{\omega} \tag{9.55}$$

$$\zeta_i = \frac{c}{2} \tag{9.56}$$

It is seen then that, if constant damping and Rayleigh damping are combined, the damping matrix is

$$\boldsymbol{C} = a\boldsymbol{M} + b\boldsymbol{K} + c\left|\frac{u}{\dot{u}}\right|\boldsymbol{K} \tag{9.57}$$

and the modal damping ratio is

$$\zeta_i = \frac{a}{2\omega_i} + \frac{b\omega_i}{2} + \frac{c}{2} \tag{9.58}$$

where a, b and c are non-negative constants.

9.7 Nonlinear earthquake response

Nonlinear* structural analysis is very important in earthquake engineering because ordinary structures will undergo nonlinear and inelastic deformation due to local or partial damage under strong earthquake excitation. The effect is usually to reduce the stiffness and increase damping for reinforced concrete, timber and steel structures. There are three effective methods for nonlinear analysis, namely step-by-step integration, response spectrum and equivalent linearization. Considering nonlinear structural deformation and strength as dual controlling design safety indices is the main achievement of the last two decades in earthquake engineering.

9.7.1 Step-by-step integration

Step-by-step integration is perhaps the most powerful and widely applicable method of dynamic analysis, especially for irregular excitation and nonlinear responses. Its fundamental procedure is similar to that described in section 9.2 for linear analysis. The important difference with the linear cases is that the structural properties, such as the stiffness matrix, are not constant, but functions of the state of deformation, described by displacement and velocity matrices $\boldsymbol{u}$ and $\dot{\boldsymbol{u}}$, and thus have to be evaluated at the beginning of each time interval (Clough and Penzien, 1975).

* The word 'nonlinear' is used here to include also inelastic, as is usual in the earthquake engineering field.

Figure 9.7 Incremental approach of nonlinear analysis.

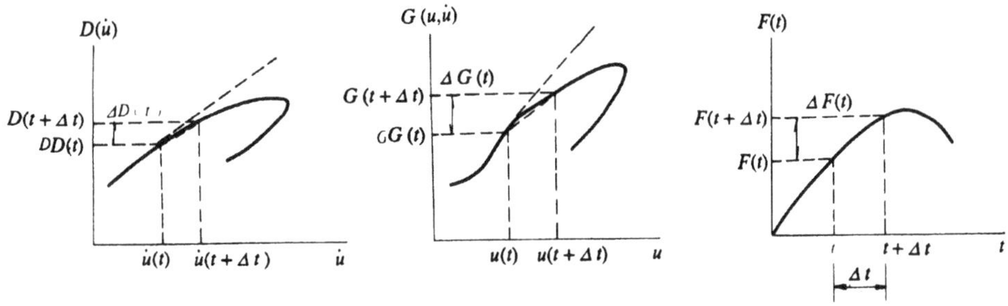

Figure 9.7 shows the hysteretic restoring force $G(u, \dot{u})$ and damping force $D(\dot{u})$ of an SDF system. The external force is $F(t)$ and the intertial force $I(t)$. The equation of motion at time t is

$$I(t)+D(t)+G(t)=F(t)$$

and at time $t+\Delta t$ is

$$I(t+\Delta t)+D(t+\Delta t)+G(t+\Delta t)=F(t+\Delta t) \tag{9.59a}$$

The difference between them is the incremental equation of motion

$$\Delta I(t)+\Delta D(t)+\Delta G(t)=\Delta F(t) \tag{9.59b}$$

where the increments of the forces are

$$\Delta I(t)=m\Delta\ddot{u}(t)$$

$$\Delta D(t)=C(t)\,\Delta\dot{u}(t)$$

$$\Delta G(t)=k(t)\,\Delta u(t)$$

$$\Delta F(t)=F(t+\Delta t)-F(t)$$

By definition, $C(t)$ and $k(t)$ are respectively the slope of the secant of damping $D(\dot{u})$ and restoring forces $G(u, \dot{u})$, as shown by the dashed lines in the figure. The slope of the secant can be found only by iteration because it is related to the quantities to be determined at the end of the current step. To avoid this complication, the following approximations are usually introduced:

$$C(t)\approx\frac{\partial D(t)}{\partial \dot{u}},\quad k(t)\approx\frac{\partial G(t)}{\partial u}$$

i.e. the use of tangent for secant. Now the incremental equation is

$$m\Delta\ddot{u}(t)+C(t)\,\Delta\dot{u}(t)+k(t)\,\Delta u(t)=\Delta F(t) \tag{9.60}$$

There are several methods available to solve this equation numerically, but only the most popular, that of linear acceleration, is given below. From equations 9.8, the velocity and displacement increments on the interval are

$$\Delta\dot{u}(t) = \ddot{u}(t)\,\Delta t + \Delta\ddot{u}(t)\frac{\Delta t}{2}$$

$$\Delta u(t) = \dot{u}(t)\Delta t + \ddot{u}(t)\frac{(\Delta t)^2}{2} + \Delta\ddot{u}(t)\frac{(\Delta t)^2}{6}$$

Now, select $\Delta u(t)$ as the basic unknown and write other unknown increments in terms of $\Delta u(t)$ and response quantities at the end of the previous step through the foregoing equations:

$$\Delta\ddot{u}(t) = \frac{6}{(\Delta t)^2}\Delta u(t) - \frac{6}{\Delta t}\dot{u}(t) - 3\ddot{u}(t) \tag{9.61a}$$

$$\Delta\dot{u}(t) = \frac{3}{\Delta t}\Delta u(t) - 3\dot{u}(t) - \frac{\Delta t}{2}\ddot{u}(t) \tag{9.61b}$$

Substituting these relations into equation 9.60, we obtain

$$\tilde{K}(t)\,\Delta u(t) = \Delta\tilde{F}(t) \tag{9.62a}$$

$$\tilde{K}(t) = k(t) + \frac{6}{(\Delta t)^2}m + \frac{3}{\Delta t}C(t) \tag{9.62b}$$

$$\Delta\tilde{F}(t) = \Delta F(t) + m\left[\frac{6}{\Delta t}\dot{u}(t) + 3\ddot{u}(t)\right] + C(t)\left[3\dot{u}(t) + \frac{\Delta t}{2}\ddot{u}(t)\right] \tag{9.62c}$$

In computation, one has to find the increment $\Delta u(t)$ at the end of the present step, and then the increment $\Delta\dot{u}(t)$. It is always a good practice to obtain from the equation of motion the acceleration

$$\ddot{u}(t) = [F(t) - D(t) - G(t)]/m \tag{9.63}$$

to avoid the accumulation of error from the approximation by using the tangent slope for the secant slope.

9.7.2 Response spectrum method

In the period 1973–76 when Newmark and Hall summarized the findings in linear and nonlinear response spectra, they suggested using them in the practice of earthquake analysis. In 1979, they modified their suggestion on the basis of a large number of additional data (Riddel and Newmark, 1979; Nau and Hall, 1982). They computed the response spectra for elasto-plastic, bilinear and stiffness degradation hysteretic systems for ten commonly used accelerograms and the resulting nonlinear response spectra were plotted on a logarithmic tripartite grid as shown in Figure 9.8, where the ordinate represents the elastic limit of the deformation capacity u_y, which the system should have in order to limit the spectral acceleration, velocity and displacement not over $\omega^2 u_y$, ωu_y and u_y respectively. Here u_y and ductility factor are requirements of the capacity for the system. Finally, four structural factors (natural period, damping ratio, type of hysteretic curve and the ductility factor μ) and three ground motion parameters (peak acceleration A, velocity V and displacement D) are obtained for analysis.

Figure 9.8 Nonlinear yield response spectra (1940 El Centro).

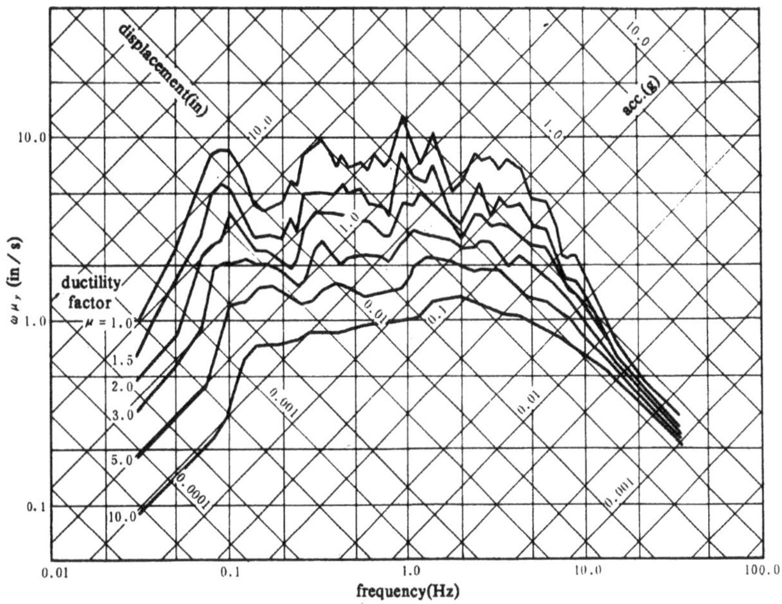

They finally suggested a method to scale both linear and nonlinear response spectra from known ground motion parameters A, V and D.

(a) Some nonlinear response spectra

In the linear case, because of the approximate relation

$$S_a = \omega S_v = \omega^2 S_d$$

all three spectra can be plotted on one tripartite grid diagram. These approximations do not hold in the nonlinear case; there are then three possible nonlinear response spectra.

1. The nonlinear yield response spectrum: here the ordinate is not displacement response, but the yield deformation which produces the displacement response $u_{max} = \mu u_y$, where $\mu = |u_{max}/u_y|$ is the ductility factor. The corresponding approximation in this case is

$$S_a = \omega S_v = \omega^2 u_y \tag{9.64}$$

and the yield strength of the structure is

$$P_y = P_{max} = k u_y = m\omega^2 u_y$$

In an ideal elasto-plastic system, P_y is the maximum internal force. In a bilinear system, the maximum internal force is

$$P_{max} = m\omega^2 u_y[1 + p(\mu - 1)] \tag{9.65}$$

where p is the ratio of slope of the second line to the first.

2. The nonlinear acceleration response spectrum: the acceleration plotted on the ordinate of the spectrum is that of the maximum internal force of the structure, P_{max}/m. The advantage of this spectrum is that it gives directly the maximum internal force; but the corresponding velocity and displacement have no meaning.
3. The nonlinear total deformation response spectrum: the displacement response of the structure, $u_{max} = \mu u_y$ is plotted on the displacement ordinate, but the corresponding velocity and acceleration ordinates have no meaning.

In conclusion, the nonlinear acceleration response spectrum gives only a meaningful acceleration ordinate, which is equal to the acceleration ordinate in the nonlinear yield response spectrum multiplied by $[1 + p(\mu - 1)]$. The nonlinear total deformation response spectrum gives only a meaningful displacement ordinate, which is equal to the displacement ordinate in the nonlinear yield response spectrum multiplied by μ. Only the nonlinear yield response spectrum gives three meaningful ordinates, which can easily be converted to obtain the other two spectra, and that is the nonlinear response spectrum used by Newmark and Hall.

(b) Elastic response spectrum, $\mu = 1$

Scaling of the elastic response spectrum is shown in Figure 9.9. Let Y be any one of the ground motion parameters (A, V, or D) and S the corresponding quantity of the response spectrum. As shown in Figure 9.9, the response quantities corresponding to A, V and D are respectively represented by the linear segments LM, KL and JK, which means that Y and S are parallel to each other. If the subscript e is used for the elastic and μ for the nonlinear response spectrum, there are then the following relations:

$$S_e = \phi_e y, \quad S_\mu = \phi_\mu S_e \tag{9.66a}$$

$$S_\mu = \phi_\mu \phi_e y = \psi_\mu y \tag{9.66b}$$

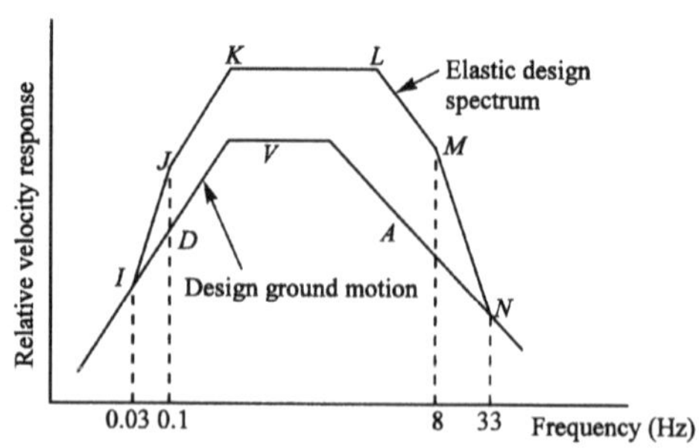

Figure 9.9 Scaling of elastic spectrum.

Table 9.2 Amplification factor ϕ_e for elastic spectrum

Damping ratio (%)	Response spectrum[a]	1979			1973		
		ϕ_e	σ	C_V	ϕ_e	σ	C_V
2	D	1.69	0.83	0.49	1.68	0.83	0.49
	V	2.03	0.85	0.42	2.06	0.92	0.45
	A	3.08	0.74	0.24	2.76	0.89	0.32
5	D	1.47	0.64	0.43	1.40	0.64	0.46
	V	1.55	0.60	0.39	1.66	0.66	0.40
	A	2.28	0.49	0.22	2.11	0.49	0.23
10	D	1.23	0.48	0.39	1.15	0.47	0.41
	V	1.20	0.44	0.36	1.34	0.47	0.35
	A	1.78	0.32	0.18	1.65	0.36	0.22

[a] D = displacement; V = velocity; A = acceleration.

Table 9.2 gives the factor ϕ_e for the elastic response spectrum for different damping. In the table, C_v is the coefficient of variation. The values of amplification factor in Table 9.2 are averages; if conservative estimates are needed,

$$\phi_{pe} = \phi_e + \delta_p \sigma_e \tag{9.67}$$

may be used for the average ϕ_e, where σ_e is the standard deviation and $\delta_p = 0$, 1 and 2 correspond to the probability of nonexceedance $P = 0.50$, 0.84 and 0.98 respectively, if the distribution is normal.

(c) Nonlinear design spectrum

Similarly, the nonlinear design response spectrum can be scaled as shown in Figure 9.10. As an example, Table 9.3 gives the result of 5% damping. Because of the complexity, the scaling of the nonlinear response spectrum is given in several bands of frequencies, whose boundaries are as follows (Figure 9.10):

Boundary of frequency band	Frequency f (Hz)
I, I′	0.03
J, J′	0.1
K, K′, L, L′	Varying
M, M′	8
N, N′	33

At a frequency $f = 33$ Hz, point N' is slightly lower than point N, with amplification factor ϕ_μ given in Table 9.4. At $f = 0.03$ Hz, point I' can be determined from equation 9.64 with $\phi_\mu = 1/\mu$.

9.7.3 *Nonlinear deformation of structures*

The maximum ductility for a given strength is a good index of a structure's earthquake resistance capacity. Several new seismic codes or guidelines, such as the API recommended practice (API, 1982) and the new Japanese Seismic Building Design Code (IAEE, 1980–), specifically require a check

Figure 9.10 Scaling of nonlinear spectrum.

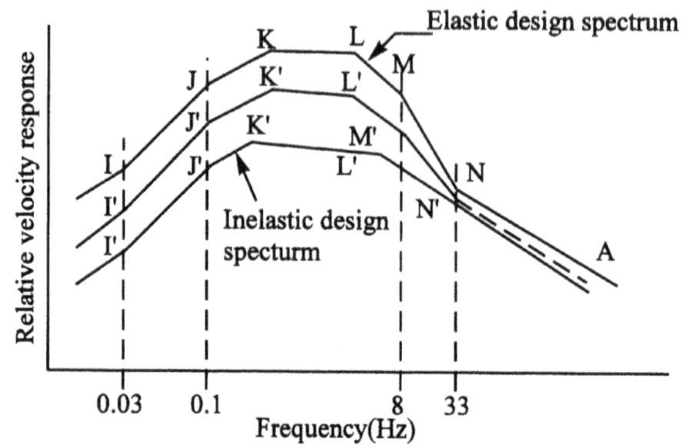

Table 9.3 Scaling factors for 5% nonlinear spectrum

Quantity	μ	Freq. band $f_l - f_u$ (Hz)	ψ_μ	σ_μ	C_v	ϕ_μ	λ_μ
Displacement	1	0.10–0.43	1.465	0.635	0.43	1.000	1.00
	1.5	0.10–0.44	0.885	0.363	0.41	0.604	0.98
	2	0.10–0.45	0.656	0.291	0.44	0.448	1.01
	3	0.10–0.47	0.439	0.194	0.44	0.299	1.01
	5	0.10–0.57	0.250	0.122	0.49	0.171	1.04
	10	0.10–0.78	0.117	0.061	0.52	0.080	1.06
Velocity	1	0.39–3.22	1.552	0.597	0.39	1.000	1.00
	1.5	0.40–3.54	0.966	0.346	0.36	0.623	0.98
	2	0.40–4.00	0.714	0.253	0.35	0.460	0.98
	3	0.42–4.42	0.508	0.176	0.35	0.328	0.97
	5	0.53–4.68	0.355	0.129	0.36	0.229	0.98
	10	0.71–5.46	0.222	0.075	0.34	0.143	0.96
Acceleration	1	2.85–8.00	2.281	0.494	0.22	1.000	1.00
	1.5	3.11–8.00	1.572	0.290	0.18	0.689	0.97
	2	3.48–8.00	1.311	0.234	0.18	0.575	0.97
	3	3.92–8.00	1.038	0.174	0.17	0.455	0.96
	5	4.22–8.00	0.781	0.134	0.17	0.342	0.96
	10	5.05–8.00	0.584	0.097	0.17	0.256	0.96

Table 9.4 Spectral acceleration amplification at $f = 33$ Hz for elasto-plastic system with 5% damping

μ	ϕ_μ	$\phi_\mu + \sigma$
1.0	1.0	1.0
1.5	0.935	0.959
2	0.913	0.943
3	0.884	0.921
5	0.846	0.891
10	0.784	0.838

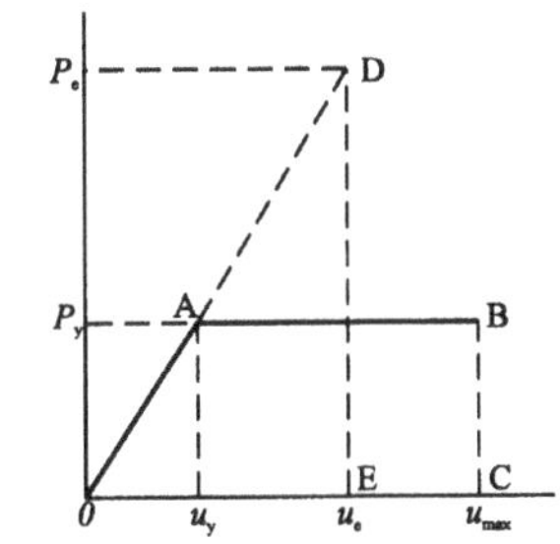

Figure 9.11 Relation between linear and nonlinear responses.

for deformation capacity to ensure the design principle of noncollapse during a strong earthquake.

Though there is still much to be desired in evaluating nonlinear deformation during very strong earthquake motion, especially that caused by nonlinear dynamic properties of structures, the following two principles are generally accepted.

(a) Approximation between linear and nonlinear deformation

Consider two systems, one linear and the other ideally elasto-plastic (Figure 9.11); point D represents the maximum linear response and point B the maximum nonlinear response. If the energy of these systems at their maximum deformations is equal, the area of triangle ODE is equal to that of the trapezoid OABC, then

$$\frac{u_e}{u_y} = (2\mu - 1)^{1/2} \tag{9.68}$$

where $\mu = u_{max}/u_y$ is the ductility factor.

(b) Concentration of nonlinear deformation

Many studies (Newmark and Rosenblueth, 1971; Kato and Akiyama, J1976; Yin *et al.*, 1981, C1981; He and Wei, 1981) show clearly that, in the case of multistory shear frames, nonlinear deformation concentrates easily at relatively weak stories. The term 'weak story' means:

1. the story stiffness is lower than that of the neighboring stories;
2. the ratio of strength to elastic response is lower than for the neighbouring stories; or
3. a weak hysteretic property gives larger deformation than for the neighboring stories.

On the basis of a large amount of numerical data for responses of shear frames, including those for systems with a stiffness degradation property (Yin *et al.*, C1981, C1983), nonlinear deformation concentrates at weak stories. When defining a yield shear coefficient $q = P_y/P_e$, their conclusions are as follows.

1. In the case of equal strengths (same q for all stories), when q decreases, the ductility factor μ of the second story does not change much but the ductility of other stories increases, especially that of the first story.
2. At a weak story, where q is less than 80% of the neighboring stories, the deformation and the ductility are much (1–5 times) greater than for the neighboring stories; the smaller the q value, the greater the deformation.
3. Where there is a weak story, the deformations or ductility factors at other stories are close.

Both Newmark and Rosenblueth and Yin *et al.* have pointed out that a system may be of nearly constant q under some ground motion but may have strong nonlinear deformation concentration under some other ground motion. One advantage of Yin's result, using a yield shear coefficient $q = P_y/P_e$, is to show that the nonlinear deformation concentration is related not only to the strength of the system P_y but also to the dynamic property of the ground motion, because P_e is a function of both.

Yin *et al.* (C1981) suggested the following empirical formula to evaluate the nonlinear deformation of multistory shear frames:

$$\bar{\mu} = \frac{1}{q^{1/2}} \exp[b(1-q)] \tag{9.69}$$

where

- $b = 1.1$ for stories of nearly the same q, except the first story in a building of uniform q with a weak story;
- $b = 1.9$ for the first story;
- $b = 2.6$ for the weak story.

Kato and Akiyama (J1978) used the following empirical formula to evaluate the nonlinear energy loss:

$$E_{qi} = \frac{E_q \alpha_i q_{\bar{i}}^{12}}{\sum \alpha_i q_{\bar{i}}^{12}}$$

$E_{qi} = W_{qi}/(mg^2/\omega_1^2)$, where ω_1 is the fundamental frequency of the structure, m the total mass, W_{qi} the nonlinear energy loss and subscript i represents the ith story; $E_q = W_q/(mg^2/\omega_1^2)$, where W_q is the nonlinear energy loss of the whole structure; $q_i = P_{yi}/P_{y0i}$, where P_{yi} is the yield shearing force and P_{y0i} the optimum yield shearing force to make all stories have the same ductility factor μ; $\alpha_i = (\sum m_i/m)(P_{y0i}/P_{y0i})^2(k_1/k_i)$, the summation from i to top story N, where k_i is the stiffness.

9.7.4 Effect of duration on structural deformation

In the last decade, the effect of strong motion duration on nonlinear deformation of structures was widely studied and accepted. Song and Zhu (C1983) pointed out that, for systems without strength degradation, when the deformation μ of the weak story is more than 4, it may increase 20% if the duration increases from 10 to 34 s. Lai and Biggs (1980) also pointed out that duration does not have a very important effect on nonlinear deformation. Xu and Weng (1982) pointed out the important effect of duration on nonlinear deformation of systems with stiffness and strength degradation. Many others (Takizawa and Jennings, 1980) have shown the important effect of duration on nonlinear deformation of various models (see also section 11.6.2).

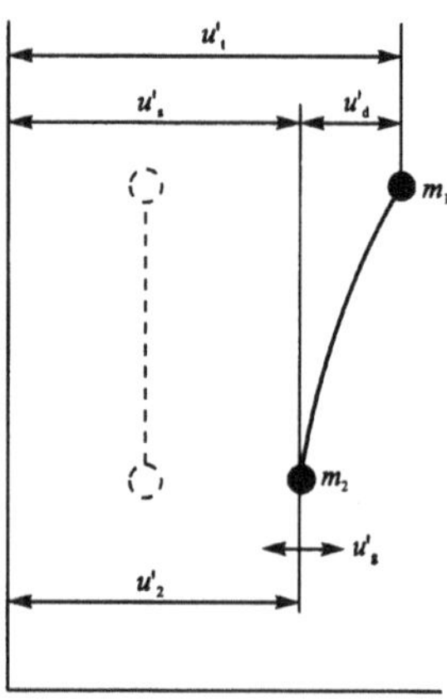

Figure 9.12 Two-mass system with no supports.

9.8 Response to multipoint and multicomponent inputs

9.8.1 General equation

To illustrate the derivation of the equation of motion of complex systems, let us study a two-mass system without support as shown in figure 9.12. The two masses m_1 and m_2 are connected by a member with stiffness matrix $\boldsymbol{K}$ and a known horizontal motion $u_g(t)$ acting on mass m_2. The equations of motion are

$$\begin{bmatrix} m_1 & 0 \\ 0 & m_2 \end{bmatrix} \begin{Bmatrix} \ddot{u}_1^t \\ \ddot{u}_2^t \end{Bmatrix} + \begin{Bmatrix} k_{11} & k_{12} \\ k_{21} & k_{22} \end{Bmatrix} \begin{Bmatrix} u_1^t \\ u_2^t \end{Bmatrix} = \begin{Bmatrix} P_1 \\ P_2 \end{Bmatrix} \tag{9.70}$$

Because there is no external force applied to m_1, $P_1 = 0$ and the first of equations 9.70 becomes

$$m_1 \ddot{u}_1^t + k_{11} u_1^t + k_{12} u_2^t = P_1 = 0 \tag{9.71}$$

Now the total displacement u^t is decomposed into two parts: the dynamic displacement u^d from vibration and the pseudo-static displacement u^s from the displacement of the input displacements $u_g(t)$. For the specific case in Figure 9.12,

$$u_1^t = u_1^d + u_1^s$$
$$u_2^t = u_2^d + u_2^s$$

Equation 9.71 then becomes

$$m_1 \ddot{u}_1^d + k_{11} u_1^d = -m_1 \ddot{u}_1^s - k_{11} u_1^s - k_{12} u_2^d - k_{12} u_2^s \tag{9.72}$$

On the basis of this definition, the pseudo-static displacement should satisfy this equation after dropping all dynamic terms, i.e.

$$k_{11} u_1^s + k_{12} u_2^s = 0$$

Because the system in Figure 9.70 has no supports, $u_1^s = u_2^s = u_g$, and

$$k_{12} + k_{11} = 0$$

which means the stiffness matrix $\boldsymbol{K}$ is singular. Moreover, because point m_2 is actually a support, $u_2^d = 0$. With these relations, equation 9.72 becomes the ordinary equation of motion

$$m_1 \ddot{u}_1^d + k_{11} u_1^d = -m_1 \ddot{u}_g \tag{9.73}$$

There are four features in the above derivation.

1. The supporting point is treated as a free point and a fixed reference coordinate system is used.
2. The total displacement is divided into two parts, a pseudo-static part and a dynamic part; there is no dynamic displacement at the support.
3. In the equation of motion of the nonsupporting points, a relation among the pseudo-static displacements is obtained by letting all dynamic displacements equal zero; and after eliminating this relation from the equation of motion, a routine equation of motion is obtained.
4. From the equation of motion of the supporting point, the supporting force or the reaction $P(t)$ at the support is obtained.

This procedure also applies to an MDF system. Consider a linear MDF system subject to multidimensional excitations $P(t)$. The controlling equation in matrix form is

$$\boldsymbol{M}\ddot{\boldsymbol{u}}^{t}+\boldsymbol{C}\dot{\boldsymbol{u}}^{t}+\boldsymbol{K}\boldsymbol{u}^{t}=\boldsymbol{P}(t) \tag{9.74}$$

The total number of degrees of freedom is $n=n_s+n_b$, including n_s degrees for displacement along which there is support, and n_b degrees of free displacement. The controlling equation is then

(9.75)

If only ground motion excitation is considered, no forces at free points are considered, or $\boldsymbol{P}_s=\boldsymbol{0}$; $P(t)$ is nonzero only in the components with ground motion input where $P(t)$ is the reaction at the support. The total response $\boldsymbol{u}^t$ can be partitioned into pseudo-static and dynamic parts,

$$\boldsymbol{u}^{t}=\{\boldsymbol{u}_s^{qs}\boldsymbol{u}_b^{qs}\}^{T}+\{\boldsymbol{u}_s^{d}\boldsymbol{0}\}^{T} \tag{9.76}$$

Let the dynamic terms be zero in the first equation of equation (9.75) to obtain a relation among the pseudo-static terms,

$$\boldsymbol{u}_s^{qs}=-\boldsymbol{K}_{ss}^{-1}\boldsymbol{K}_{sb}\boldsymbol{u}_b^{t} \tag{9.77}$$

After eliminating the pseudo-static terms from the first of equations 9.75, the relation among the dynamic terms is

$$\boldsymbol{m}_{ss}\ddot{\boldsymbol{u}}_s^{d}+\boldsymbol{C}_{ss}\dot{\boldsymbol{u}}_s^{d}+\boldsymbol{K}_{ss}\boldsymbol{u}_s^{d}=[\boldsymbol{m}_{ss}\boldsymbol{K}_{ss}^{-1}\boldsymbol{K}_{sb}-\boldsymbol{m}_{sb}]\ddot{\boldsymbol{u}}_b^{t}+[\boldsymbol{C}_{ss}\boldsymbol{K}_{ss}^{-1}\boldsymbol{K}_{sb}-\boldsymbol{C}_{sb}]\dot{\boldsymbol{u}}_b^{t} \tag{9.78}$$

In the case where damping is proportional to stiffness, the last term is zero; in the case of other kinds of small damping, the last term can be neglected, and the following approximation is obtained:

$$\boldsymbol{m}_{ss}\ddot{\boldsymbol{u}}_s^{d}+\boldsymbol{C}_{ss}\dot{\boldsymbol{u}}_s^{d}+\boldsymbol{K}_{ss}\boldsymbol{u}_s^{d}\approx[\boldsymbol{m}_{ss}\boldsymbol{K}_{ss}^{-1}\boldsymbol{K}_{sb}-\boldsymbol{m}_{sb}]\ddot{\boldsymbol{u}}_b^{t} \tag{9.79}$$

Compared with equations 9.75, this equation has one obvious advantage in that it is only required to know the acceleration $\ddot{\boldsymbol{u}}_b^t$, but not the velocity $\dot{\boldsymbol{u}}_b^t$ or the displacement process $\boldsymbol{u}_b^t$.

The modes of natural vibration $\boldsymbol{\Phi}_s$ for fixed supports can be calculated from the following equation obtained from equation 9.78:

$$\boldsymbol{m}_{ss}\ddot{\boldsymbol{u}}_s^{d}+\boldsymbol{K}_{ss}\boldsymbol{u}_s^{t}=\boldsymbol{0}$$

If $\boldsymbol{\Phi}_s$ is used to express the dynamic response,

$$\boldsymbol{u}_s^{d}=\boldsymbol{\Phi}_s\boldsymbol{y}_s \tag{9.80}$$

the equation of motion (equation 9.79) may be written as follows:

$$\boldsymbol{M}_s\ddot{\boldsymbol{y}}_s+\boldsymbol{C}_s\dot{\boldsymbol{y}}_s+\boldsymbol{K}_s\boldsymbol{y}_s=\boldsymbol{\Phi}^{T}[\boldsymbol{m}_{ss}\boldsymbol{K}_{ss}^{-1}\boldsymbol{K}_{sb}-\boldsymbol{m}_{sb}]\ddot{\boldsymbol{u}}_b^{t} \tag{9.81}$$

where $\boldsymbol{M}_s$, $\boldsymbol{C}_s$, and $\boldsymbol{K}_s$ are all $n_s\times n_s$ diagonal matrices.

$$M_s \equiv \Phi_s^T m_{ss} \Phi_s$$

$$K_s \equiv \Phi_s^T K_{ss} \Phi_{ss} = \omega_s^2 M_s$$

If C_{ss} also satisfies orthogonality, it is then obtained that

$$C_s = \Phi_s^T C_{ss} \Phi_s = \frac{M_s}{2\omega_s \zeta_s}$$

If C_{ss} does not satisfy the condition of orthogonality, C_s is not a diagonal matrix and equation 9.81 is still coupled in damping.

From the second of equations 9.75, the forces at the supports can be found.

9.8.2 A special case: $n_b = 2$ and $n_s = 1$

As shown in Figure 9.13, points A and B are supports and only the motion along the x-direction is considered. The total displacement is

$$u^t = \begin{Bmatrix} u_1^t(t) \\ u_{gA}(t) \\ u_{gB}(t) \end{Bmatrix} = \begin{Bmatrix} u_1^{qs}(t) \\ u_{gA}(t) \\ u_{gB}(t) \end{Bmatrix} + \begin{Bmatrix} u^d(t) \\ 0 \\ 0 \end{Bmatrix}$$

where $u_{gA}(t)$ and $u_{gB}(t)$ are respectively the input motion at supports A and B.

Pseudo-static displacement is, from equation 9.78

$$u_1^{qs} = -\frac{k_{12}}{k_{11}} u_{gA} - \frac{k_{13}}{k_{11}} u_{gB}$$

and the dynamic displacement is, from equation 9.79

$$\ddot{u}_1^d + 2\zeta_1 \omega_1 \dot{u}_1^d + \omega_1^2 u_1^d = \left(\frac{k_{12}}{k_{11}} - \frac{m_{12}}{m_{11}}\right)\ddot{u}_{gA} + \left(\frac{k_{13}}{k_{11}} - \frac{m_{13}}{m_{11}}\right)\ddot{u}_{gB} = -A\ddot{u}_{gA} - B\ddot{u}_{gB} \quad (9.82)$$

where

$$\omega_1^2 = \frac{k_{11}}{m_{11}}, \quad \zeta_1 = \frac{C_{11}}{2m_{11}\omega_1}$$

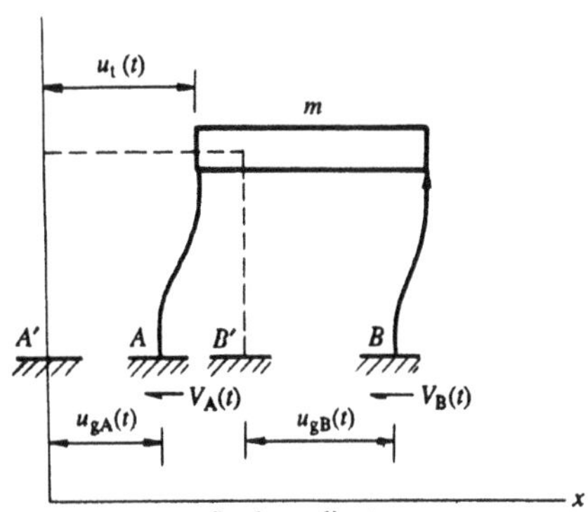

Figure 9.13 Horizontal vibration of symmetric frame.

$$A \equiv -\left(\frac{k_{12}}{k_{11}} - \frac{m_{12}}{m_{11}}\right), \quad B \equiv -\left(\frac{k_{13}}{k_{11}} - \frac{m_{13}}{m_{11}}\right)$$

Let $r \equiv A/B$ and equation 9.82 will be

$$\ddot{u}_1^{\mathrm{d}} + 2\zeta_1\omega_1\dot{u}_1^{\mathrm{d}} + \omega_1^2 u_1^{\mathrm{d}} = -B(r\ddot{u}_{\mathrm{gA}} + \ddot{u}_{\mathrm{gB}})$$

$r = +1$ means that the motion of the two supports is in phase, $r = -1$ means that the motion is in antiphase, or opposite.

The equation of motion of a symmetric frame, where $r = 1$ and $\omega_1 = k/m$ but the ground motion at the two supports is not the same, may be decomposed into two, one in phase and one in antiphase, i.e.

$$u_1^{\mathrm{qs}} = \frac{u_{\mathrm{gA}} + u_{\mathrm{gB}}}{2} \tag{9.83a}$$

$$\ddot{u}_1^{\mathrm{d}} + 2\zeta_1\omega_1\dot{u}_1^{\mathrm{d}} + \omega_1^2 u_1^{\mathrm{d}} = -\frac{\ddot{u}_{\mathrm{gA}} + \ddot{u}_{\mathrm{gB}}}{2} \tag{9.83b}$$

and the shear forces in the columns are

$$V_{\mathrm{A}}(t) = \frac{k}{2}\left(u_1^{\mathrm{d}} + \frac{u_{\mathrm{gB}} - u_{\mathrm{gA}}}{2}\right)$$

$$V_{\mathrm{B}}(t) = \frac{k}{2}\left(u_1^{\mathrm{d}} - \frac{u_{\mathrm{gB}} - u_{\mathrm{gA}}}{2}\right)$$

It is clear then that the in-phase part of the ground motion $(u_{\mathrm{gA}} + u_{\mathrm{gB}})/2$ produces dynamic responses, and the antiphase part $(u_{\mathrm{gA}} - u_{\mathrm{gB}})/2$ produces only pseudo-static responses; the internal forces in the columns consist of both dynamic and pseudo-static parts.

9.8.3 Structural response for space-correlated ground motion

Spatial correlation of ground motion studies the statistical relation between motion in different directions or at different locations nearby.

The simplest representation of the relation of motion between two points of distance r apart is a stationary wave with a time delay τ; for surface waves traveling with a speed r/τ, it is

$$a_1(t) = a_2(t + \tau)$$

with no attenuation, or

$$a_1(t) = Ca_2(t + \tau) \qquad (C < 1)$$

with attenuation. The latter relation in the frequency domain is

$$A_1(\omega) = C(\omega)\,A_2(\omega)\exp(\mathrm{i}\rho\omega) \quad (C(\omega) \leqslant 1) \tag{9.84}$$

where $\rho\omega$ is the phase difference of these two waves, and it means that the attenuation is frequency-dependent.

In statistical considerations, the correlation of motion should be represented by a cross-correlation function or cross-spectrum density function, such as the normalized cross-spectrum density

$$R_{12}(i\omega)=\frac{S_{12}(i\omega)}{[S_1(\omega)\,S_2(\omega)]^{1/2}} \tag{9.85}$$

where $S_1(\omega)$, $S_2(\omega)$ and $S_{12}(\omega)$ are respectively the spectrum density of motion at points 1 and 2, and the cross-spectrum density between points 1 and 2. Equation 9.85 includes relations both of amplitude and phase and can be written in the following form:

$$R_{12}(i\omega)=\exp[-(\rho_1\omega+\rho_2)\Delta+i\rho_3\omega\Delta] \tag{9.86}$$

where the real part represents the amplitude relation and the imaginary part the phase difference; Δ is the distance between the two points. Constants ρ_1, ρ_2 and ρ_3 show the relation between the motion at point 1 and that at point 2. On the basis of limited data, ρ_1 and ρ_2 are roughly in the following ranges: $\rho_1=(0.4-2)\times10^{-5}$ l/rad – m; $\rho_2=(20–90)\times10^{-4}$ l/m; ρ_3^{-1} is the apparent shear-wave velocity.

With given spatial correlation of ground motion, the responses of structures on multiple supports can be found. Yamamura and Tanaka (1990) suggested a deterministic method of analysis in both time-history and response spectrum approaches.

9.9 Soil–structure interaction

Soil–structure interaction (SSI) was first studied a long time ago in analysis of the vibration of machinery foundations, but has greatly developed in the past two decades because of the rapid advances in computer technology. It may be classified into time and frequency domains in methodology, or total structure, substructure and hybrid structure in structural decomposition. There are many good reviews on the subject (Lysmer, 1978; Idriss and Kennedy, 1980; Rosenblueth, 1980; Luco, 1982; Roesset and Tassoulas, 1982). Most investigators have adopted frequency-domain analysis because:

1. it is convenient to consider the frequency-dependent foundation impedance when the whole system is divided into structure and ground;
2. the transmitting boundary for a finite element model or wave propagation theory is frequency-dependent.

The time-domain SSI method has also been developed (Bayo and Wilson, 1983), which is good for nonlinear SSI analysis.

Generally speaking, SSI analysis consists of four fundamental problems.

1. The free-field problem: the rock (ground) motion is found from other given motions.
2. The scattering problem: the response of the ground is studied with the structure removed, at the contacts (points c in Figure 9.14(c)) of the structure and ground. The difference between this and the free-field

problem is the absence of soil, which is replaced by structure in the free-field problem. Therefore, for problems with foundations on free ground, these two steps reduce to one. In the hybrid structure model, the motion of the contact points of the hybrid structure is studied, including the foundation soil and the remaining half space.

3. Impedance: the displacements $U_c(\omega)\exp(i\omega t)$ at the contact points are found for forces applied at the contact points $F_c(\omega)\exp(i\omega t)$. The impedance matrix $I_s(\omega)=C^{-1}(\omega)$ or the flexibility matrix $C(\omega)$ of the ground is defined in the following relation:

$$U_c(\omega)=C(\omega)F_c(\omega)$$

If the foundation of the structure is on the ground surface, the ground substructure is a half space; for the case with only one horizontal layer on top of a homogeneous space, there is an analytical solution for $C(\omega)$.

4. There is finally the soil–structure interaction (SSI) system.

In the following, the substructure method in the frequency domain and the total structure method in the time domain is used to illustrate these steps.

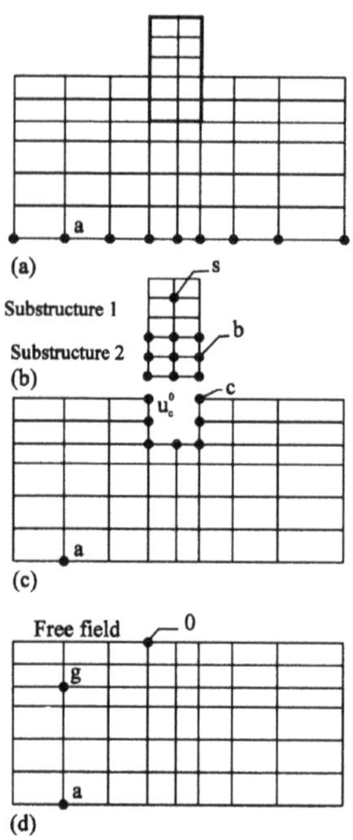

Figure 9.14 Substructures in SSI.

9.9.1 Substructure method in the frequency domain

This method is also called the method of impedance, which divides the system into two substructures (substructures 1 and 2) at the contact surfaces as shown in Figures 9.14(b), (c), studies their responses separately first, and combines them later to satisfy the continuity condition. It has been used in both the time and frequency domains.

Let the known ground motion be a general three-dimensional time history $u_g(t)$, whose frequency domain expression is $U_g^0(\omega)$, at some depth below the surface. The four primary problems are as follows.

1. Free-field response: The input motion $u_a(t)$ or $U_a(\omega)$ is evaluated at bedrock points 'a' in Figure 9.14(d), the lower boundary of the ground substructure, for given motion at point g. Let the analysis use an operator L_1

$$L_1:\ u_a(t),\ U_a(\omega)\Leftarrow u_g(t),\ U_g^0(\omega)$$

2. Scattering problem: the response of substructure 2 is studied (Figure 9.14(c)). Let there be an operator L_2

$$L_2:\ u_c^0(t),\ U_c^0(\omega)\Leftarrow u_a(t),\ U_a(\omega)$$

3. Impedance problem: the response of substructure 2 is studied (Figure 9.14 (c)), under an external excitation $f_c(t)=F_c(\omega)\exp(i\omega t)$ acting along one direction at any point c. The response at point c, $u_c(t)=U_c(\omega)\exp(i\omega t)$, for a given force $f_c(t)=F_c(\omega)\exp(i\omega t)$ at point c is

$$U_c(\omega)=C(\omega)F_c(\omega)\quad\text{or}\quad F_c(\omega)=I_s(\omega)U_c(\omega) \tag{9.87}$$

4. The response $\boldsymbol{u}_b(t)$ or $\boldsymbol{U}_b(\omega)$ of substructure 1 under external forces $\boldsymbol{f}_b(t)=\boldsymbol{F}_b(\omega)\exp(iwt)$ is studied at contact points b

$$\boldsymbol{F}_b(\omega)=\boldsymbol{M}_b(\omega)\,\omega^2\,\boldsymbol{U}_b(\omega) \qquad (9.88)$$

where $\boldsymbol{M}_b(\omega)$ is the equivalent, frequency-dependent, mass matrix of the superstructure.

The four primary problems are four steps to find the operators L_1, L_2, $\boldsymbol{C}(\omega)$ or $\boldsymbol{I}_s(\omega)$ and $\boldsymbol{M}_b(\omega)$. The final solution is only an assemblage of the four to satisfy the compatibility requirement at the boundary points b and c.

The continuity condition at boundary points b and c is that the displacement $\boldsymbol{u}_b(t)$ or $\boldsymbol{U}_b(\omega)$ is equal to the sum of displacements $\boldsymbol{u}_c^0(t)$ or $\boldsymbol{U}_c^0(\omega)$ of substructure 2 and the displacements $\boldsymbol{u}_c(t)$ or $\boldsymbol{U}_c(\omega)$ due to the SSI effect of substructure 1 on substructure 2,

$$\boldsymbol{u}_b(t)=\boldsymbol{u}_c^0(t)+\boldsymbol{u}_c(t) \text{ or } \boldsymbol{U}_b(\omega)=\boldsymbol{U}_c^0(\omega)+\boldsymbol{U}_c(\omega) \qquad (9.89)$$

The equilibrium condition requires that the sum of the forces $\boldsymbol{f}_c(t)$ or $\boldsymbol{F}_c(\omega)$ between the ground and foundation and the forces $\boldsymbol{f}_b(t)$ or $\boldsymbol{F}_b(\omega)$ between the structure and foundation resist the inertial force $\boldsymbol{f}_i(t)$ or $\boldsymbol{F}_{i(\omega)}$ of the rigid foundation, i.e.

$$\boldsymbol{f}_c(t)=\boldsymbol{f}_i(t)+\boldsymbol{f}_b(t) \text{ or } \boldsymbol{F}_c(\omega)=\boldsymbol{M}_0\omega^2\boldsymbol{U}_b(\omega)+\boldsymbol{F}_b(\omega) \qquad (9.90)$$

where $\boldsymbol{M}_0\omega^2\boldsymbol{U}_b(\omega)$ is the Fourier transform of the inertial force $\boldsymbol{f}_i(t)$ and $\boldsymbol{M}_0$ the mass matrix of the rigid foundation.

From these four equations, it is found finally that

$$\boldsymbol{U}_b(\omega)=\{\boldsymbol{I}-\omega^2\boldsymbol{C}(\omega)[\boldsymbol{M}_0+\boldsymbol{M}_b(\omega)]\}^{-1}\boldsymbol{U}_c^0(\omega) \qquad (9.91)$$

or

$$\boldsymbol{U}_b(\omega)=\{\boldsymbol{I}_s(\omega)-\omega^2[\boldsymbol{M}_0+\boldsymbol{M}_b(\omega)]\}^{-1}\boldsymbol{F}_c^0(\omega) \qquad (9.92a)$$

$$\boldsymbol{F}_c^0(\omega)=\boldsymbol{I}_s(\omega)\,\boldsymbol{U}_c^0(\omega) \qquad (9.92b)$$

where $\mathbf{I}$ is the unit matrix. If three-dimensional ground motion is considered, a rigid foundation has six components, three translational and three rotational, and $\mathbf{I}$ is 6×6. Equation 9.91 shows clearly the interactions. The effect of input motion $\boldsymbol{u}_g(t)$ is represented by $\boldsymbol{U}_c^0(\omega)$, the foundation input motion, obtained by operators L_1 and L_2; the interaction of the superstructure, foundation and ground by the term $\boldsymbol{C}(\omega)[\boldsymbol{M}_0+\boldsymbol{M}_b(\omega)]$; and $\boldsymbol{U}_b(\omega)$ is the result of the input motion on the soil–structure system, from which the structural response is obtained.

In equations 9.92, $\boldsymbol{F}_c^0(\omega)=\boldsymbol{I}_s(\omega)\,\boldsymbol{U}_c^0(\omega)$ is the driving force vector, which includes the forces and moments from the rigid foundation on the ground, when points c of the substructure 2 tend to deform and the foundation resists the deformation. Equations 9.92 are easier to use than equation 9.91.

9.9.2 Substructure method in the time domain

In the substructure method in the time domain, there are only three primary problems; the impedance problem is not needed.

Categories of displacements of different points are shown in Figure 9.15; the superscript 0 indicates quantities in the foundation substructure in Figure 9.15(b), and superscript i indicates those due to interaction effects in Figure 9.15(c). Matrices and vectors are as follows:

$$u^{\mathrm{i}} = \begin{Bmatrix} u_{\mathrm{s}}^{\mathrm{i}} \\ u_{\mathrm{b}} \\ u_{\mathrm{g}} \end{Bmatrix}, \qquad u^0 = \begin{Bmatrix} 0 \\ u_{\mathrm{b}}^0 \\ u_{\mathrm{g}}^0 \end{Bmatrix} \tag{a}$$

$$m = \begin{bmatrix} m_{\mathrm{s}} & m_{\mathrm{b}} & 0 \\ m_{\mathrm{b}}^{\mathrm{T}} & m_{\mathrm{bb}} & 0 \\ 0 & 0 & 0 \end{bmatrix}, \qquad m^0 = \begin{bmatrix} 0 & 0 & 0 \\ 0 & m_{\mathrm{bb}}^0 & m_{\mathrm{bg}}^0 \\ 0 & m_{\mathrm{gb}}^0 & m_{\mathrm{bb}}^0 \end{bmatrix} \tag{b}$$

Stiffness and damping matrices are similar. Notice that the displacement matrices include two parts, one for contact points b and another for the remaining points g in the ground. The equation for the ground is

$$\begin{bmatrix} m^0 & m_{\mathrm{a}}^0 \\ m_{\mathrm{a}}^{0\mathrm{T}} & m_{\mathrm{aa}}^0 \end{bmatrix} \begin{Bmatrix} \ddot{u}^0 \\ \ddot{u}_{\mathrm{a}}^0 \end{Bmatrix} + \begin{bmatrix} c^0 & c_{\mathrm{a}}^0 \\ c_{\mathrm{a}}^{0\mathrm{T}} & c_{\mathrm{aa}}^0 \end{bmatrix} \begin{Bmatrix} \dot{u}^0 \\ \dot{u}_{\mathrm{a}}^0 \end{Bmatrix} + \begin{bmatrix} k^0 & k_{\mathrm{a}}^0 \\ k_{\mathrm{a}}^{0\mathrm{T}} & k_{\mathrm{aa}}^0 \end{bmatrix} \begin{Bmatrix} u^0 \\ u_{\mathrm{a}}^0 \end{Bmatrix} = \begin{Bmatrix} 0 \\ 0 \end{Bmatrix} \tag{9.93}$$

The first of equations 9.93 is

$$m^0 \ddot{u}^0 + C^0 \dot{u}^0 + k^0 u^0 = - m_{\mathrm{a}}^0 \ddot{u}_{\mathrm{a}}^0 - c_{\mathrm{a}}^0 \dot{u}_{\mathrm{a}}^0 - k_{\mathrm{a}}^0 u_{\mathrm{a}}^0 \tag{9.94}$$

where the right-hand side represents the input bedrock motion. Putting the superstructure back on its foundation, the system studied is then shown in Figure 9.15(c). The left-hand side of the equation will be changed but the right-hand side is assumed to be unchanged because the superstructure is far from the bedrock. Now the equation becomes

$$[m^0 + m]\{\ddot{u}^0 + \ddot{u}^{\mathrm{i}}\} + [c^0 + c]\{\dot{u}^0 + \dot{u}^{\mathrm{i}}\} + [k^0 + k]\{u^0 + u^{\mathrm{i}}\} = - m_{\mathrm{a}}^0 \ddot{u}_{\mathrm{a}}^0 - c_{\mathrm{a}}^0 \dot{u}_{\mathrm{a}}^0 - k_{\mathrm{a}}^0 u_{\mathrm{a}}^0 \tag{9.95}$$

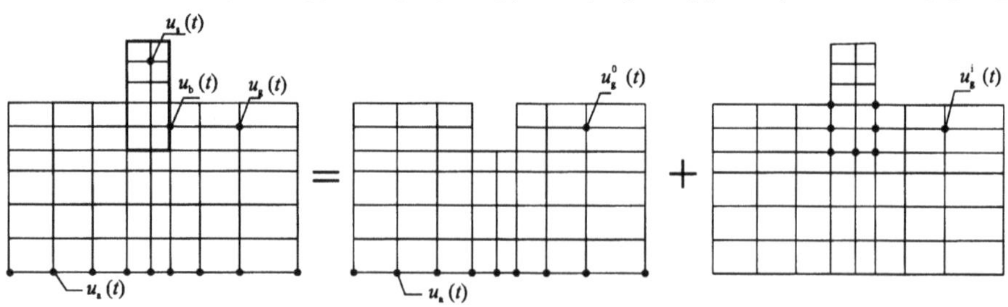

(a) (b) (c)

Figure 9.15 Steps of the substructure method in the time domain: (a) interaction problem $u^i(t) + u^0(t)$; (b) foundation problem $u^0(t)$; (c) additional motion $u^i(t)$.

where $\boldsymbol{m}$, $\boldsymbol{c}$ and $\boldsymbol{k}$ are those of the superstructure. Combining these two equations gives

$$[\boldsymbol{m}^0 + \boldsymbol{m}]\,\ddot{\boldsymbol{u}}^{\mathrm{i}} + [\boldsymbol{c}^0 + \boldsymbol{c}]\,\dot{\boldsymbol{u}}^{\mathrm{i}} + [\boldsymbol{k}^0 + \boldsymbol{k}]\,\boldsymbol{u}^{\mathrm{i}} = -\boldsymbol{m}\ddot{\boldsymbol{u}}^0 - \boldsymbol{c}\dot{\boldsymbol{u}}^0 - \boldsymbol{k}\boldsymbol{u}^0 \tag{9.96}$$

which is the solution of the additional motion problem shown in Figure 9.15(c). Substituting equations (a) and (b) into equation 9.94 gives

$$[\boldsymbol{m}^0 + \boldsymbol{m}]\,\ddot{\boldsymbol{u}}^{\mathrm{i}} + [\boldsymbol{c}^0 + \boldsymbol{c}]\,\dot{\boldsymbol{u}}^{\mathrm{i}} + [\boldsymbol{k}^0 + \boldsymbol{k}]\,\boldsymbol{u}^{\mathrm{i}} = - \{\boldsymbol{m}_{\mathrm{b}}\boldsymbol{m}_{\mathrm{bb}}\boldsymbol{0}\}^{\mathrm{T}}\ddot{\boldsymbol{u}}_{\mathrm{b}}^0 - \{\boldsymbol{c}_{\mathrm{b}}\boldsymbol{c}_{\mathrm{bb}}\boldsymbol{0}\}^{\mathrm{T}}\dot{\boldsymbol{u}}_{\mathrm{b}}^0 - \{\boldsymbol{k}_{\mathrm{b}}\boldsymbol{k}_{\mathrm{bb}}\boldsymbol{0}\}^{\mathrm{T}}\boldsymbol{u}_{\mathrm{b}}^0 \tag{9.97}$$

It should be noted that the right-hand side represents the input only at the contact points in Figure 9.15(c); the additional motion in the superstructure is the total displacements, where nonlinearity of the structure can be considered. For structures on the ground surface, the input motion $\boldsymbol{u}_{\mathrm{b}}^0$ to the additional motion problem is just the free-field motion; for an embedded structure, the foundation substructure problem has to be solved for input motion $\boldsymbol{u}_{\mathrm{b}}^0$ at the contact points.

Equation 9.97 may be simplified by following the procedure in section 9.8. Let the total displacement be decomposed into

$$\boldsymbol{u}^{\mathrm{i}} = \boldsymbol{u}^{\mathrm{d}} + \boldsymbol{u}^{\mathrm{s}}$$

Setting all dynamic parts to zero, the pseudo-static displacement is

$$\boldsymbol{u}^0 = -[\boldsymbol{k}^0 + \boldsymbol{k}]^{-1} \begin{Bmatrix} \boldsymbol{k}_{\mathrm{b}} \\ \boldsymbol{k}_{\mathrm{bb}} \\ \boldsymbol{0} \end{Bmatrix} \boldsymbol{u}_{\mathrm{b}}^0 = \alpha \boldsymbol{u}^0 \tag{9.98}$$

Then equation 9.97 may be written as follows:

$$[\boldsymbol{m}^0 + \boldsymbol{m}]\,\ddot{\boldsymbol{u}}^{\mathrm{d}} + [\boldsymbol{c}^0 + \boldsymbol{c}]\,\dot{\boldsymbol{u}}^{\mathrm{d}} + [\boldsymbol{k}^0 + \boldsymbol{k}]\,\boldsymbol{u}^{\mathrm{d}} = \left[(\boldsymbol{m}^0 + \boldsymbol{m})\,\alpha + \begin{Bmatrix} \boldsymbol{m}_{\mathrm{b}} \\ \boldsymbol{m}_{\mathrm{bb}} \\ \boldsymbol{0} \end{Bmatrix} \right] \ddot{\boldsymbol{u}}_{\mathrm{b}}^0 - \left[(\boldsymbol{c}^0 + \boldsymbol{c})\,\alpha + \begin{Bmatrix} \boldsymbol{c}_{\mathrm{b}} \\ \boldsymbol{c}_{\mathrm{bb}} \\ \boldsymbol{0} \end{Bmatrix} \right] \dot{\boldsymbol{u}}_{\mathrm{b}}^0 \tag{9.99}$$

In the case of small damping, the second term on the right side may be neglected and an approximate relation is obtained

$$[\boldsymbol{m}^0 + \boldsymbol{m}]\,\ddot{\boldsymbol{u}}^{\mathrm{d}} + [\boldsymbol{c}^0 + \boldsymbol{c}]\,\dot{\boldsymbol{u}}^{\mathrm{d}} + [\boldsymbol{k}^0 + \boldsymbol{k}]\,\boldsymbol{u}^{\mathrm{d}} = -\left[(\boldsymbol{m}^0 + \boldsymbol{m})\,\alpha + \begin{Bmatrix} \boldsymbol{m}_{\mathrm{b}} \\ \boldsymbol{m}_{\mathrm{bb}} \\ 0 \end{Bmatrix} \right] \ddot{\boldsymbol{u}}_{\mathrm{b}}^0 \tag{9.100}$$

After obtaining the dynamic displacement of the SSI system, the internal forces are

$$\boldsymbol{f} = [\boldsymbol{k}^0 + \boldsymbol{k}](\boldsymbol{u}^0 + \boldsymbol{u}^{\mathrm{d}} + \boldsymbol{u}^{\mathrm{s}}) = [\boldsymbol{k}^0 + \boldsymbol{k}] \left[\begin{Bmatrix} \boldsymbol{0} \\ \boldsymbol{u}_{\mathrm{b}}^0 \\ \boldsymbol{u}_{\mathrm{g}}^0 \end{Bmatrix} + \begin{Bmatrix} \boldsymbol{u}_{\mathrm{s}}^{\mathrm{d}} \\ \boldsymbol{u}_{\mathrm{b}}^{\mathrm{d}} \\ \boldsymbol{u}_{\mathrm{g}}^{\mathrm{d}} \end{Bmatrix} + \alpha \boldsymbol{u}_{\mathrm{b}}^{\mathrm{d}} \right] = [\boldsymbol{k}^0 + \boldsymbol{k}] \left[\begin{Bmatrix} 0 \\ \boldsymbol{u}_{\mathrm{b}}^0 \\ \boldsymbol{u}_{\mathrm{g}}^0 \end{Bmatrix} + \begin{Bmatrix} \boldsymbol{u}_{\mathrm{s}}^{\mathrm{d}} \\ \boldsymbol{u}_{\mathrm{b}}^{\mathrm{d}} \\ \boldsymbol{u}_{\mathrm{g}}^{\mathrm{d}} \end{Bmatrix} \right] - \begin{Bmatrix} \boldsymbol{k}_{\mathrm{b}} \\ \boldsymbol{k}_{\mathrm{bb}} \\ \boldsymbol{0} \end{Bmatrix} \boldsymbol{u}_{\mathrm{b}}^0 \tag{9.101}$$

The internal forces in the superstructure are

$$f_s = \boldsymbol{k}\boldsymbol{u}_s^d + \boldsymbol{k}_b\boldsymbol{u}_b^d \tag{9.102}$$

It is clear that the internal forces in the superstructure are determined only by the dynamic displacements $\boldsymbol{u}_s^d$ and $\boldsymbol{u}_b^d$.

The above mentioned method has two shortcomings. Firstly, in the ground substructure, the transmission of energy through the boundaries is not considered. To reduce this effect, the outside boundaries of substructure 2 should be far away from the superstructure, or an energy transmitting boundary should be introduced. Secondly, the number of degrees of freedom is usually very large, which makes the solution of the eigenvalue problem time-consuming; and because the damping matrix is not orthogonal, modal superposition may not be used, but the Ritz vector method may be used instead (Bayo and Wilson, 1983).

9.9.3 *Transmitting boundary*

A transmitting boundary is created by adding some requirement at the boundary to absorb the energy propagating to the boundary to simulate the effect of the infinite ground. Because this geometric damping effect is usually very large, the transmitting boundary has attracted both theoretical and practical attention.

(a) One-dimensional wave propagation

In one-dimensional media, the energy-absorbing effect of infinite media may be represented by frequency-independent viscous damping. Considering a harmonic P-wave propagating from the end of an infinite half bar (x direction); the displacement, velocity and stress in the bar will be

$$u = A\exp[\mathrm{i}(\omega t - kx)]$$

$$\dot{u} = A\mathrm{i}\omega\exp[\mathrm{i}(\omega t - kx)]$$

$$\sigma_x = -EA\mathrm{i}k\exp[\mathrm{i}(\omega t - kx)] = -\frac{E\dot{u}k}{\omega} = -v_P\rho\dot{u} \tag{9.103}$$

because the P-wave velocity is $v_P = (E/\rho)^{1/2} = \omega/k$. This result shows that, if a piece is cut from the bar and replaced by a dashpot with a damping coefficient $v_P\rho$, when the p-wave hits the dashpot, a damping force $-v_P\rho\dot{u}$ is produced in the dashpot which simulates exactly the action of the piece cut from the bar. For a shear wave, the dashpot damping coefficient is $v_s\rho$. Because the damping of the dashpot is independent of frequency, this result can be used in both time- and frequency-domain analysis.

Similarly, if the principle is also used for the two-dimensional case, there should be two dashpots at the boundary, one with damping ρv_P perpendicular to the boundary and one with ρv_s parallel to the boundary. For the three-dimensional case, a third dashpot with ωv_s is added in the plane of boundary but perpendicular to the other two. This is only approximate because, in a real case, waves may propagate to the boundary from any direction.

(b) Frequency-dependent transmitting boundary

Theoretically speaking, artificial damping of the transmitting boundary should be frequency-dependent, but available data (Lysmer and Kuhlemeyer, 1969) show that constant damping corresponding to the natural frequency gives acceptable results; in the case of a larger model and with internal damping considered, the error is smaller.

(c) Iteration transmitting boundary (Liao and Wong, 1984)

Because the incident angle of waves to the boundary may vary over a wide range and is unknown, the apparent wave velocity is thus also unknown. Liao suggested in a numerical method to take some adequate value of the apparent wave velocity v_a and the error, he proved, will be reduced to $v(\Delta t)^n$ upon the nth iteration at the boundary, if the time interval Δt, space interval Δx and the apparent velocity v_a satisfy the following relations:

$$v_a \Delta t < \Delta x \tag{9.104a}$$

$$v_a \leqslant 2v_{min} \tag{9.104b}$$

where v_{min} is the minimum velocity of waves considered. Experience has shown that $n = 3$ gives good results.

10 Earthquake response of earthworks and foundations

Earthquake response of earthworks and foundations is an important problem in earthquake engineering because damage of structures is related closely to ground or soil. Study in this branch of earthquake engineering has been active internationally for several decades. This branch of science also includes dynamic properties of soils, wave propagation in soil, soil–structure interaction, pile foundations, underground structures and soil liquefaction.

10.1 Field experience on earthworks and foundations

Since the earliest earthquake studies earthquake damage to ground and foundations has been used as an index for intensity scales. In general, ground damage may include: earthquake faulting, landslides and foundation failure.

Earthquake faulting occurs only in a narrow region of the earthquake origin, and landslide and mud–rock flow only in mountain areas. Although they can do great harm, there is not much that can be done to reduce them. Therefore only foundation failures will be discussed in the following.

10.1.1 Ground deformation

Ground and earthworks often deform during strong earthquakes. Deformation may be classified in two kinds, according to its cause, namely that from slope slide, and that from ground vibration. Here slope slide is defined as the motion of an earth body under gravity with or without additional vibrational action, and deformation related to slope slide is called landslide deformation. A bird's eye view of a landslide deformation shows in general a curved deformation band, with its top portion, sometimes very long, a straight line, parallel to the edge of the slope and curved downward to the slope at its ends. It is easy to distinguish on steep slopes, but not easy on very gentle slopes, especially when part of the ground surface is flat. The second kind of deformation means only soil deformation, especially unequal deformation or differential settlement of inhomogeneous ground, such as the compressional deformation shown by the bending of rails. Large deformations are visible; smaller ones can be found indirectly from behavior of structures on the ground. Sinking or cracks on a building may help identify them. For example, cracks of the form shown in Figure 10.1 indicate relative settlement of the middle portion of the building and a reversed pattern near the top indicates relative settlement of the ends of the building. This can be found on the filled lands of old rivers or lakes.

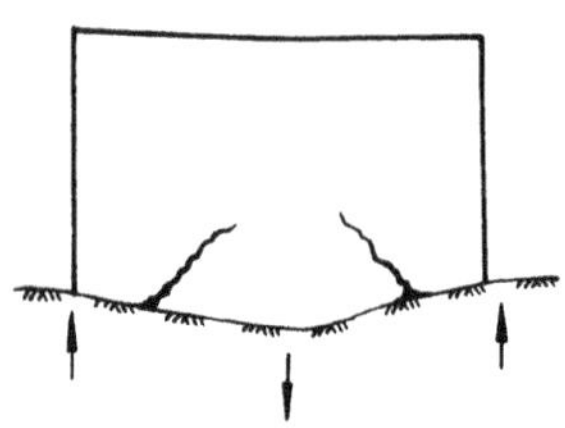

Figure 10.1 Crack pattern on wall from ground.

10.1.2 Soil liquefaction

Soil liquefaction is very common on lands of loose sandy soils with a high water table subjected to strong earthquakes. Typical phenomena of lique-

Figure 10.2 Inclination and overturning of apartments.

faction are sandblows and waterspouts, where a liquid mixture of sand and water is ejected into the air under pressure from a depth of a few to about 15m below the surface. They may occur in the process of strong shaking or many minutes or hours after, because it takes time for the liquefied sand to break through the overburdening deposit above it. The ejection may continue for a few hours or just a few minutes. Sand liquefaction may bring damage to superstructures, and may cause the inclination or overturning of them. Figure 10.2 shows a sketch of apartments inclined or overturned immediately after the 1964 Niigata earthquake in Japan. Although settlement, uniform or differential, is a common damage pattern induced by liquefaction, floating-up can also be found as a result of liquefaction. For example, immediately after the strong shaking of the 1975 Haicheng earthquake, an irrigation and drainage pumping station with an underground basement floated up and ceased to function. Figure 10.3 shows the deformation and cracking of an earth dam over the Dou river after the Tangshan earthquake. Analysis showed that the damage was due to liquefaction of the sand layer underneath the dam.

Field identification of liquefaction of a sand layer underground can be made only from sandblows and waterspouts, although sand layers may liquefy without their occurence. Possible liquefaction may be indirectly

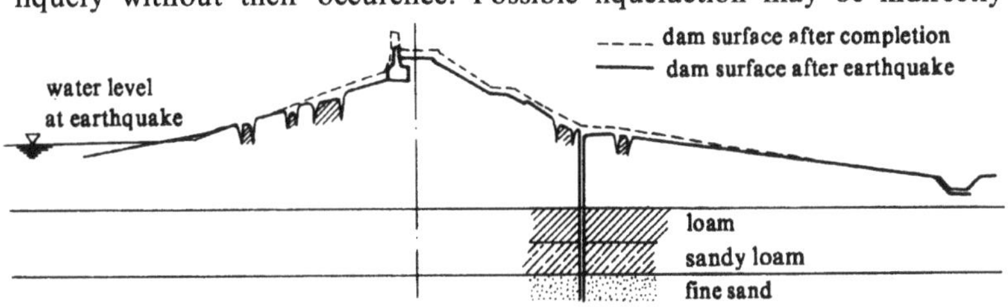

Figure 10.3 Earth dam over the Dou River damaged in 1976 Tangshan earthquake.

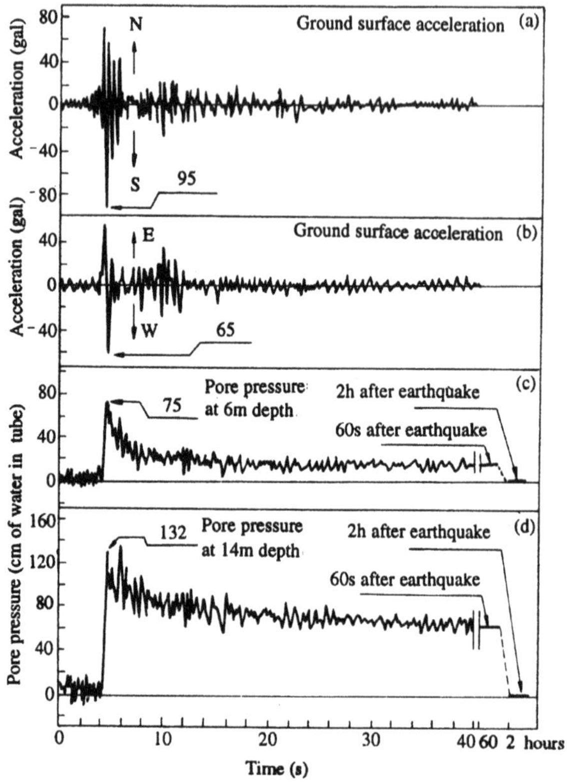

Figure 10.4 Observed pore pressure of sand liquefaction.

estimated from a landslide, but not with certainty. Field measurement of pore pressure of a sand layer is a direct indicator of liquefaction, but such measurements are rare. Figure 10.4 shows such a record of pore pressure together with acceleration time history records obtained on a fill ground with very loose fine sand, which shows that the sand liquefied almost immediately after the maximum peak acceleration. For sands which are not so loose, liquefaction will take place after some repetition of strong ground motion.

10.2 Dynamic properties of soil

10.2.1 Soil material

Soil is a complex natural material of engineering interest. Its complexity includes the following three aspects.

1. Soil particles are quite different in size and shape, and in physical and chemical properties.
2. Particles of different kinds may mix in quite different proportions.

3. The voids between particles may be filled with water or air; the mixture is then a multiphase medium of solid, liquid and gas.

Soils can be classified into two kinds: cohesive and cohesionless. Cohesionless soil consists of sand and gravel, which resists shear by friction only; cohesive soil consists of clay and clayey soil, which resists shear by both cohesion and friction. The shear strength of soil may be written as

$$\tau=(\sigma-u)\tan\phi+c \tag{10.1}$$

where τ is the shear stress and σ the normal stress on the failure surface; u is the pore pressure; ϕ is the internal friction angle; c is the cohesion strength. $(\sigma-u)$ is usually called the effective stress and σ the total stress. The first term on the right-hand side is friction and the second term is cohesion. For clay, c is very large and the second term dominates; for sand, c is very small and the first term dominates. Pore water may take compression when confined. The strength of soil is controlled by the strength of both the particles and the binding between them. The binding includes cohesion, interlocking and water content. For different kinds of soil, the relative importance of these factors is different.

(a) Sand

Although the shear strength correlates with the smoothness of the sand particle surface, it is controlled primarily by the void ratio or the density and the water content. Ordinary sand has a void ratio, the ratio of the void volume to the volume of particles, in the range 40%–100%. The shear deformation of sand is not reversible and there is volume change during shear deformation. There is a critical density for a certain sandy soil; under vibration, sand with density lower than critical will be compacted, and sand with density higher than critical will loosen.

When the voids in sand are completely filled with water, the sand is saturated. For a load acting on a saturated sand deposit, both the sand skeleton and the water in the void will be stressed to maintain the equilibrium. Under normal load and confinement, the skeleton will deform much more than the water because water is usually considered incompressible, and the load is finally taken entirely by the water; if the water is not confined, the stress distribution between solid and liquid depends on the permeability of the soil. Generally speaking, the larger the diameter of the soil particles, the larger is the permeability. Under shear, the saturated sand will deform with no stress. For loose sands, especially saturated, liquefaction may take place (section 10.3).

(b) Clayey soil

Clay particles are very small and can show solid, semi-solid, plastic or flow behavior, depending on water content. Under forces, the pore pressure will increase, and the soil will be condensed if water can flow out through the boundary. Because the permeability of clay is very small, the flow of water takes a long time to reach equilibrium, referred to as consolidation.

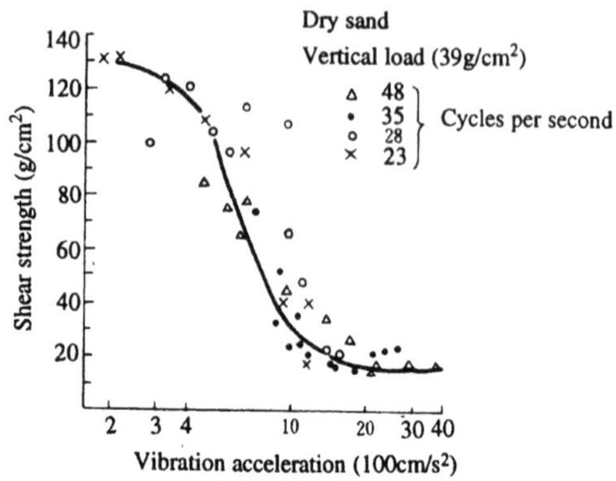

Figure 10.5 Shear strength of sand subject to vibration.

10.2.2 Dynamic strength of sand

By subjecting dry sand in a container to harmonic vertical vibration of various accelerations a, Mogami and Kubo (1953) obtained the result shown in Figure 10.5. It is seen from the figure that the simple shear strength of the sand quickly decreases when $a > 200\,\text{cm/s}^2$.

10.2.3 Dynamic strength of clay

Seed and Chan (1966) studied this problem using a specially designed instrument and their results for undisturbed samples of silty clay are shown in Figure 10.6. They applied first a certain constant load p to the sample and then a vibrational load with amplitude q and measured the deformation of the sample under the vibration. They define a certain amount of deformation unacceptable by engineers as failure. In Figure 10.6 σ is the static strength of the soil.

It is seen from their results that a vibrational load has two kinds of influence on the deformation strength of clay, depending on the number of repetitions. The sum $[(p+q)/\sigma]$ % of the abscissa and ordinate at any point on the curve represents the strength of clay. For $N=1$, it is smaller (127%) for large p/σ (80%), and greater (157%) for small p/σ (20%); on the contrary, for $N=100$, it is greater (96%) for large p/σ (80%) and smaller (78%) for small p/σ (20%). This fact means that the dynamic effect increases the strength of clay for small N but decreases it for large N.

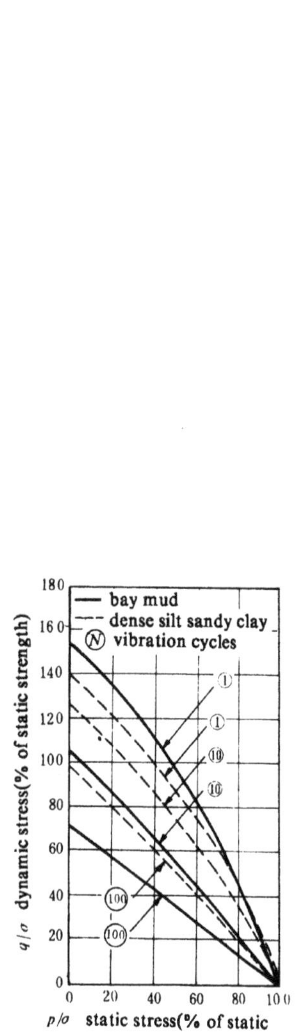

Figure 10.6 Strength of clay subject to vibration.

10.2.4 Dynamic deformation characteristics of soil

Through laboratory tests and field measurements, it is generally realized (Ishihara, 1981; Ohsaki, 1982a, b, c) that ordinary soil shows very strong nonlinearity, i.e. the important mechanical properties, such as the modulus and damping, are strong functions of strain. It is then necessary to have a general understanding of the range of possible variation of the important parameters of soil. Table 10.1 gives the probable ranges of deformation of various methods of tests or source of vibration.

There are two ways to represent the nonlinear properties of the soil: one is by giving the nonlinear stress–strain relationship, or the constitution law, of the soil, such as shown in Figure 10.7, and the other, simplified, one is

Table 10.1 Deformation ranges of soil

Max. deformation (%)	10^{-6}	10^{-5}	10^{-4}	10^{-3}	10^{-2}	10^{-1}	10^{-0}
Seismic prospecting	—	—	—	—	—		
Laboratory waves	—	—	—				
Resonance (solid)		—	—	—	—		
Resonance (cylindrical)					—	—	—
Dynamic simple shear					—	—	—
Mechanical vibration		—	—	—			
Forced vibration				—	—	—	—
Nuclear test (near)							—
Strong earthquake							—

by giving the deformation-related equivalent modulus and damping; both are commonly used.

Figure 10.7 is one of the commonly used nonlinear dynamic models, which is simple, and if the parameters are adequately selected, it will be consistent with test data. This model assumes that the stress–strain relation consists of two parts; the skeleton curve A′OA and the hysteretic loop ABA′B′A. The slope of the straight line OA, G_{eq}, may be taken as the secant modulus or equivalent modulus, and ζ_{eq} = (area of loop ABA′B′A)/(area of triangle OAD)/(4π) is the equivalent damping ratio, and both G_{eq} and ζ_{eq} are functions of strain γ. Near the origin O, the strain is very small, and the secant OA is the tangent of the skeleton curve OA at O; the equivalent modulus is $G_{eq}=G_{max}$ and damping $\zeta=0$. As the strain increases, G_{eq} decreases and ζ increases.

(a) Hyperbolic skeleton curve

On the basis of a large amount of test data, Hardin and Drnevich (1972) suggested a hyperbolic skeleton curve for the stress–strain relations as follows:

$$\tau=\frac{\gamma}{(1/G_{max}+\gamma/\tau_{max})} \tag{10.2}$$

Their results show that the damping ratio decreases as the confining pressure $\bar{\sigma}_0$ decreases. When defining a reference strain $\gamma_r=\tau_{max}/G_{max}$ (Figure

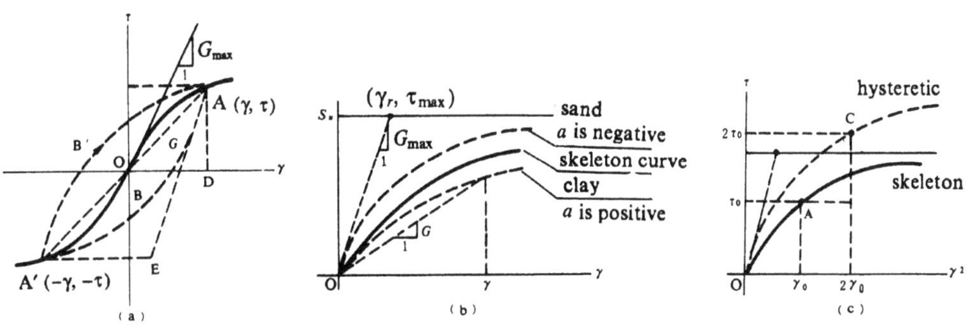

Figure 10.7 Nonlinear model of stress–strain relation of soil.

10.7(b)), they found that the damping will be represented by the same curve for the normalized strain γ/γ_r.

Reference strain γ They defined first G_{max} as the modulus for $\gamma=0.25\times 10^{-4}$ from test data and, for undisturbed clay and sand,

$$G_{max}=326\frac{(2.973-e)^2}{1+e}(\mathrm{OCR})^{\alpha}\bar{\sigma}_0^{1/2} \tag{10.3}$$

where e is the void ratio and OCR the overconsolidation ratio; the shear modulus G_{max} and confining pressure $\bar{\sigma}_0$ are in kg/cm^2. The result given by this equation may be small for G_{max} when $e>2$. The parameter α is related to plastic index PI and takes values of 0, 0.18, 0.30, 0.41, 0.48 and 0.50 for PI=0, 20, 40, 60, 80 and $\geqslant 100$ respectively. The maximum stress τ_{max} is taken as

$$\tau_{max}=\left[\left(\frac{1+K_0}{2}\bar{\sigma}_v\sin\bar{\phi}+\bar{c}\cos\bar{\phi}\right)^2-\left(\frac{1-K_0}{2}\bar{\sigma}_v\right)^2\right]^{1/2} \tag{10.4}$$

where K_0 is the static lateral stress coefficient, $\bar{\sigma}_v$ is the vertical effective stress and $\bar{c}$ and $\bar{\sigma}$ are strength parameters expressed in the effective stress equation 10.1.

Estimating K_0 as a function of OCR and ϕ and substituting the expressions of G_{max} and τ_{max} in the definition of the reference strain γ_r, one obtains the normalized reference strain $\gamma_r/\bar{\sigma}^{1/2}$ of sand; or estimating $\bar{\phi}$ as a function of PI and K_0 as a function of OCR and $\bar{c}$, one obtains the normalized reference strain of clayey soil. Numerical study shows that the void ratio has the strongest effect on strain. Hardin and Drnevich explain this in detail.

Shear modulus G and damping ζ. Simple expressions may be obtained for G and ζ under the following two assumptions. Firstly, the slope of the reverse curve AB at A is equal to G_{max} and secondly, the ratio of the loop area ABA′B′A to the triangle area AEA′ is a constant, say C_1. Then

$$\zeta=\frac{C_1}{\pi}1-\frac{G}{G_{max}}=\zeta_{max}\left(\frac{1-G}{G_{max}}\right) \tag{10.5}$$

which matches the test data well. The final results are then

$$\frac{G}{G_{max}}=\frac{1}{1+\gamma_h} \tag{10.6a}$$

$$\frac{\zeta}{\zeta_{max}}=\frac{\gamma_h}{1+\gamma_h} \tag{10.6b}$$

where $\gamma_h=\gamma/\gamma_r$.

Modified hyperbolic relation. Because measurement data show that the real stress–strain relation does not follow the assumption of a hyperbolic law and the hysteretic loop does not follow exactly the two assumptions, Hardin and Drnevich (1972) suggested modifying the value of γ_h by the following formula on the basis of test data:

$$\gamma_h = \frac{\gamma}{\gamma_r}\left[1 + a\exp\left(-\frac{b\gamma}{\gamma_r}\right)\right] \tag{10.7}$$

where constants a, b and ζ_{max} are given functions of the number of cycles and other parameters.

(b) Empirical relation of Seed *et al.*

Seed *et al.* (1969, 1976) have studied the problem of equivalent modulus and damping of soils and they believe that the most important factors are relative density D_r and confining pressure. Their results are consistent with strong motion records and damage, and are therefore widely used. Figures 10.8 and 10.9 show some of their results.

The results in Figure 10.8 are for medium sand with $D_r = 80\%$, where σ_0' is the effective confining pressure, in lbf/ft^2 (1 lbf/ft^2 = 47.8 Pa), the same units as the modulus G. The same results may be used for other sandy soils, if multiplied by a modifying factor. For example, for silt, sandy silt and silty sand, this factor is in the range of 0.5–0.9, depending on the silt content; for gravel it is 1.2–2.0, depending on the gravel content; for sands of different density, it is

$$\frac{G_{D_r}}{G_{80}} = \frac{5}{9}\left(1 + \frac{D_r}{100}\right) \tag{10.8}$$

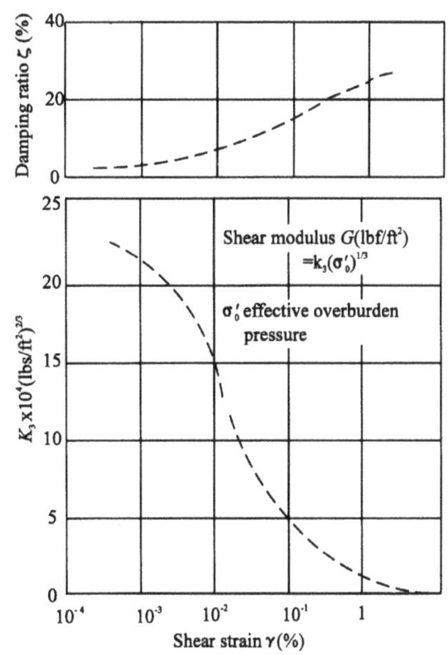

Figure 10.8 Equivalent shear modulus and damping of medium sand. (1 lbf/in^2 = 6.89 kPa).

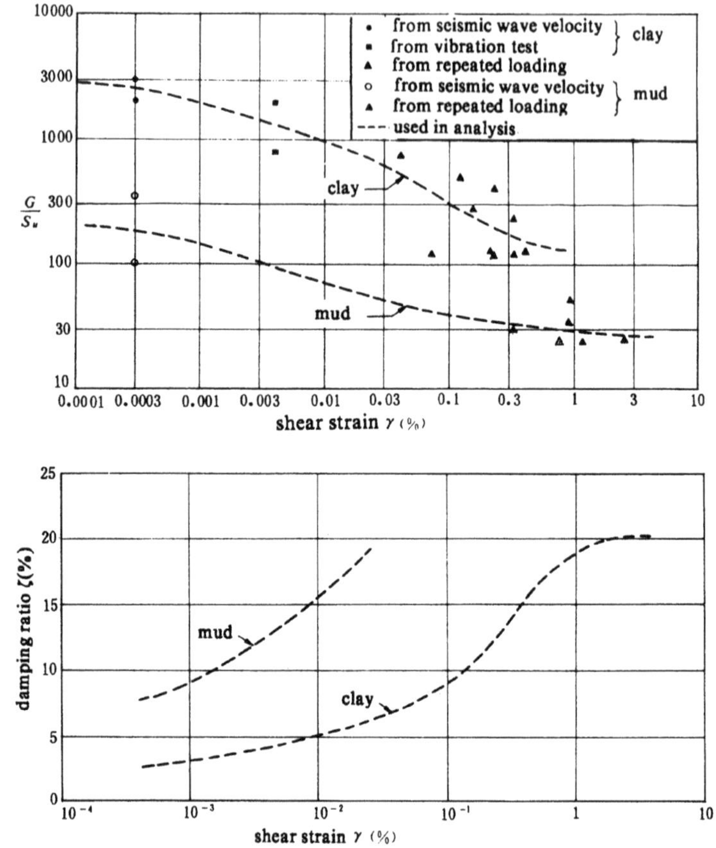

Figure 10.9 Equivalent shear modulus and damping for saturated clay and mud.

The damping ratio is not affected by D_r. Results in Figure 10.9 are for saturated clay and mud, where S_u is the undrained shear strength.

10.3 Sand liquefaction

10.3.1 Field experience on sand liquefaction

In section 10.1.2, typical phenomena of liquefaction are described. Analysis of such phenomena provides the following information. Firstly, when the ground liquefies, superstructures are mostly damaged indirectly by ground failure and seldom by structural vibration (Ambraseys and Sarma, 1969; Hu *et al.*, C 1980), because the liquefied soil layer prevents the shear wave from propagating through and thus limits the structural vibration. Provided that there is a sufficiently thick and strong surface layer on top of the liquefied layer to prevent the liquefied materials from breaking through, the superstructures will be damaged less than those on ground without that liquefied layer. Experience from the 1975 Haicheng earthquake shows this clearly

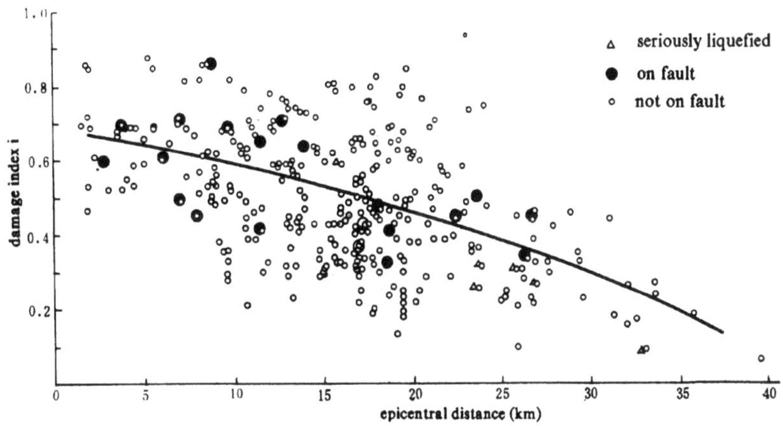

Figure 10.10 Effect of liquefaction on damage.

(Hu and Sun, 1982). Figure 10.10 shows such a case during the 1975 Haicheng earthquake.

Secondly, sand liquefaction occurs usually in the saturated loose fine sand layer near the ground surface, mostly less than 10 m from the top, sometimes 10–17 m deep, but not more than 20 m. There are perhaps four reasons as follows.

1. A higher pore pressure is needed for liquefaction to overcome the higher confining pressure at greater depth.
2. Sand at greater depth usually has higher density and larger grains and is more difficult to liquefy.
3. Ground motion is smaller at greater depth.
4. It is more difficult for the liquefied sand to break through to the surface and not easy to identify if the sand is liquefied.

Thirdly, time is required for the sand to liquefy and to find its way to come to surface. There are at least three steps: the initial and local liquefaction, the liquefaction of the whole layer and arrival at the surface. The time needed to complete these steps depends on ground motion velocity and duration, the mechanical and geometrical properties of the sand and those of its top layers. Ambraseys and Sarma (1969) estimated the time required for the bearing capacity of the top layer to decrease to a certain amount and found that it may require from a few tens of seconds to a few minutes for a top layer of a few tens of meters.

Fourthly, strong ground motion may condense loose sand and loosen a dense sand layer. Field evidence shows this clearly (Ohsaki, 1966).

10.3.2 Evaluation of sand liquefaction

Figure 10.11 shows liquefaction test results for saturated fine sand under cyclic loading. It can be seen that, after a certain repetition of vibration, the pore pressure and the dynamic strain suddenly increase and the sand

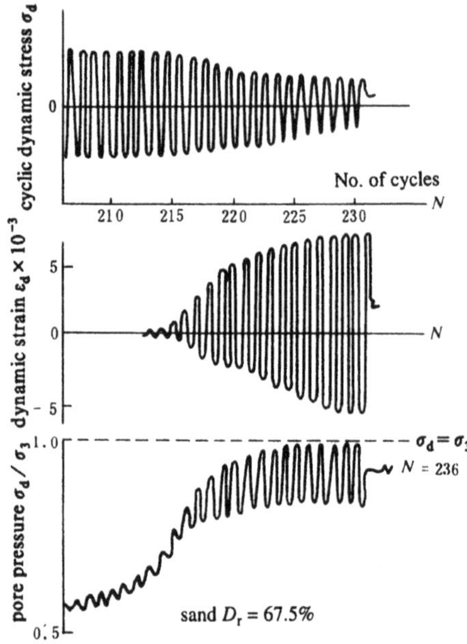

Figure 10.11 Records of liquefaction test of sand.

quickly acts like a liquid when pore pressure reaches the confining pressure or when the total dynamic deformation is 0.05 or 0.1.

Evaluation of sand liquefaction is an important and practical problem. The following methods are available.

(a) Simple method (Seed and Idriss, 1971)

This method simplifies the ground motion to an equivalent harmonic shear stress τ_{av} of N cycles with the amplitude

$$\tau_{av} = 0.65\gamma h r_d a_{max} \tag{10.9}$$

where γ is the weight per unit volume and h the depth of deposit above the considered soil; the product γh is then the overburden pressure, a_{max} is the maximum horizontal acceleration over g; r_d is a depth-reduction coefficient to consider the effect of soil response, which may vary linearly in the range from 1.0 at ground surface to 0.9 at depth 10 m. A coefficient of 0.65 is included to consider the ratio of average to maximum dynamic stress. The number of cycles N is related to earthquake magnitude M, and Seed *et al.* suggested $N = 10$, 20 and 30 for $M = 7$, 7.5 and 8 (Velera and Donovan, 1977).

The field shear strength τ_R of soil is then estimated. Their laboratory result is

$$\frac{\tau_R}{\sigma_v'} = \left(\frac{\sigma_{dc}}{2\sigma_a}\right)_{D_{50}} C_r \frac{D_r}{50} \tag{10.10}$$

where σ_v' is the effective stress in the soil; σ_{dc} is the cyclic deviator stress; σ_a is the original consolidation compressional stress; D_r the relative density

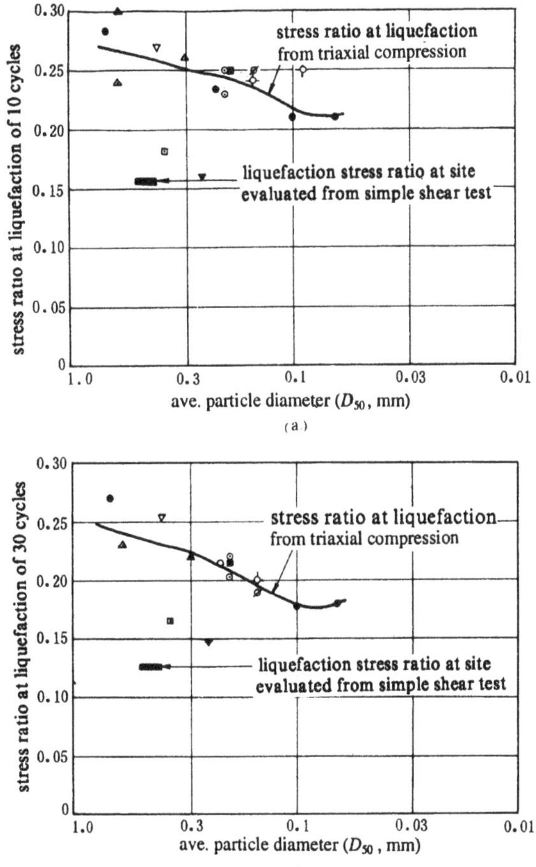

Figure 10.12 Stress ratio $\sigma_{dc}/(2\sigma_a)$.

as a percentage. C_r is a correction factor for using the laboratory result in the field; it is 0.55, 0.55, 0.55, 0.60, 0.69 or 0.74 for $D_r=0$, 20, 40, 60, 80 or 90. $\sigma_{dc}/(2\sigma_a)$ is the ratio of stresses of liquefaction in a laboratory test at N cycles, which is a function of the average diameter of grains, d_{50}. Figure 10.13 gives their relationship. From a comparison of the results, τ_{av} and τ_R, of equations 10.9 and 10.10, there will be a possibility of liquefaction if $\tau_{av}>\tau_R$. The result thus obtained is shown in Figure 10.13. This result was modified later by Seed (1981) using the earthquake magnitude M for N, as shown in Figure 10.16.

It is seen that there are two groups of parameters considered in this method, which are ground motion (including peak acceleration a_{max} and number of cycles N) and the soil properties (including the grain size d_{50}, relative density D_r and effective stress or depth to the surface).

(b) Empirical method

The difficulty in applying the above method is in the determination of the relative density of loose sand without disturbance. It is easier to use the

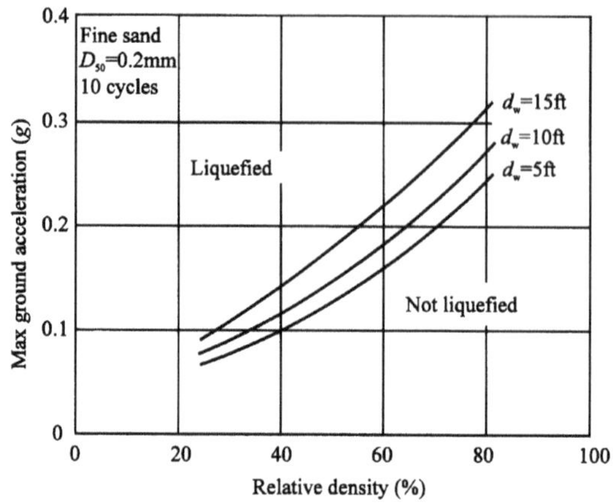

Figure 10.13 Liquefaction evaluation.

standard penetration blow counts instead because relative density is usually determined through blow counts in practice. Ohsaki (1966) was the first to apply this method to analyze the liquefaction cases during the 1964 Niigata earthquake. On the basis of data available at Niigata City, he found various depths where the blow count N_{cr} did not change before and after the Niigata earthquake and obtained the curve $N_{cr}-H$, as shown in Figure 10.14.

In China, a large number of field liquefaction data since 1960 have been analyzed to find a rule for liquefaction evaluation, and such a relation has

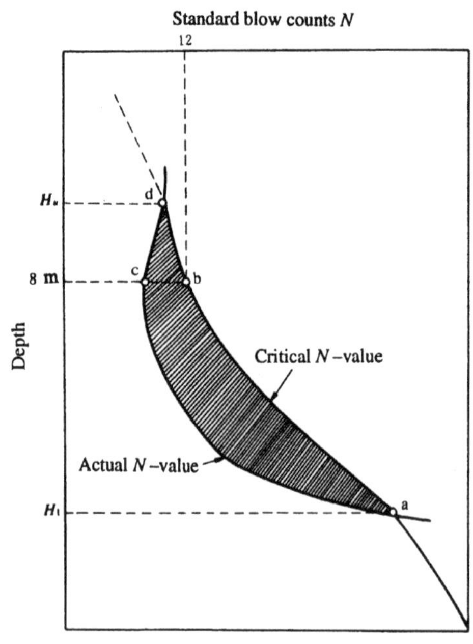

Figure 10.14 Critical blow count.

been adopted in the China Code on Seismic Design of Buildings since 1974. The early specification was

$$N_{cr} = \bar{N}\ [1 + 0.125(d_s - 3) + 0.05(d_w - 2)] \qquad (10.11)$$

where d_s is the depth of the sand layer and d_w the water table depth, both in meters; $\bar{N}$ is the critical N-value at $d_s = 3$ and $d_w = 2$, which is 6, 10 and 16 for intensity VII, VIII and IX respectively. In the current Building Seismic Design Code GBJ 11–89, this formula is modified to include both the silt and the distance–magnitude effect, as discussed later (equation 14.14).

It is obvious that the specification in the current China code can easily be further improved by specifying directly the magnitude and distance instead of indirectly through intensity, because the duration of strong motion or number of load repetition cycles is a direct function of earthquake magnitude and distance but not the intensity.

In the 1970s, progress was made in seismic hazard assessment, seismic slope design and damage reduction; Seed's (1981) achievements were parallel to this development and he suggested quite a complete and simple empirical method of liquefaction evaluation, of which the main steps are as follows.

1. Determine the standard penetration blow count N (blow count for penetration of 1 ft (30.5 cm)) of the considered horizontal layer of sand and convert it to the blow count N_1 at a confining pressure of 0.975 kgf/cm^2 as follows:

$$N_1 = C_N N \qquad (10.12)$$

   The coefficient C_N is given in Figure 10.15.
2. From N_1 and magnitude M of earthquake, find the necessary cyclic stress ratio τ/σ_v' for a pore pressure ratio of 100% under static stress ratio $\alpha = \tau_{fc}/\sigma_{fc}$ from Figure 10.16, where the magnitude M replaces the

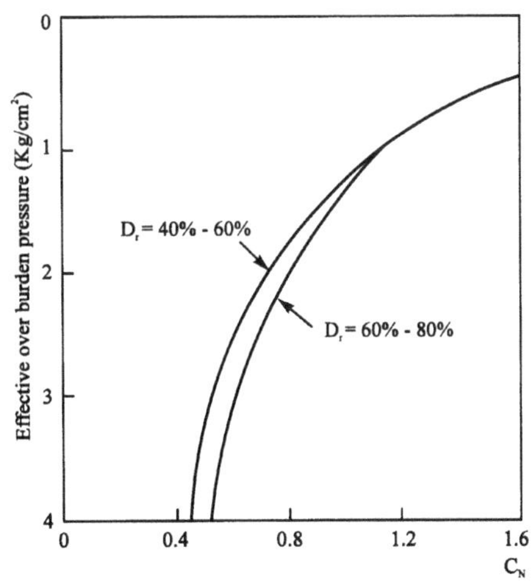

Figure 10.15 Effect of confining pressure on blow count.

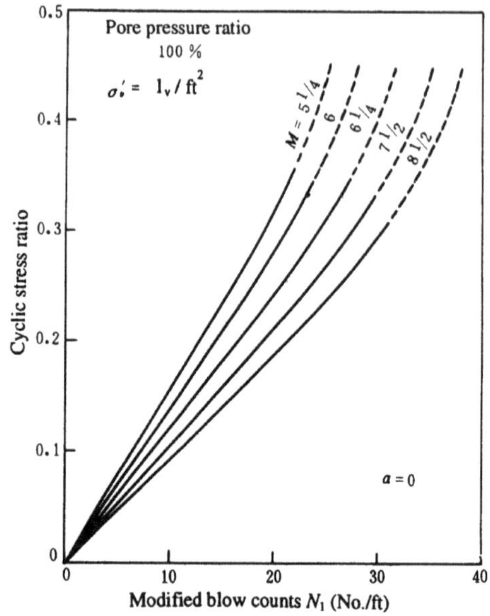

Figure 10.16 Liquefaction under horizontal ground.

number of cycles previously used by Seed *et al.*; τ_{fc} is the initial horizontal shear stress and σ_{fc} the initial vertical normal stress before earthquake; for a horizontal ground surface, $\alpha=0$.

3. From the static stress ratio α and Figure 10.17, find the correction factor K_α for the liquefaction cyclic stress ratio

$$\left(\frac{\tau_c}{\sigma_c'}\right)_{\ell,\alpha}=\left(\frac{\tau_c}{\sigma_v'}\right)_{\ell,\alpha=0}K_\alpha \tag{10.13}$$

If the effective vertical stress σ_0' is greater than 1.37 kgf/cm^2, another correction factor K_σ from Figure 10.18 is required:

$$\left(\frac{\tau_c}{\sigma_v'}\right)_{\ell}=\left(\frac{\tau_c}{\sigma_v'}\right)_{\ell,\alpha}K_\sigma \tag{10.14}$$

4. Finally obtain the cyclic shear stress required for 100% pore pressure

$$\tau_c=\left(\frac{\tau_c}{\sigma_v'}\right)_{\ell}\sigma_v' \tag{10.15}$$

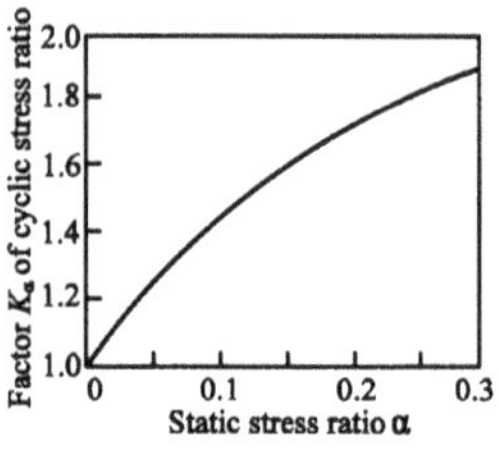

Figure 10.17 Effect of initial shear stress on liquefaction.

This method is derived mainly from earthquake field experience in the same way as the China code method; the results are thus similar, as shown in Figure 10.19. But Seed's method considers more factors and has the following advantages, when compared with equations 10.9 and 10.10.

1. It is better than methods based on a limited number of tests.

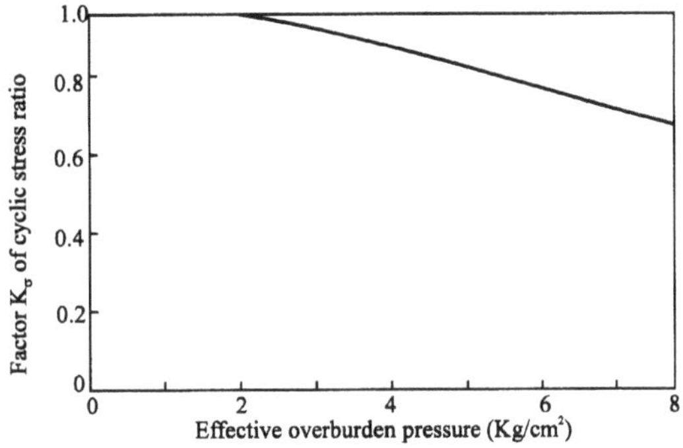

Figure 10.18 Correction factor for initial shear stress.

2. It is not affected by disturbance from soil sampling, transportation and processing.

The main disadvantage is the difficulty in estimating the blow count.

(c) Other methods

Other methods have been applied in some cases, such as the energy method by Davis and Berrill (1982), Mostaghel and Habibagahi (1979); method of discrimination (Tanimoto and Noda, 1976; Yu and Shi, C1980), and probability method (Haldar and Tang, 1979).

10.3.4 Liquefaction risk

The relation between liquefaction and risk is not simple. Phenomena have been mentioned above from many case histories where damage of structures has been intensified due to foundation liquefaction, but also some cases where the damage to the superstructure has been reduced. There are many cases which indicate the importance of the depth of liquefaction on damage:

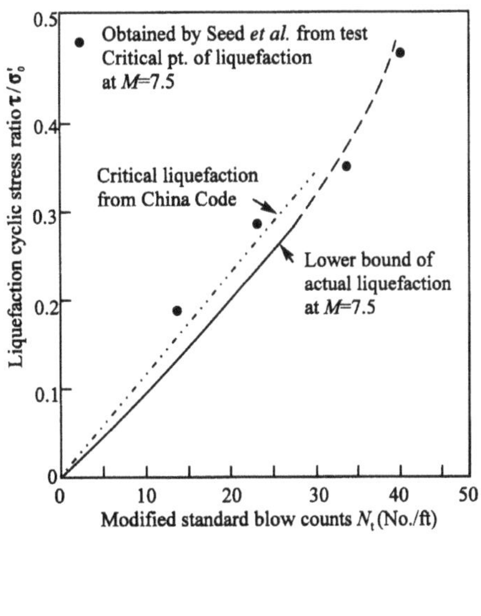

Figure 10.19 Comparison of China specification and Seed's result.

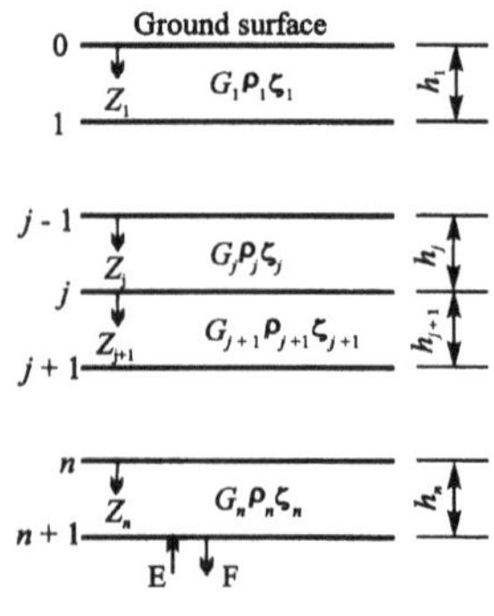

Figure 10.20 Waves in a layered medium.

the deeper the liquefied layer, the lighter the damage. Moreover, it is clear that there is a transition period before liquefaction. Therefore, a liquefaction risk index P_L is defined as follows (Iwasaki *et al.*, J1980; Qiao and Liu, C1984):

$$P_L = \int_0^H F \cdot w(z) \mathrm{d}z \tag{10.16}$$

where H is the total depth of the ground considered; z a depth variable, measured from surface, $w(z)$ a weighting function, varying linearly from the surface $w(z=0)=k$ to the deepest position considered $w(z=H)=0$; F is assigned as follows. Let $F_L = R/L$ be a liquefaction index, where R is the liquefaction resistance and L the ground motion at the same depth; then $F=1-F_L$ if $F_L<1$ and $F=0$ if $F_L \geqslant 1$. Iwasaki *et al.* used $H=20$ m, $k=10$ and corresponding stresses for R and L; Qiao and Liu used $H=15$ m, $k=10$ and corresponding blow counts for R and L. They compared their results with data from Japan and China (Haicheng and Tangshan) respectively and found the relations between damage and P_L shown in Table 10.2.

It is better to take the constant $k=w(z=0)=2/H$ to make $\int_0^H w(z)\,\mathrm{d}z=1$ or to use the normalized $P_L/(\int w(z)\,\mathrm{d}z)$ instead of equation 10.16.

The idea of a liquefaction risk index considers the ratio of capacity and load as well as depth effects. However, the effect of isolation of shear wave propagation is not considered at all.

10.4 Earthquake response analysis of sites

There are two aims in studying the seismic response of sites: first, to know the relation between ground surface and 'bedrock', for example to find the motion at the structural foundation in order to study the seismic structural behavior; and secondly, to know the seismic behavior of the ground itself, for example if the ground will liquefy.

The ground or soil has two features compared with structures:

1. strong nonlinearity;
2. an infinite or no boundary, or half space, which has the important property of geometric damping or high energy-absorbing capacity.

There are two groups of methods for analysis of site response. One is for horizontally layered sites, which may be treated as a one-dimensional

Table 10.2 Liquefaction risk index P_L

Damage	Iwasaki *et al.*	Qiao and Liu
Very light	0 (0)	0 (0)
Light	1–5 (0.01–0.05)	1–3 (0.013–0.04)
High	5–15 (0.05–0.15)	3–6 (0.04–0.08)
Very high	> 15 (> 0.15)	> 6 (> 0.08)

problem; this includes the shear beam, lumped mass, wave propagation and characteristic line methods. The other is applicable to complex sites, including two- or three-dimensional problems, such as the finite element method.

10.4.1 Method of equivalent linearization for horizontally layered sites

(d) Wave propagation in layered medium

Consider a site of n horizontal layers of viscous elastic soil in a half space as shown in Figure 10.20, subject to a shear wave propagating vertically from a half space. With coordinates and material properties as shown, each layer should satisfy the equation (section 2.5)

$$\rho_j \frac{\partial^2 u_j}{\partial t^2} = G_j \frac{\partial^2 u_j}{\partial z_j^2} + \eta_j \frac{\partial^3 u_j}{\partial z_j^2 \partial t} \qquad \text{(a)}$$

where u_j is the displacement at depth z of the jth layer in the horizontal direction; and G, ρ and η are respectively the shear modulus, dry density and damping constant. The stationary solution of this differential equation is

$$u_j(z_j, t) = U_j(z_j)\exp(i\omega t) \qquad \text{(b)}$$

$$U_j(z_j) = E_j \exp(\mathrm{i}k_j z_j) + F_j \exp(-\mathrm{i}k_j z_j) \qquad \text{(c)}$$

The E term is the upward propagating wave and the F term the downward propagating wave; $k_j = \omega / V_j$; U, E and F are all functions of frequency ω; the shear wave velocity V_j of the jth layer is

$$V_j^2 = \frac{G_j + \mathrm{i}\omega\eta_j}{\rho_j} = \frac{G_j}{\rho_j}(1 + \mathrm{i}2\zeta_j) = \frac{G_j^*}{\rho_j} \qquad \text{(d)}$$

where ζ_j is the damping ratio of the jth layer and G_j^* the complex shear modulus (section 9.6.1).

The shear stress is

$$\tau = G\frac{\partial u}{\partial z} + \eta \frac{\partial^2 u}{\partial z \partial t} = G^* \frac{\mathrm{d}u}{\mathrm{d}z}$$

The continuity conditions between two neighboring layers of force is $\tau_j|_{z=h_j} = \tau_{j+1}|_{z=0}$, of displacement is $u_j|_{z=h_j} = u_{j+1}|_{z=0}$, which yield

$$E_{j+1} = \frac{1}{2}[E_j(1+\alpha_j)\exp(\mathrm{i}k_j h_j) + F_j(1-\alpha_j)\exp(-\mathrm{i}k_j h_j)] \qquad (10.17\text{a})$$

$$F_{j+1} = \frac{1}{2}[E_j(1-\alpha_j)\exp(\mathrm{i}k_j h_j) + F_j(1+\alpha_j)\exp(-\mathrm{i}k_j h_j)] \qquad (10.17\text{b})$$

where $\alpha_j = \rho_j V_j / (\rho_{j+1} V_{j+1})$.

It is obtained from equation (c) that the amplitude of ground motion at top of the jth layer is

$$A_j = U_j(z=0) = E_j + F_j \tag{10.18}$$

At the free surface or the top of the first layer, shear $\tau_1|_{z=0} = 0$, and $E_1 = F_1$, therefore

$$E_1 = F_1 = \frac{A_0}{2} \tag{10.19}$$

where $A_0 = U_1(0)$ is the amplitude of displacement at the ground surface. It is important to know that, from equation (c), $E_k + F_k$ is the ground motion at the top of the kth layer.

On the basis of these relations, all E_j and F_j ($j = 1, 2, \ldots, n+1$) may be found from the given motion A_0 at the surface. For example, if A_0 is known, E_1 and F_1 may be found from equation 10.19; then E_j and F_j for $j = 2$, iterating for $j = 3$ and so on, up to $j = n+1$, from equations 10.17.

If any ground motion x_m, A_m or E_m, is known at the top of the mth layer, proceed as in the foregoing example for $A_0 = 1$, and find E_k and F_k for all values of $k = 2 \ldots, n+1$, and then the computed motion x_m'. Define then a modification factor

$$r_{x_m} = \frac{x_m}{x_m'} \tag{10.20a}$$

and then multiply all E and F by the factor r_{x_m} and the problem is solved. In the case where the upward propagating wave E_{m0} at the top of the mth layer is known, the modification factor is then

$$r_{E_m} = \frac{E_m}{E_{m'}} \tag{10.20b}$$

If the incident wave amplitude E_{n+1} is known at the bottom of the nth layer, the modification factor is then

$$r_{E_{n+1}} = \frac{E_{n+1}}{E_{n+1}'} \tag{10.20c}$$

(b) Amplification of a single horizontal layer

From equations 10.17, the ground motion at the surface and at the top of the half space are respectively

$$A_0 = A_1 = E_1 + F_1 = 2E_1$$

$$A_2 = E_2 + F_2 = E_1 \exp(\mathrm{i}k_1 h_1) + F_1 \exp(-\mathrm{i}k_1 h_1)$$

The amplification or the transfer function from the bottom to the top of the surface layer is then

$$H(\omega) = \frac{A_0}{A_2} = \frac{2}{\exp(\mathrm{i}k_1 h_1) + \exp(-\mathrm{i}k_1 h_1)} = \frac{1}{\cos(\omega h_1 / V_1)}$$

If complex damping is considered, it is only necessary to substitute V_1 by $V_1(1 + 2\zeta_1 \mathrm{i})^{1/2}$, and the transfer function with damping is

$$H(\omega)=\frac{1}{\cos[\omega h_1/V_1(1+2\zeta_1 \mathrm{i})^{1/2}]}=\frac{2(A-B\mathrm{i})}{A^2+B^2} \tag{10.21}$$

where

$$A=[\exp(-kd)+\exp(kd)]\cos kc,\ B=[\exp(-kd)-\exp(kd)]\sin kc \tag{10.22a}$$

$$c=\left(\frac{(1+4\zeta^2)^{1/2}+1}{2(1+4\zeta^2)}\right)^{1/2},\ d=\left(\frac{(1+4\zeta^2)^{1/2}-1}{2(1+4\zeta^2)}\right)^{1/2},\ k=\frac{\omega h_1}{V_1} \tag{10.22b}$$

(c) Computational steps

When considering the nonlinearity of soil, strictly speaking, the method of analysis in the frequency domain is not applicable for an exact solution; but it is commonly used as an approximate method together with the method of equivalent linearization. The computational steps are as follows.

1. Divide the ground into horizontal layers. The half space surface is selected first. If the bedrock surface of the site is not too deep, it is selected as the top of the half space; if it is, or if the bedrock surface is not known, a sufficiently deep and hard (say, $v_s \geqslant 700\,\mathrm{m/s}$) surface is selected, where the soil below has a suddenly increased ρV value (for example, the ρV value suddenly increases by a factor more than 2) and there is no layer underneath of lower ρV value. For division of layers, soils of quite different properties are divided into different layers; and one layer cannot be too thick when the nonlinear properties of soil are considered. In the method of equivalent linearization, one layer has the same dynamic properties, such as equivalent shear modulus and damping. If the layer is very thick, the stress and strain in the layer may vary by a large amount and an equivalent mean will not be a good approximation. The point where ground motion is known, specified or required, is taken as the layer boundary.
2. Assume the equivalent shear modulus and damping for each layer on the basis of their elastic values and the given ground motion, or just the elastic values in the absence of ground motion data.
3. Find the Fourier transforms $U_m(\omega, z_m)$ of the given motion $u(t, z_m)$ with quiescent motion $\ddot{u}=0$ added after the given motion to make the time interval Δt and the total number of data points N acceptable for the frequency band requirement (section 9.4.1).
4. Assume $E_1(\mathrm{i}\omega)=F_1(\mathrm{i}\omega)=0.5U_1\ (\mathrm{i}\omega, z_1=0)=1$, and calculate all other $E_k(\mathrm{i}\omega)$ and $F_k(\mathrm{i}\omega)$ for $k=2, 3, \ldots, n+1$. Although it is possible to carry out the above computations for all $\omega=k\cdot\Delta\omega$ ($k=1, 2, \ldots, (N/2)+1$), where $\Delta\omega=2\pi/(N\Delta t)$, in practice, only a small fraction is computed directly for the following two reasons. Firstly, filtering is usually used to remove very high and very low frequencies; and secondly, at those frequencies away from the resonance band, the variation of transfer functions is slow, and interpolation may give good results, reducing computer time.
5. Compute the shear strain γ at the middle point of each layer in the frequency domain

$$\Gamma(i\omega, z_k = 0.5h_k) = ik_k\,[E_k \exp(ik_k z_k) - F_k \exp(-ik_k z_k)]$$

and then find the strain time history $\gamma_k(t)$. Find the equivalent amplitude of the simple harmonic motion $\bar{\gamma}_k = C\gamma_{k,\max}$ where C is an empirical constant, usually taken as 0.65 as suggested by Seed (Martin and Seed, 1982; Schnabel *et al.*, 1972).

6. Compare the equivalent modulus and damping of each layer corresponding to the computed shear strain $\bar{\gamma}_k$ with those assumed in step 2; if the error is too large for some, then modify them and repeat the whole process from steps 3–6 until errors between the assumed and the computed shear moduli G and damping ζ are smaller than required.

(d) Convergence of iteration

Although there is still no exact proof of the convergence of this iteration procedure, numerical examples have demonstrated it. The authors' experience has shown that any reasonable initial guess will come to the same result after less than 10, or 20 at most, iterations for most of the practical problems.

10.4.2 Finite element method for arbitrary site

The plane finite element method has been widely applied to the study of nonhorizontal soil-layered sites, such as inclined surfaces and inclined bedrock and soil with soft soil lenses, as well as slopes and dams. The equivalent linear method in the frequency domain is popular because of its simplicity, at least in formulation. The following procedure applies also for three-dimensional finite element problems in principle.

An earthwork, such as a dam or slope, may sometimes be treated as a plane strain problem. The cross-section of a long earthwork is divided into many finite elements and its equation of motion in matrix form is

$$\boldsymbol{M}\ddot{\boldsymbol{u}} + \boldsymbol{K}\boldsymbol{u} = \boldsymbol{m}_1 \ddot{u}_g - \boldsymbol{m}_2 \ddot{v}_g$$

where $\boldsymbol{u}$ is the displacement matrix of the nodes, including a horizontal and a vertical component in the cross-section of the earthwork; $\boldsymbol{M}$ is the mass matrix; $\boldsymbol{K}$ is the complex stiffness matrix which is a function of strain in the case of nonlinear soils. $\boldsymbol{Ku}$ may be replaced by $\boldsymbol{C}\dot{\boldsymbol{u}} + \boldsymbol{Ku}$ where $\mathbf{K}$ and $\boldsymbol{C}$ are real stiffness and damping matrices. $\ddot{u}_g$ and $\ddot{v}_g$ are respectively the horizontal and vertical input motion acting at the base of the model; $\boldsymbol{m}_1$ and $\boldsymbol{m}_2$ are related mass vectors. The following problems are important in establishing this equation. Firstly, there are some effective boundary models available to consider the requirement of no reflection at ground boundary (section 9.9.3), or the dimensions of the foundation model may be taken to be much larger than the dimensions of the base of the superstructure. Secondly, no input motion is applied to the vertical boundaries of the ground except when considering the waves propagating along the ground surface, such as surface waves. Thirdly, the finite element cannot be too large or it will prevent the transmission of waves of high frequency or short wavelength. Experience shows that the size of finite element in the direction of wave propagation should not be greater than an eighth or a quarter of the wavelength of the highest frequency component considered. Fourthly, it should be noted that higher modal responses are much reduced by the assumption of the damping ratios of the lower modes when using Rayleigh damping (section 9.6.3)

The following methods may be used to solve this equation: step-by-step integration in the time domain, especially for nonlinear response; method of modal superposition; and equivalent linearization in the frequency domain.

10.5 Earthquake stability of earth dams and slopes

10.5.1 Experience from earthquake damage and field measurement

There have been numerous case histories of slope and dam failure during strong earthquakes. Their detailed analysis reveals the cause of failure and the weak points in design. In addition to examples given in the following sections, two cases from the 1975 Tangshan earthquake (Section 6.4) are described as follows.

(a) Miyun reservoir earth dam

The dam lies about 70 km to the north of Beijing and about 150 km from the epicenter of the 1975 Tangshan earthquake (magnitude 7.8). The dam is 65.3 m high and about 1 km long and built on the deposit of sandy gravel and gravel layers of maximum depth more than 40 m. The dam body is made of gravelly sand with slightly more than 60% of grains of diameter greater than 5 mm with an inclined thin clay core wall. The cross-section of the dam is shown in Figure 10.21.

During the earthquake, the dam was in a region of intensity VI. The upstream protection outside of the core wall and below the water level slid down the slope for a distance of 40 to 100 m. The accelerographs on and near the dam recorded the main shock with peak accelerations of $0.05g$ and $0.14g$ respectively at the foundation and top of the dam, with a total duration of 140 s. Results of analysis show that the slide is due to the liquefaction of the gravelly sand of the protection layer because there is not enough gravel to form a gravel skeleton and the protection material acts just like loose sand with gravel floating between the sand particles.

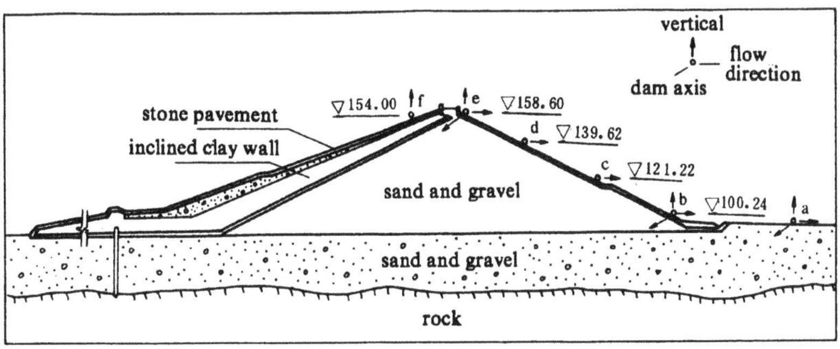

Figure 10.21 Cross-section of an earth dam at Miyun reservoir.

(b) Earth dam on the Dou River

The dam, located 15 km to the northeast of Tangshan City, is made of compacted sandy clay, of a height 22 m and about 6 km long at the top; the foundation is sandy clay and clayey sand, with a fine sand layer of 20 m thickness about 6 to 8 m below surface, and the hard gravel sand layer is about 50 m below surface. A covering layer of clay of 1.5 m thickness and 100 m long was placed under the dam foundation. During the earthquake, the intensity near the dam was rated IX, and the dam was badly damaged, but fortunately did not collapse because the water depth was only 6 m during that time. There were several open cracks on both the upstream and downstream slopes and through the cross-section of the dam. The largest crack was 1.0–1.5 m wide on the upstream slope, with visible depth 1.5 m, and 1.5 m wide on the downstream slope, with visible depth 2 m. There were more than 90 transverse cracks on the top of the dam but they were not very deep; the maximum settlement of the top was about 1.5 m, and the top road was in the form of a wave (Figure 10.3). Water and mud spouts continued for more than one day at the downstream side and fine sands from liquefaction were found on the downstream slope. Many liquefaction holes were found downstream. The water level raised 4–5 m on the water pressure gauges on both sides of the dam. The results of analysis show that the damage of the dam was primarily due to the liquefaction of the fine sand layer under the dam foundation.

10.5.2 Pseudo-static method of analysis

The pseudo-static method has been used to analyze the safety against sliding of earth dams for the last 40 years. According to this method, the effect of horizontal action of an earthquake is considered as an equivalent static horizontal inertial force, equal to the product of the earthquake coefficient $k=a/g$ and the weight W of the sliding mass, as shown in Figure 10.22. The aim of the analysis is to find a sliding arc adc on the slope where the resistance, or the stability coefficient, against slide along the arc is

$$G_s=\frac{SLR}{EW+kFW} \tag{10.23}$$

where L is the length of the arc adc and S the shear strength of the slope material. The denominator is the sliding moment and the numerator the

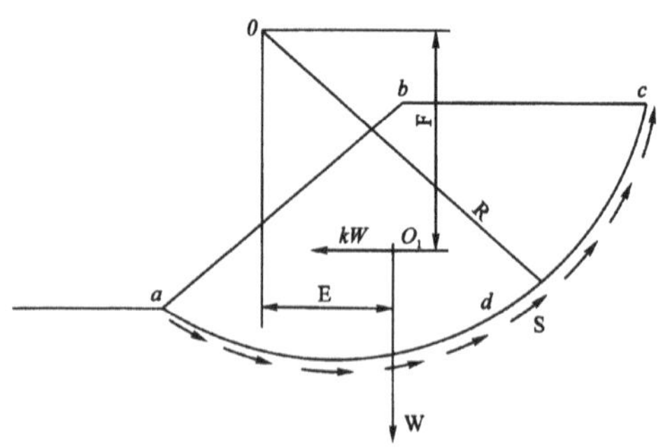

Figure 10.22 Pseudo-static analysis of earth slope.

resisting moment, both about the center O of the circular arc adc. If there is an arc where $G_s \leq 1$, the wedge abcd of the slope may slide along the arc adc; if no such arc can be found, the slope is considered stable, and the smallest coefficient G_s is the sliding stability safety factor of the slope under the earthquake motion k.

The earthquake coefficient k depends on the strength of the ground motion. Even before 1950, it was suggested to take $k=0.1$ (Rossi–Forel intensity IX), 0.25 (Rossi–Forel intensity X) and 0.5 (catastrophic). In 1950, Terzaghi pointed out clearly that: because equation 10.23 assumes a permanent action in the same direction of a horizontal acceleration on the slope material, the earthquake action on the slope so evaluated is at best highly inaccurate. Theoretically $G_s=1$ means slope slide; but actually the slope may be stable even when $G_s<1$ and may be damaged when $G_s>1$, depending on the property of the slope material. The most stable material is clay, which is insensitive under plastic conditions, dense sand and loose sand above water level; and the most unstable materials are soils of fine grain with weak cohesion, such as loess and silty sand, with high water content.

Most of Terzaghi's achievements in soil mechanics have had great influence in practice, but his warning on the seismic design of earth dams went unnoticed, and earth dams were still designed according to the pseudo-static method with earthquake coefficients k even smaller than those given above. For example, in the most severe seismic area of California, k is taken as 0.05–0.15; in Japan, it is often less than 0.2, and similarly in severe seismic regions in other parts of the world. It seems that engineers believe these low k values together with the pseudo-static method to be suitable for ensuring the safety of their dams.

Research shows that the pseudo-static method of design cannot guarantee the safety of sandy dams built on sandy foundations. Newmark (1965) suggested a method to evaluate the permanent deformation of earth dams, applicable to materials whose strength does not decrease suddenly, such as clayey soil, dry sand and very dense cohesionless soil. It can be proved that, when the acceleration at dam top is less than $0.75g$, the deformation of the dam is usually small and acceptable if the pseudo-static method for $k=0.15$ gives $G_s \geq 1.15$. Newmark's method assumes the dam material to be elasto-plastic or rigid–plastic, and assumes that permanent deformation occurs only when the acceleration aids downward slide and is greater than k_y, the earthquake coefficient for when the material yields; permanent deformation is only unidirectional because of the gravitational effect of the slope. Therefore, as shown in Figure 10.23, deformation comes only from the shaded portion when $k>k_y$; a twice integration of the shaded portion gives the deformation. It seems that, for cases where the pseudo-static method guarantees the safety of the dam, it controls the degree of permanent deformation.

10.5.3 Earthquake response of dams of saturated cohesionless soil

Loose and medium dense sands differ from cohesive soil and dense sand, and have no clear elastic limit strength; the seismic response of this material

Figure 10.23 Permanent deformation of slope slide.

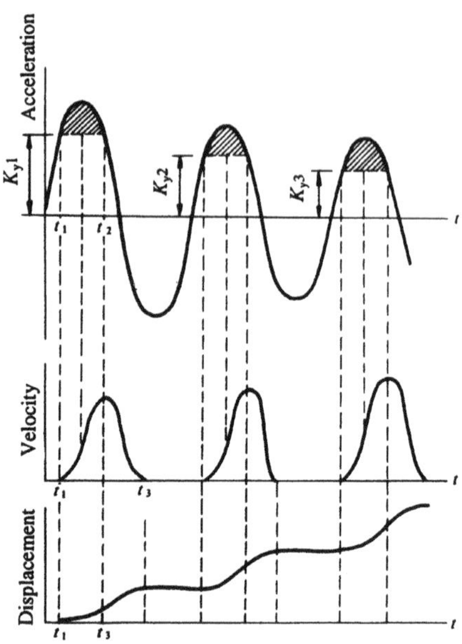

is very complicated because of the redistribution of the water pressure caused by the high pore pressure introduced by vibration in the soil. The stress–strain relation, before and after the application of cyclic seismic load, may be quite different; and the seismic design of dams of this material should consider the induced pore pressure and the related effect.

Seed (1966) suggested, and later improved, a method to estimate the deformation of dams of this kind of soil. The fundamental principles are as follows.

1. Determine the following items: the cross-section of the dam or slope including the foundation; the maximum earthquake input time history; the initial static stress in the dam before the earthquake; dynamic properties of all the kinds of soil in the earthwork considered, and their strain relationships; and an adequate method of analysis, such as the finite element method.
2. Find the dynamic stresses in the earthwork under the given earthquake input time history, with the nonlinear behavior considered.
3. Test some typical elements of soil under the static and dynamic stresses found in the preceding steps and determine the increase of pore pressure and the development of strain. Test enough elements so that dynamic responses of all elements can be found by interpolation.
4. On the basis of the pore pressure, soil deformation and strength properties induced by the earthquake, evaluate the safety factor of the earthwork during and after the earthquake. If it is safe, evaluate the total deformation of the earthwork on the basis of the strain under the joint action of the static and dynamic loads.

5. Judgement may be needed in the preceding four steps, which are based on the understanding of typical soil properties, details of the finite element method and earthquake response of earthworks.

Because this method is rather complicated, several simplified methods have been suggested. The simplest one is to rely entirely on empirical assessment, excluding computations; this is, of course, on the basis of the understanding of the soil property, results of past analyses and experience of field observation and investigation of similar systems.

The best application of this dynamic method is perhaps to calculate on a computer the formation and dissipation of pore pressure during and after the earthquake (Booker *et al.*, 1976), and use the finite element technique to coordinate the strain between elements to evaluate the final strain distribution (Serff *et al.*, 1976). This method has been applied to eight earth dams, of which two showed very large slope slides, one very large deformation, one small deformation and the other four no damage. The results of analysis are consistent with real damage.

10.6 Earthquake response of pile foundations

Pile foundations are necessary for structures on soft ground. For shallow soft deposits, bearing piles are used which transfer the load of the structure to the hard ground at the tip of the pile; but for very deep soft soils, frictional piles are usually used which transfer the load by friction to the soil surrounding the piles. This section discusses only the latter, because they are most sensitive to earthquakes.

10.6.1 Earthquake damage and field observation

(a) Damage

Experience shows that pile foundations behave far better than those with no piles on similar ground, although they have also been damaged by earthquakes. For example, during the 1948 Fukui and 1952 Tokachioki earthquakes in Japan, the pile foundations on weak grounds were usually seriously damaged due to excessive differential settlement of the superstructure and the supporting soil. Experience from 1964 Niigata, 1964 Alaska and 1968 Tokachioki earthquakes shows further that permanent displacement due to liquefaction produces damage in pile foundations and thus serious damage to the superstructure. For example, during the Niigata earthquake, a five-span bridge over the Shinano River collapsed; field investigation (Takata *et al.*, J1965) after the earthquake revealed that liquefaction of a sand layer about 10 m below the surface had caused serious deformation of the piles and the corresponding large deformation of the piers caused the collapse of the bridge. Figure 10.24 shows the deformation of a steel pile from the foundations.

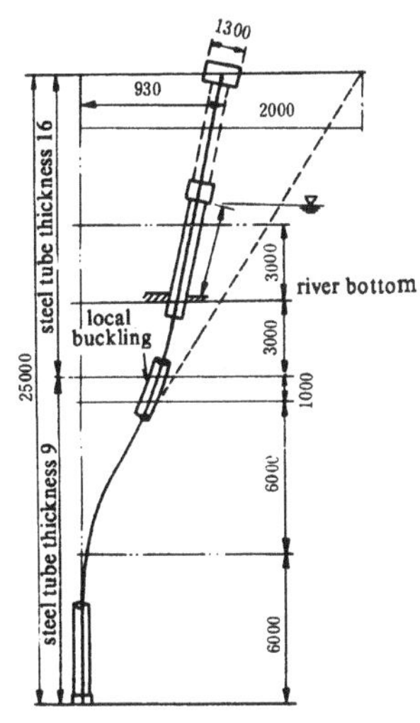

Figure 10.24 Deformation of pile (in mm).

During the 1975 Haicheng earthquake in China, a steel tube inside a drill hole on the bank of the Shuangtaizi River in the Panjin district bent significantly at a depth 7–14 m below the surface, where sand had liquefied nearby. The Liaohe Bridge at Tianzhuangtai was under construction during the earthquake; after the earthquake, the abutments and piers near banks slid toward the river channel or inclined. The most severely damaged abutment slid 6 m and inclined 8°. All abutments and piers showed obvious settlement; ground fissures appeared on two banks and flooded land, with maximum differential settlement of roughly 1 m. In a field inspection after the earthquake, removal of the surface soil under some abutments revealed that cracks had appeared on the river side of piles just below the foundation, and boreholes in the piles at the piers revealed that cracks had appeared in the piles at the level of the liquefaction layer and had filled with thin layers of fine sand.

During the 1978 Miyagioki earthquake, several case histories showed that piles under tall buildings can also fail in an earthquake. Table 10.3 shows a few cases of pile damage identified by excavation, where, except in the last case, the buildings were on flat ground with horizontal layers of deposits, and no sliding was found. The excavations were made only because the settlement of the buildings continued after the earthquake. Most of the cracks were found near the top; but, except in the first case, where the building collapsed, the buildings were not or only slightly damaged, and the more severe the damage of the pile, the less the damage of the building. Investigators believe that pile damage was due to vibration.

The evidence from these cases shows that a liquefied sand layer may induce slope slides, break piles and in turn damage superstructures.

(b) Strong motion observation

Many earthquake records have been obtained on structures with pile foundations in Japan. Sugimura (1977) mentioned three of them, which are: a seven-story reinforced concrete (RC) apartment block with piles 12 m long; three tower-type structures of height 50, 30 and 15 m on a common foundation, with steel piles of 40 m long; a 14-story RC building of height 39 m. Sugimura's conclusions are as follows.

1. Predominant periods and amplification factors are consistent with those calculated.

Table 10.3 Damage to buildings with pile foundations in the 1978 Miyagioki Earthquake[a]

Building		Pile				Building Damage	Ground	
Type	N	Type	L(m)	D(m)	Damage		Topology	Soil
BoxRC	3	RC	5	0.25	Bd; Ck	Collapse	Flat	Mud–clay
BoxRC	4	PRC	5	0.35	Bd–Sh; Ck	DS	Flat	Clay–sand
StRC	11	PRC	12	0.60	Bd–Sh; Ck	DS; small Ck	Flat	Silt–sand
StRC	14	PRC	24	0.5	Bd–Sh; Ck	DS; small Ck	Flat	Mud–sand
RC	4	PRC	10	0.30	Bd–Sh; Ck	DS; small Ck	Slope	Loam

[a] N = number of stories; RC = reinforced concrete; P = prestressed; St = steel; Bd = bending; Sh = shear; Ck = cracks; DS = differential settlement.

2. From the ratio of spectra of the pile top to pile tip, two predominant periods were found which are close to the first two natural periods of the site, and the valley of the spectral ratio is at the natural period of the structure.
3. The vibration of the piles was controlled by the vibrational property of the site; but near the pile top, the effect of superstructure was large because of the soil–pile–structure interaction.

10.6.2 Analysis of structures with pile foundations

This is an SSI problem and the only differences lie in the modeling of the piles. There are three kinds of method for this:

1. lumped system;
2. beam on elastic foundation;
3. finite element.

(a) Lumped system (Wiegel, 1970, ch. 14)

This method is also known as Mindlin static model, which consists of the following two steps. In the first step, the free-field problem is solved, i.e. the half space with no structure and no piles, to find the ground motion process $u_i(t)$ for all points i in the half space where the piles will be placed, under the given earthquake motion $u_g(t)$ at the bedrock. In the second step, the soil–pile–structure system is solved for the given ground motion $u_g(t)$ at the bedrock and the ground motion $u_i(t)$ on the piles. Nonlinear properties of the soil are usually considered for strong motion either by equivalent linear or by other nonlinear models.

(b) Beam model (Kagawa and Kraft, 1980a, b, 1981)

p–y curve. The key step in this method is to find the p–y curve, i.e. the displacement y at the point of the action of a force p in the foundation. The horizontal force p is the reaction of the soil on the pile per unit length due to the horizontal force $F\exp(\mathrm{i}\omega t)$ and the moment $M\exp(\mathrm{i}\omega t)$ at the top of the pile induced by input bedrock motion $u_g\exp(\mathrm{i}\omega t)$.

For simplicity, the piles are considered as a Winkler beam in a uniform foundation, by replacing the varying foundation reaction on the beam by a constant reaction coefficient. The nondimensional form of the p–y curve may be written as

$$\frac{p}{E^*(u_p-u_f)}=\frac{p}{E_s^* y}=\delta_1+\mathrm{i}\delta_2=\delta \tag{10.24}$$

where E_s^* is the complex Young's modulus of the soil, u_f the free-field displacement, u_p the relative displacement of the pile to the bedrock, $y=u_p-u_f$ the relative displacement of pile to the free field, δ_1 the spring effect between pile and soil and δ_2 the energy dissipation of the soil and radiation damping.

Factors influencing p–y curve. The foundation and pile are under the action of three excitations: $u_g\exp(\mathrm{i}\omega t)$, $M\exp(\mathrm{i}\omega t)$ and $F\exp(\mathrm{i}\omega t)$. Let subscripts h and f represent the quantity induced by forces at the pile head and foundation motion respectively; the *p–y* curve, equation 10.24, can be written as

$$\frac{p}{E_s^* y}=\frac{1}{E_s^*}\frac{p_h-p_f}{y_h+y_f}=\frac{\delta_h y_h+\delta_f y_f}{y_h+y_f}=\delta \tag{10.25}$$

It is seen here that the three *p–y* relations δ_h, δ_f and δ are related respectively to three loadings: the load at the pile head only, $u_g=0$; bedrock motion only, $M=F=0$; and both. They are all functions of frequency. The radiation damping δ_2 increases very fast when frequency increases.

Physical quantities involved in the *p–y* curve are: modulus G_s or E_s, density ρ_s, damping ratio ζ and Poisson ratio ν of soil; bending rigidity EI, radius r_0 and length H, density ρ_p of pile; or ρ_s/ρ_p, slenderness ratio $H/(2r_0)$, rigidity coefficient of pile $K_r=EI/(E_s H^4)$ and ν and ζ in nondimensional form.

Average p–y curve. A good simplified approach is to use an average $\bar{\delta}$ to replace the depth-varying $\bar{\delta}$ and to find the response in the Winkler beam. Consider two systems. One is the system used above with soil reaction p and displacement y; the other is a Winkler beam with soil response $\bar{p}=E_s^*\bar{\delta}\bar{y}$ and displacement $\bar{y}$. Equating the work done by the two systems,

$$W_1=\int_0^H \text{p}y\,\text{d}z=W_2=\int_0^H E_s^*\bar{\delta}\bar{y}\cdot\bar{y}\,\text{d}z$$

and requiring the two systems to have the same displacement $\bar{y}=y$, then

$$\bar{\delta}=\int_0^H py\,\text{d}z\Big/\int_0^H E_s^*\,y^2\text{d}z=\bar{\delta}_1+\text{i}\bar{\delta}_2 \tag{10.26}$$

p–y curve for inhomogeneous soil. If E_s^* is varying with depth, equation 10.26 may still be used as an approximation but an average modulus should be used for E_s^*:

$$\bar{E}_s^*=\int_0^H E_s^*(z)\,y^2\text{d}z \tag{10.27}$$

(c) FEM model (Berger *et al.*, 1975)

The piles and the surrounding soil may be modeled by finite ring elements.

11 Vibrational testing and dynamic properties of structures

11.1 Purpose of vibrational testing

Field investigation and observation of strong earthquakes, laboratory tests on structures and theoretical studies of structural vibration are three fundamental activities in earthquake engineering. Understanding of the actual behavior of structures in a strong earthquake is the ultimate purpose of these activities; but such behavior is so complex that the understanding is usually only qualitative, not quantitative, because there are so many uncertain factors involved in both the ground motion and the structure. It is also limited to what information existing structures can provide.

A laboratory test is controlled; a new structure or part of the structure can be tested under known conditions. A laboratory test differs from field investigation in the following aspects.

1. In a test, structural conditions are controlled, while in the field, although the structural types and properties are actual examples, they cannot be controlled, and the properties can be known only within certain limits.
2. In a test, the input motion is also controlled, while in the field it is usually unknown.
3. In a test, factors for investigation can be isolated, such as effects of vertical input motion or the type of lateral bracing.

In order to understand the behavior of a structure, it is better to treat its parts separately. When carrying out tests on a member or a part of a structure, care should be taken to simulate the dynamic properties of earthquake loads, which are:

1. of a very short duration, say a few or a few tens of seconds;
2. very irregular and dynamic;
3. very strong so that the material may become nonlinear.

11.2 Development of structural testing

Tests are carried out following certain theoretical ideas; the test technique and content are linked and develop along with theory. Theory in earthquake engineering has had three stages of development: static, response spectrum and dynamic (section 13.1); it is important to see the parallel development in structural testing.

11.2.1 Early structural testing

In the era of static theory (roughly 1920–60), structural testing was performed to find the natural vibrational properties (period, damping and modes) of structures. The Japanese started this kind of testing in the early 1920s. Field testing of existing structures was started in the US Coast and Geodetic Survey (USCGS, forerunner of the US Geological Survey, USGS) in the United States about ten years later (Carder, 1936, 1966); from 1934 to 1962, about 400 buildings, 44 tall towers and many bridges and dams were tested. The Chinese performed similar tests in the 1960s (Yin and Peng, C1962). The purpose of these tests on natural vibrational properties was to follow the requirements of the response spectrum theory.

Laboratory testing in this period was done under linear theory on small scale models with emphasis on natural properties, mostly by applying simple harmonic motion or impact.

At the same time, some inelastic and strength testing of materials and members was also done but the static method was mostly used; it has since been realized that, after strong earthquakes, the natural periods elongate and damping increases.

11.2.2 Recent vibration tests on structures

Starting in the 1960s, test methodology made two developments: firstly, elastic and inelastic response testing under one- or two-dimensional earthquake action; and secondly, pseudo-static testing of members and parts of structure up to failure.

(a) Structural response under earthquake motion

Only after the construction of large earthquake-simulation shaking tables in the early 1970s, could large scale models of structures be tested under earthquake motion (Clough *et al.*, 1973; Clough and Hidalgo, 1973; Rea and Penzien, 1973). This kind of shaking table differs from the early ones in the following features.

1. The table can simulate the irregular earthquake motion in one or more directions;
2. The table size is large, usually about 5 m × 5 m, and the motion may be strong, up to 0.5*g*–1.0*g*, so that a large scale structural model or part of real structure can be shaken to failure;
3. Data processing is done digitally on computers.

Many types of structure have been tested on earthquake-simulation shaking tables. The following may provide a general idea:

1. effect of abrupt change of stiffness in multistory reinforced concrete frameshear wall systems (Sozen and Moehle, 1980);
2. damage mechanism and strengthening of liquid-storage tanks;
3. overturning of rigid bodies (Yim *et al.*, 1980; Psycharis and Jennings, 1983);
4. base-isolation systems (Kelly and Beucke, 1983);
5. ductility of offshore platforms of X-braced pile trusses (Ghanaat and Clough, 1983);
6. spatial piping systems and their supports (Schneider *et al.*, 1982; Stiemer and Chow, 1984);

7. horizontal and torsional coupled vibration of offshore platforms and damage detection;
8. active mass dampers controlled in two horizontal directions (Aizawa *et al.*, J1990);
9. rigid blocks on saturated dense sand layers (Fukutake *et al.*, 1990).

In addition to the shaking table test, a pseudo-dynamic test has been developed since the 1970s, in which the forces or displacements simulating the earthquake responses are calculated and applied at very short time steps to the structure. In the calculation, the actual nonlinear dynamic properties observed during the test in the previous time steps are used to find the earthquake responses and forces to be applied to the structure in the next step. Pseudo-dynamic tests and earthquake shaking tables have both become widely applied important laboratory test methods.

(b) Inelastic properties of members and parts of structures

The response characteristics are determined by the properties of ground motion and structures. Earthquake motion differs from other dynamic loads in the following aspects.

1. The earthquake motion considered for design may be strong enough to introduce inelastic behavior in the structure.
2. Earthquake motion is three-dimensional with three components in translation and three in rotation; because of inelasticity, superposition is not allowed and members subject simultaneously to forces in two or three directions must be considered.
3. The repetition of motion in a few to a few tens of cycles requires consideration of low-cycle fatigue.
4. There are some important features of earthquake motion not yet considered in design, for example duration of strong motion and rotational motion; some special requirements in detailing and constructional measures are important to take care of unexpectedly strong motion.

In design and construction, and in laboratory testing of members and joints, these special features should be kept in mind; remember also that members and parts of structures should be tested according to their function in a structure under earthquake action.

Because of the pronounced nonlinear behavior which occurs under real earthquake motion, the model scale is usually large, such as one-third to a half, and sometimes full scale. Modern testing differs from static testing as follows.

1. Tests range from small amplitude to large amplitude and well into the plastic or failure stage.
2. Because it is difficult to determine dynamic nonlinear behavioral laws of materials and structures, large scale models and thus large shaking tables are necessary.

3. because the superposition principle cannot be used, the input motion is earthquake simulated, not simple harmonic.
4. because of the limitation of the large shaking table size, full structures cannot be tested on shaking tables, and pseudo-dynamic testing is used.

All these developments are directly related to the design principle of 'no collapse in a very strong earthquake'.

11.3 Methods of structural testing

11.3.1 Pseudo-static testing

Pseudo-static testing is a form of cyclic static testing which loads the structure in one direction, increases from zero to a certain value, say P, then decreases gradually back to zero, and then loads in the opposite direction, usually also to P; this loading and reverse loading process is continued to simulate the earthquake vibrational motion.

The pseudo-static test system consists of three parts:

1. the test frame to support the test specimen;
2. a loading system to apply loads or displacement to the specimen;
3. a set of measuring instruments.

The test frame, usually made of steel, should be rigid enough to provide the required boundary conditions for the specimen. The loading system consists usually of several synchronized hydraulic jacks, with reversible direction of action, to apply force or displacement, and of large working range of displacement (several centimeters in each direction) and force (several tens of tons). The pseudo-static test system may be quite small, to test a member or a joint, or very large, to test a multistory building. The latter requires a very rigid reaction wall to apply horizontal loads to the model. For example, the reinforced concrete reaction wall of the Building Research Institute in Japan is 25 m high, 20 m wide and 6.6 m thick, with 4000 Tf ultimate shear capacity and $(4\text{–}7)\times 10^4$ tf·m bending capacity, and may be loaded at several levels.

The important problem in pseudo-static testing is how to load the specimen. The usual way is to apply a gradually reversible load in the range $+P$ to $-P$ or displacement in the range $+d$ to $-d$. The load P or d is small at the beginning, and repeats a few cycles of $(+P, -P)$ or $(+d, -d)$; if the result after a complete cycle is nearly the same, the amplitude P or d is then increased. The process continues until the specimen behaves nonlinearly; then the increase of load will be reduced, and the test continues until the specimen fails at a very large displacement.

Pseudo-static testing can also be carried out to simulate earthquake response in the following way. The structure is analyzed first with assumed nonlinear dynamic properties, which may be obtained or guessed from linear testing of the specimen, for some given earthquake motion. The time histories $P(t)$ or $d(t)$ of the forces and displacements on the part of the structure to be tested are then found, and applied to the specimen statically as one step in the pseudo-static test. All instruments may be controlled by computers.

11.3.2 Pseudo-dynamic testing

Pseudo-dynamic testing is a wise extension of pseudo-static testing; it differs from advanced earthquake-simulated pseudo-static testing mentioned above only in the way assumptions are made about the nonlinear properties of the structure under test. This method will be discussed in Section 11.8.

11.3.3 Vibrational testing of linear systems

The free vibration, resonance, ambient vibration and wind- or machine-excited vibration tests are the most common and easy methods of dynamic testing, and are very efficient for linear structures; the differences are only in the way the structure is excited, and in the method of analysis. The first three are still used widely, especially the ambient method where an actual structure is tested *in-situ*, because no interruption of the functioning of the structure is required and a large amount of natural vibrational information can be obtained using modern computer and instrumental technology.

11.3.4 Earthquake-simulation shaking tables

The use of large earthquake-simulation shaking tables is very popular. It started first in the United States in the early 1970s; the table at the University of California at Berkeley is perhaps the most well known; but the Japanese are leading in both the size and number of tables, with two tables of 15 m × 15 m. In China, the best table is of size 5 m × 5 m, controlled in six degrees of freedom; and there are about ten tables of size about 5 m × 5 m.

11.4 Dynamic properties of materials

In an earthquake, although the loading rate is not as high as in an explosion, the load is applied repeatedly at a high amplitude; and though much is known on the dynamic properties of materials, there is still great uncertainty.

11.4.1 Effect of loading rate

For some structural materials, such as concrete (Figure 11.1) and brick, although the dynamic strength may be higher than the static strength by 5–15%, brittle failure occurs more easily, which decreases its earthquake resistance greatly. Materials, including soils, which show high rheological effect, also show strong nonlinear dynamic effects, and both of them can be modeled by physical models using springs and dashpots, such as the Kelvin, Voigt and Maxwell models.

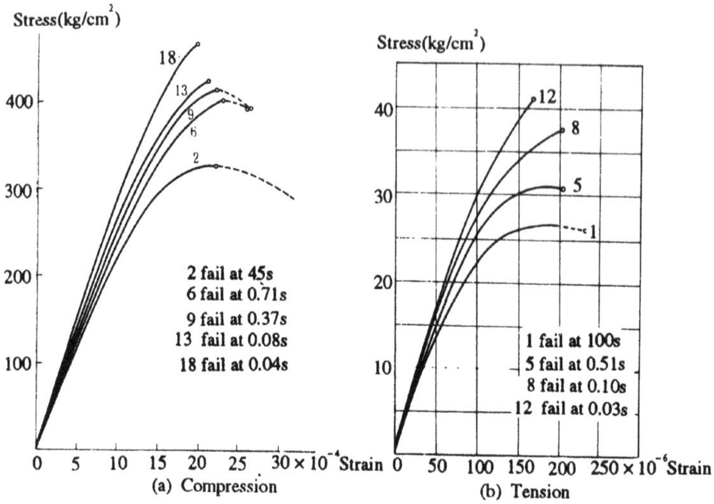

Figure 11.1 Effect of loading rate on stress–strain relations.

11.4.2 Low-cycle fatigue

For ordinary fatigue strength, the Miner criterion assumes on the basis of test data that

$$\sum \frac{n_i}{N_i} = 1 \tag{11.1}$$

where n_i and N_i are respectively the actual and critical number of cycles in the ith narrow range of stress or deformation, and the summation is over all possible ranges of i. This assumption assumes that the damage can be accumulated by linear superposition.

Ordinary fatigue occurs when the number of cycles reaches more than 10 000; but low-cycle fatigue deals with repetition of only a few or a few hundred cycles at most.

Yao and Munse (1962) gave the low-cycle fatigue strength criterion as follows:

$$\sum \left(\frac{\Delta p_i}{\varepsilon_{ui}}\right)_{\bar{z}} = 1 \tag{11.2a}$$

$$z = 1 - 0.86 \frac{\Delta n_i}{\Delta p_i} \tag{11.2b}$$

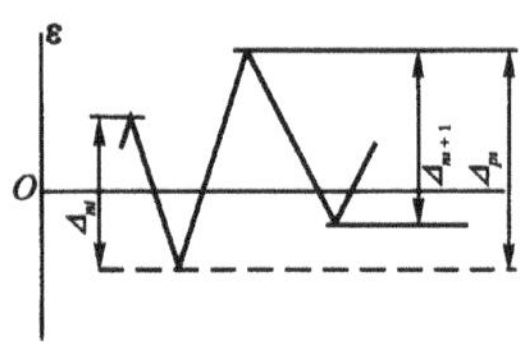

Figure 11.2 Definition of plastic strain. (After Yao and Munse, 1969.)

where Δp_i is the difference (or the sum of the absolute values) of the maximum positive and minimum negative plastic strains of the ith reversed plastic deformation; ε_{ui} is the ratio of the pre-compressed plastic strain of this cycle to the ultimate tensile strain (for low-carbon steel $\varepsilon_{ui} = 0.85$); and Δn_i is the pre-compressed plastic strain occurring before the tensile plastic strain, as shown in Figure 11.2. Yao *et al.* applied this criterion to the response analysis of SDF frames.

The stiffness and strength degradation phenomena are closely related to low-cycle fatigue, and any of them may be related to the microfissures occurring in the material under repeated loading; the development of fissures may finally result in failure.

11.5 Dynamic properties of members

11.5.1 Reinforced concrete beam

When a member bends, test data show that, once a clear diagonal tension crack occurs, the stiffness of the beam reduces fast, and deformation increases as the load repeats. The phenomena of stiffness degradation and deformation increase also happen when the main reinforcement slides due to failure in bonding. If bending reverses, the main reinforcement may elongate first in tension and then buckle in compression after spalling of the concrete. Increasing the number of stirrups, or decreasing the spacing between stirrups, may help to prevent this failure.

Some design criteria suggested by Blume *et al.* (1961) for reinforced concrete beams should be followed. They assumed that the moment – curvature relation might be simplified to ideal elasto-plastic and with ultimate curvature α_u, as shown in Figure 11.3; the maximum bending moment M_{max} might be taken as the yield moment M_y. For beams of rectangular cross-section and with no negative reinforcement,

$$\frac{\alpha_u}{\alpha_y} = \frac{\varepsilon_{cu}}{\varepsilon_y}\frac{1-k}{1+k_u} \leqslant 20 \tag{11.3}$$

where ε_{cu} is the limit strain of the concrete; ε_y is the yield strain of the reinforcement in tension; $k=[(np)^2+2np]-np$; $k_u=pf_y/f_{cu}$; n is the ratio of moduli of reinforcement to concrete; p is the reinforcement ratio; f_y is the elastic limit stress; and f_{cu} is the ultimate average compressive stress. Additional negative reinforcement increases ductility by decreasing k_u:

$$k_u = \frac{(p-p')f_y}{f_{cu}}, \quad (P' < P) \tag{11.4}$$

where p' is the compressive reinforcement ratio. A conservative simplified relation may replace the first two equations:

$$\frac{\alpha_u}{\alpha_y} = \min\left(\frac{10}{p-p'}, 20\right) \tag{11.5}$$

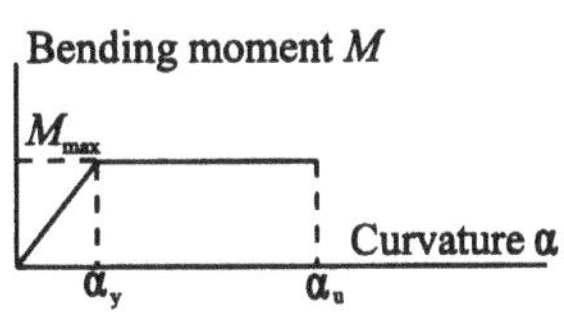

Figure 11.3 Moment–curvature relation of reinforced concrete beam.

For a beam with known bending moment diagram, the curvature may be estimated as shown in Figure 11.4, where B is the support and A the inflection point. The actual curvature of the beam may be a complex curve as shown in the lower diagram and the shadow zone is in inelastic deformation, which may be simplified to a rotation of a plastic hinge shown by the dashed line at point B. The width of the plastic zone x_0 may be taken as half the effective height of the beam, but not more than AB/2. Under this

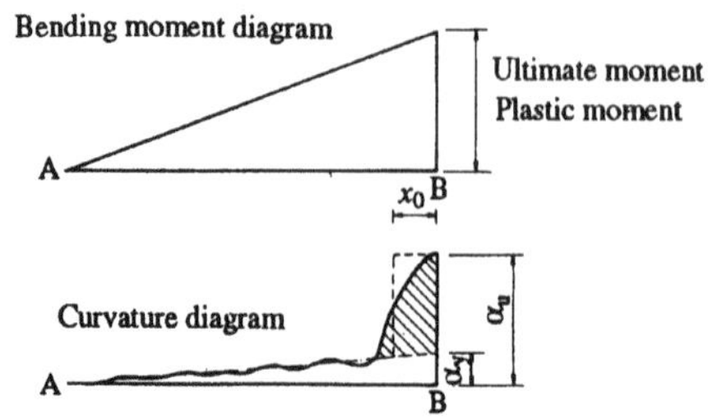

Figure 11.4 Computation of curvature in reinforced concrete beam.

assumption, the curvature consists of two parts, one from the elastic deformation from A to B, varying linearly from zero at A to α_u at B, and the other from the rotation of the plastic hinge or $(\alpha_u - \alpha_y)\,x_0$.

The ratio α_u/α_y is the local ductility factor of a beam in bending; the global ductility factor is usually smaller than the local factor.

Experience shows that a beam of traditional design cannot provide sufficient ductility to develop its full strength. The weak points of design include:

- early failure in slip of the reinforcement at anchorage or splice;
- crushing and spalling of concrete in compression;
- widening of cracks;
- buckling of unconfined main reinforcement due to inadequate consideration of bond, or other strength degradation;
- confinement stirrups.

A prestressed beam has a much lower energy absorption capacity than ordinary concrete but lower degradation under repeated load.

11.5.2 Reinforced concrete beam-column joint

The joints of members are vulnerable to repeated loads because of stress concentration. Test results show that concrete inside the joint panel may crack under a few load cycles at quite low amplitudes, and that the best measure is to add confinement to the concrete. Figures 11.5 (a) and (b) show inadequate joint detail; Figure 11.5 (c) is much better, where additional stirrups are provided inside the joint panel. In the case of three-dimensional frames, a beam in the perpendicular direction of the joint helps to provide lateral confinement.

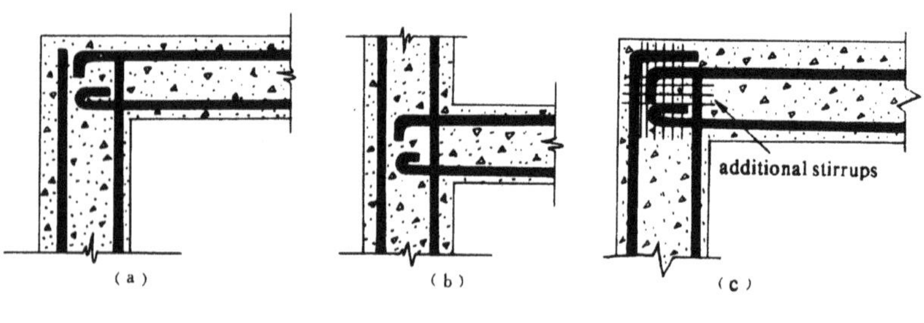

Figure 11.5 Joint of reinforced concrete beam and column.

11.5.3 Dynamic property of part of a frame

The simplest frame consists of two columns and a beam, with or without an infill wall or lateral bracing, whose dynamic properties are controlled by those of its components: the beam, columns, joints and lateral system if any, but emphasis should be given to the stress redistribution in the inelastic stage. Umemura and Aoyama (1969) suggested a detailed procedure to evaluate the nonlinear displacement of a reinforced concrete frame from those of its components, including shear deformation of members and joints. The models commonly used for a member, part of a structure or a whole structure are discussed in the following.

A nonlinear model consists generally of a skeleton curve and a hysteretic loop, with the former applicable to the first loading and the latter to unloading and reloading.

(a) Bilinear model

The bilinear model, shown in Figure 11.6, includes the ideal elasto-plastic model as a special case where the second straight line is horizontal or $p=0$; it is often used for shear beam models of a building. The second line represents the effect of strain hardening. Because stiffness or strength degradation is not considered, it can be used only for loads with just a few cycles or no repetition. When $p=0$ and $\Delta_y=0$, it degenerates to the rigid-plastic model, which has been used by Newmark (1965) to estimate the permanent displacement of earth dams.

(b) Masing model (Ohsaki, 1982a)

Suppose that the skeleton curve follows the Ramberg–Osgood model

$$\gamma=\frac{\tau}{G_{max}}\left(1+a\left|\frac{\tau}{S_u}\right|^b\right) \quad (11.6)$$

where $S_u=G_{max}\gamma_r$ is the undrained shear strength; $a=(0.01\,G_{max}/S_u)-1$; $b=1.4$ (clay) or 1.6 (sand). The Masing model is used for the hysteretic loop (Figure 10.7(c)); if the skeleton curve is $\tau=f(\gamma)$, the hysteretic curve is then

$$\frac{\tau-\tau_0}{2}=f\left(\frac{\gamma-\gamma_0}{2}\right) \quad (11.7)$$

where (γ_0, τ_0) are the coordinates of the initial point A of the loop (Figure 10.7(a)) (Newmark and Rosenblueth, 1971, section 5.1). The tangent and secant moduli and damping ratio from equation 10.8 are

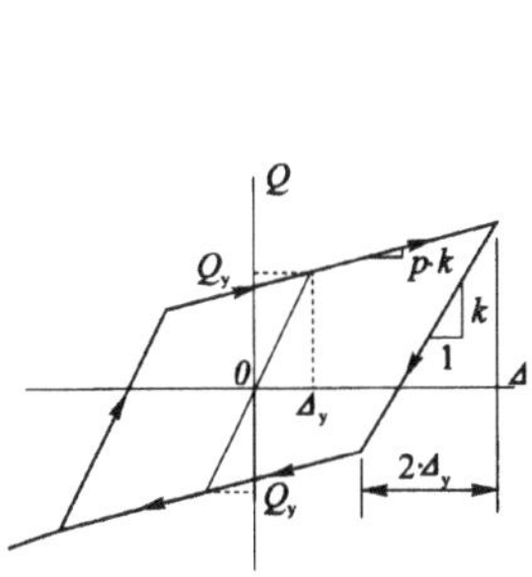

Figure 11.6 Bilinear model.

$$G=\frac{d\tau}{d\gamma}=\frac{G_0}{1+a(b+1)|\tau/S_u|^b} \quad (11.8a)$$

$$\bar{G}=\frac{\tau}{\gamma}=\frac{G_0}{1+a|\tau/S_u|^b} \quad (11.8b)$$

$$\frac{\bar{G}}{G_{max}}=\frac{1}{1+a|\tau/S_u|^b} \quad (11.8c)$$

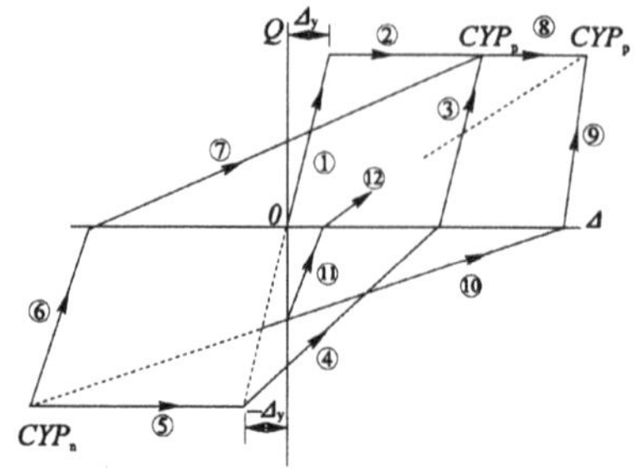

Figure 11.7 Clough model of stiffness degradation.

$$\zeta=\frac{2}{\pi}\left(\frac{b}{b+2}\right)\left(1-\frac{\bar{G}}{G_{\max}}\right) \tag{11.9}$$

Test data for clay and sand show that this model is good in the range $\gamma=0.001\%$–1%.

(c) Clough model of stiffness degradation

This model considers stiffness degradation under repeated loads, which has not been considered by the earlier models. This model consists of six main steps (steps 1–6 in Figure 11.7); step 7 is quite similar to step 4 and steps 8 and 9 are similar to steps 2 and 3. This model has been applied to shear- and bending-framed buildings.

(d) Model of X-braced frame

This is also called the sliding, reverse S or Z model. Consider a frame with a cross-bracing of slender bars, which can take only tension, as shown in Figure 11.8. Its hysteretic curve is shown in the same figure. The shear stiffness of the columns is negligible when compared with the bracing. The model consists of the following steps.

- OA (loading): H increases from zero to H_y which is the horizontal component of the yield strength F_y of the bracing; bracing 2 buckles and takes no force; force F_1 in bracing 1 increases from zero to its yield strength F_y; the deflection Δ at the top of the frame increases from zero to Δ_y.

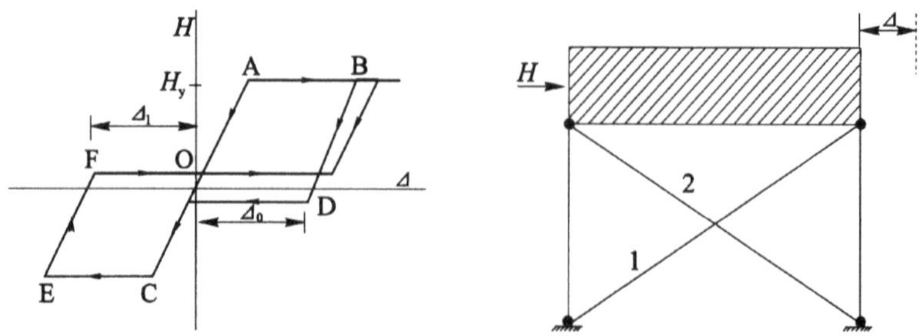

Figure 11.8 (a) An X-braced frame; (b) its hysteretic model.

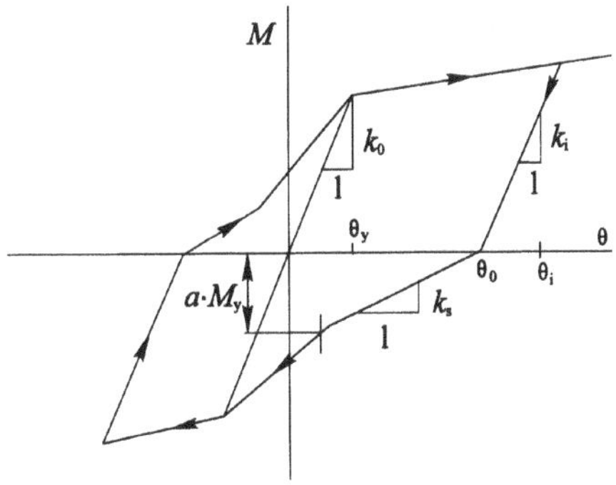

Figure 11.9 A pinching model.

- AB (loading): $H=H_y$; $F_1=F_y$; $F_2=0$; Δ increases from Δ_y to $(\Delta_y+\Delta_0)$, where Δ_0 is equal to the horizontal component of the permanent elongation of bracing 1.
- BD (unloading): H decreases from H_y to zero; F_1 decreases from F_y to zero; $F_2=0$; Δ decreases from $(\Delta_y+\Delta_0)$ to Δ_0.
- DO (unloading): H tries to unload by a very small amount, i.e. remain zero; bracing 1 in its elongated state is in compression now and buckles, therefore $F_1=0$; bracing 2 is still in the buckled state, therefore $F_2=0$; the top of the frame will move freely to the left until it goes back to the original position, or Δ decreases from Δ_0 to zero.
- DC (reverse loading): $H=0\rightarrow -H_y$; bracing 1 buckles and $F_1=0$; bracing 2 in tension and $F_2=0\rightarrow F_y$, $\Delta=0\rightarrow -\Delta_y$.
- CE (reverse loading): similar to step AB.
- EF (reverse unloading): similar to BD.
- FOD (reverse unloading): at position F, both bracings are elongated with a horizontal component of elongation equal to Δ_0. When the reverse unloading starts, bracing 2 is in compression and $F_2=0$; but bracing 1 remains in the buckled position and $F_1=0$; both bracings cannot function and the top moves freely back to the original position, point O. When H starts to increase from zero to the right just as at the first loading step, bracing 2 is in compression and takes no force, but bracing 1, now elongated compared with step OA, remains in the buckled position and cannot take any force, therefore $F_1=0$; the free movement continues until $\Delta=+\Delta_0$ again, and at this time, bracing 1 comes back to a straight line position and is ready to take tension force.
- DB (loading): $H=0\rightarrow H_y$; $F_2=0$; $F_1=\rightarrow F_y$; $\Delta=\Delta_0\rightarrow(\Delta_0+\Delta_y)$.

(e) Pinching model

For reinforced concrete members, if the effect of shear, axial compression, bond degradation or slip is significant, the hysteretic curve will show a pinching effect near the origin, as shown in Figures 11.7–11.9, where $k_0=M_y/\theta_y$, $k_i=(\theta_y/\theta_i)^r k_0$, and $k_s=\lambda M_y/(\theta_y+\theta_0)$ with parameters r, α and λ determined from test data. The pinching effect occurs both in members and in complete buildings.

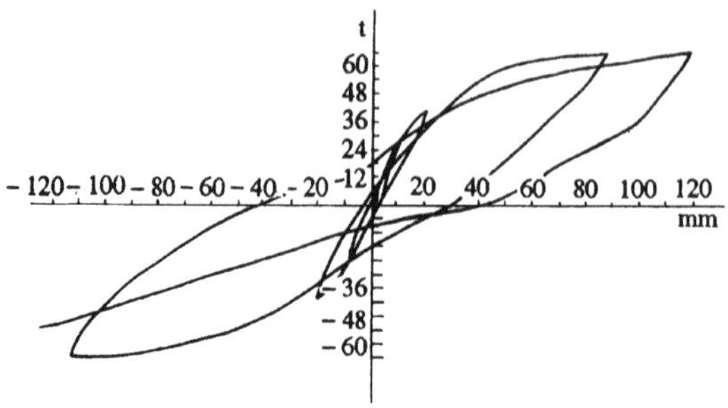

Figure 11.10 Hysteretic curve of a reinforced-concrete-framed building.

11.6 Dynamic properties of structures

11.6.1 Nonlinearity

The most significant and important property of a structure is its nonlinearity which comes mostly from the nonlinearity of materials and local failure or microfailure of its components.

At small amplitudes, a structure may usually be considered linear. Dynamic theory provides sufficiently good evaluation of the natural period and mode, but the effect of nonstructural members should be included. However, damping can be determined only from experiment. Theoretically, damping is also a kind of nonlinear behavior, but fortunately it can be treated as linear when small.

At large amplitudes, structure behaves nonlinearly as shown in Figure 11.10 and Tables 11.1–11.3. For simplicity, the nonlinear property is often described by the bilinear model in Figure 11.6, for which the equivalent linear parameters are

Table 11.1 Change in fundamental period of Beijing Hotel

Date of measurement	Source of excitation	Period (s)	
		Long.	Trans.
October 1974	Ambient vibration	1.00	0.90
4 February 1975	Haicheng earthquake	1.10	0.95
28 July 1976	Tangshan earthquake	1.30	1.40

Table 11.2 Change in natural period of Marunouchi Building

Date measured	State of building	Period (s)	
		NS	EW
December 1921	Under construction	1.11	1.14
February 1922	Just completed	0.89	0.94
May 1922	After an earthquake	1.01	1.09
October 1922	After retrofit	0.67	0.71
December 1923	After Kanto earthquake	1.11	1.18
May 1926	After retrofit	0.50	0.48

Table 11.3 Change in period and damping of Union Bank

Date measured	Excitation	Period (s)				Damping (%)			
		1st	2nd	3rd	4th	1st	2nd	3rd	4th
–1971	Environment	3.1	1.1	0.6	0.4	1.7	1.5	–	2.0
9 February 1971	Earthquake	4.61	1.49	1.00	0.66	4.2	5.8	–	6.6
1971–	Environment	3.8	1.3	0.8	0.5	–	–	–	–

$$\omega_e = \omega_0 \left(p + \frac{(1-p)}{\mu} \right)^{1/2} \tag{11.10a}$$

$$\zeta_e = \frac{2}{\pi} \frac{(1-p)[1-(1/\mu)]}{1-p+p\mu} \quad \left(\mu = \frac{u_{max}}{u_y} \geqslant 1 \right) \tag{11.10b}$$

11.6.2 Low-cycle fatigue

Low-cycle fatigue is possible for engineering works under strong earthquakes, as shown in structural members (Figure 13.20) and soil liquefaction. Dual parameters of ground motion (amplitude and duration) and structural responses (deformation and number of repetitions) are needed for low-cycle fatigue evaluation (Yao and Munse, 1962; Yamada, 1969; Ang and Hu, 1983; Park and Ang, 1983; Ang *et al.*, 1984; Hu and Tao, 1984).

11.7 Ambient vibration

In Japan, ambient vibration was known in the early days of earthquake engineering as tremor or microtremor, which refers to vibration induced by sources not well defined, or vibration induced by a possible combination of sources such as wind, traffic, machinery, human-produced excitations and very small earthquakes. The maximum amplitude of displacement on ground surface in a busy city may reach a few millimeters, and may be only a few microns on rock sites in quiet countryside; its measurement requires large amplification of a few thousands to a few hundred thousands. Its frequency band is wide, usually from 0.05 to 10 Hz, just about the bandwidth of earthquake motion, which is the best example of a natural white-noise motion (Yin and Peng, C1962; Zhou, Hu and Chen, C1965)

11.7.1 Ambient vibration of ground

(a) Primary features

The primary feature of ambient vibration of ground and structures is its randomness, which nearly satisfies all the conditions of a stationary random process, such as: numerous factors of influence but with none controlling, and no repetition, even over a very long record. Figure 11.11 shows some typical samples given by Kanai's group; the records a–d are respectively from grounds of bedrock, diluvium, alluvium and recent fill.

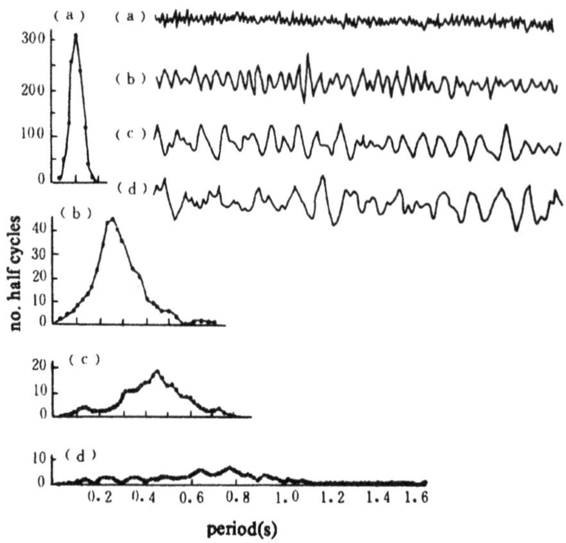

Figure 11.11 Ambient vibration of ground.

(b) Kanai's work (Kanai *et al.* J1954–68)

Ambient vibration or microtremor was connected with Kanai's name in the 1950s and 1960s; his studies cover almost all aspects of ambient vibration, especially ground motion. He found that spectral shape does not change with time of day or the seasons, but the amplitude is smaller at night; similar site conditions show similar spectra. The Japanese seismic design code adopted his suggestion by classifying site conditions according to the amplitude and spectrum of ground motion. It has been developed into a widely used method of microzonation (Sherif, 1981; Anon., 1973). He concluded that the spectrum of ambient vibration is similar to that of earthquake motion on the same ground. This conclusion is good for linear response, but it is questionable in general, because the amplitudes of ambient and earthquake motions are different by a few orders of magnitude, and soil is highly nonlinear.

11.7.2 Ambient vibration of structures

The ambient vibration of structures may be considered a filtered narrow-band random vibration and can therefore be used to find the natural vibrational properties of the structure at the elastic stage of response, because structures usually show clearer elastic properties before they are damaged. Table 11.4 gives a comparison of the results by ambient and forced vibrations.

It is not difficult to obtain the vibrational property of the first few modes in each direction by careful ambient vibrational measurements (Zhou, Hu and Chen, C1965; Bao and Lai, C1981). For lower modes, vibrations may be measured in terms of displacement; for higher modes, accelerations are recommended.

Table 11.4 Comparison of natural vibrational property of structures measured by ambient and forced vibrations

Structure	*Height (story or m)*	*Period (s)*		*Damping ratio*	
		Forced	Ambient	Forced	Ambient
Brick building	3	0.34	0.31	0.06	0.042
RC frame	13	0.53	0.53	0.032	0.032
Auditorium	40 m	0.555	0.556	0.045	0.058
Industrial	1	0.675	0.665	0.03	0.026
Furnace	–	0.318	0.318	0.03	0.015
Railway bridge	120 m span	0.50	0.51	0.026	0.028
Concrete dam	–	0.189	0.193	0.05	0.05

11.7.3 Analysis of ambient vibration

Standard procedures and computer programs are available for the analysis of ambient vibration records with the assumption of stationarity and ergodicity, which include the autospectrum $S(\omega)$, cross-spectrum $S_{xy}(i\omega)$ and coherence spectrum

$$\gamma_{xy} = \frac{S_{xy}(i\omega)\, S^*_{xy}(i\omega)}{S_x(\omega)\, S_y(\omega)} \tag{11.11}$$

To obtain better results, it is recommended that the averages of the spectra be used from five to 50 segments of records measured under nearly the same environment, instead of only one.

To obtain damping ratios for higher modes, a technique of filtering out of lower modes and other higher modes away from the one considered (Rainer, 1979) may be used.

11.8 Failure testing of structures

11.8.1 Pseudo-dynamic 'testing'

Pseudo-dynamic testing started in 1975 (Takanashi *et al.*, 1980) in Japan and became popular worldwide in the 1980s.

Pseudo-dynamic testing combines the advantages of pseudo-static testing (simplicity and possibility of testing large components to failure) and the earthquake simulation shaking table (realistic dynamic and earthquake simulation), and combines testing, analysis and simulation or experimental and computer techniques. It is possible to use the pseudo-dynamic method for:

1. components too large, massive or strong to be tested on available shaking tables;
2. problems involving geometric nonlinearity, three-dimensional and multiple support excitations, and soil–structure interaction;
3. real structural sub-assemblies.

The main advantage of pseudo-dynamic testing is its simplicity, ease and economy to implement and its capacity to simulate the earthquake behavior of a large structure both in the linear and the nonlinear stage up to failure. It has become a necessary test facility for any important institution of earthquake engineering research.

A pseudo-dynamic test consists of the following important components:

1. pseudo-static test equipment, but with several servo-controlled electro-hydraulic actuators;
2. computers on-line with the test for data acquisition and processing, response analysis and actuator control;
3. nonlinear dynamic analysis software to evaluate the restoring force of each member of the structure tested for each step of the test, which will be used by the software for nonlinear dynamic analysis of the whole structure;
4. software to evaluate the loading histories of each controlled point of the structure and to control the actuators to apply the loads or the displacements to the controlled points with sufficient accuracy.

11.8.2 Difficulties of pseudo-dynamic test

There are still some difficulties in pseudo-dynamic testing compared with other test methods (Mahin *et al.*, 1989).

(a) Damping effect

The pseudo-dynamic method accounts explicitly for Coulomb and hysteretic damping but not for viscous damping. Fortunately, it is not large and its relative importance becomes smaller when structural response enters the inelastic range because hysteretic damping is usually much larger than elastic damping.

(b) Loading rate effect

This effect has been studied recently. It should be considered for short-period structures and for impulsive-type loading, but this is not necessary for steel structures.

(c) Test control

Test control for heavy and stiff MDF systems is still an unsolved problem. Hydraulic and control system requirements for reliable pseudo-dynamic tests remain to be established.

11.8.3 Important steps in pseudo-dynamic testing

The reliability of a test depends heavily on careful implementation of the test, in which the following are important.

1. A good time integration algorithm: this is necessary to reduce the errors of step-by-step integration. An explicit algorithm is usually used because implicit algorithms may require solution schemes that rely on the tangent stiffness of the structures under test, which is difficult to measure during a test. But explicit algorithms are conditionally stable

and with less artificial damping to suppress the unnecessary higher mode responses induced by experimental error. Shing and Manivannan (1990) have shown that an implicit algorithm may give better results.

2. Care to reduce experimental errors: there are systematic and random errors in measurements and control. Systematic errors will propagate and lead to a failure of the test. Careful choices and arrangement of the instruments are vitally important to limit this kind of error.

11.9 Structural identification

Structural or system identification studies the model and parameters of the system from its measured inputs and outputs. It was introduced from modern control theory to earthquake engineering in the 1970s (Beck, 1982; Hart and Yao, 1976).

11.9.1 Problem statement

For linear problems, the Fourier transformation of the equation of motion in the frequency domain is equation 9.36 or 9.37, i.e.

$$Y(\mathrm{i}\omega) = H(\mathrm{i}\omega) * X(\mathrm{i}\omega)$$

where $X(\mathrm{i}\omega)$ and $Y(\mathrm{i}\omega)$ are respectively the Fourier transformations of the input motion $x(t)$ and output motion $y(t)$; and $H(\mathrm{i}\omega)$ is the transfer function. For this system, three kinds of problem may be defined as follows.

1. Direct analysis: to find $y(t)$ from given $x(t)$ and $H(\mathrm{i}\omega)$,

$$y(t) \leftarrow Y(\mathrm{i}\omega) = H(\mathrm{i}\omega) * X(\mathrm{i}\omega) \leftarrow X(t) \tag{11.12a}$$

2. Input identification: to find $u(t)$ for given $H(\mathrm{i}\omega)$ and $y(t)$,

$$x(t) \leftarrow X(\mathrm{i}\omega) = H^{-1}(\mathrm{i}\omega) * Y(\mathrm{i}\omega) \leftarrow y(t) \tag{11.12b}$$

3. Structural identification: to find $H(\mathrm{i}\omega)$ for given $y(t)$ and $x(t)$,

$$H(\mathrm{i}\omega) = \frac{Y(\mathrm{i}\omega) \leftarrow y(t)}{X(\mathrm{i}\omega) \leftarrow x(t)} \tag{11.12c}$$

Structural identification is useful to improve the structural model (Ibanez, 1977; Pilkey and Cohen, 1972; Beck, 1978, 1982), and input identification to estimate input when it is not measured or is difficult to measure (Hu, 1980, C1982).

11.9.2 Principle of structural identification

The principle of structural identification is to minimize the error between the calculated and the given output. Let $x(t)$ and $y(t)$ be the measured or given input and output time histories and suppose that the model of the structure is known, but with some parameters $\boldsymbol{\alpha}$ to be determined by structural identification. They are all vectors, i.e. with multiple inputs and

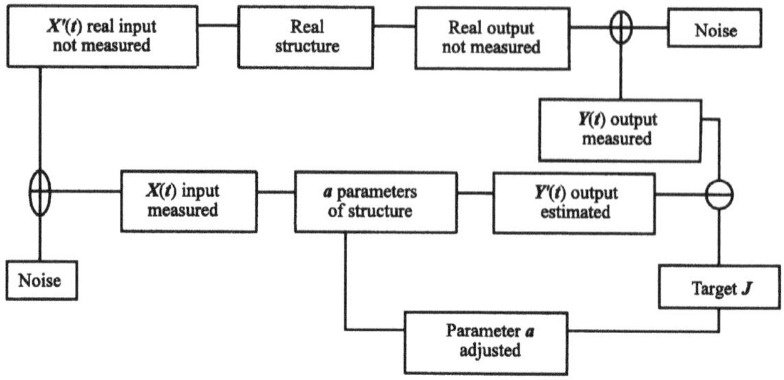

Figure 11.12 Principle of structural identification method of minimizing the error in output.

outputs, and they may contain measurement errors. Suppose the initial conditions are $y^0=y(t=0)$ and $\boldsymbol{a}=\{y^0\alpha\}^{\mathrm{T}}$ are the estimated values of initial condition and parameters of the structure. From Figure 11.12 the error between the estimated and the measured output is

$$v(t, \boldsymbol{a})=y(t)=y'(t, \boldsymbol{a}, \boldsymbol{x}) \tag{11.13}$$

which is a function of both t and a. Let $J(a)$ be a certain selected function of error $v(t, \boldsymbol{a})$, or the target function which should be minimized by selection of $\boldsymbol{a}$, i.e.

$$J(\boldsymbol{a})=\frac{\int_0^T |v(t, \boldsymbol{a})|^2 \mathrm{d}t}{\int_0^T |y(t)|^2 \mathrm{d}t}+\left|\frac{(\boldsymbol{a}_0-\boldsymbol{a})}{\boldsymbol{a}_0}\right|^2=\min \tag{11.14}$$

where T is the duration of vibration; and $\boldsymbol{a}_0$ is the assumed value of $\boldsymbol{a}$. The final equations may be linear or nonlinear, whose solution can be obtained by routine methods (Eykhoff, 1974, p. 113 and p. 446); the Gauss–Newton method of minimization is usually used.

Hu (C1982) suggested a method to minimize the error of the sum of input and output,

$$J(\boldsymbol{a})=\frac{\int_0^T |v(t, \boldsymbol{a})|^2 \mathrm{d}t}{\int_0^T |y(t)|^2 \mathrm{d}t}+\frac{\int_0^T |x-x'|^2 \mathrm{d}t}{\int_0^T |x(t)|^2 \mathrm{d}t}+|a_0-a|^2=\min \tag{11.15a}$$

$$v(t, \boldsymbol{a})=y(t)-y'(t; \boldsymbol{a}, \boldsymbol{x}') \tag{11.15b}$$

where $x'(t)$ is the modified input. This principle considers that both output and input are subject to measurement error.

11.9.3 Structural identification from strong motion records

Many studies have used the output error method in the frequency domain; among these, many used the model superposition method in which parameters may be estimated through the transfer function (Udwadia and Trifunac, 1974; McVerry, 1979; Pauschke *et al.*, 1981). When the transfer function (equation 11.12c) is used, care should be taken in the case where the input spectrum has nearly zero amplitude. If the output error approach (equations 11.14 or 11.15), in the time or frequency domain, is used, multiplication (equation 11.12a) is used instead of division and there will be no trouble of this kind. The advantages of this method should include also the following.

1. Parameters may be estimated when two neighboring modes are coupled.
2. Multiple responses may be used to estimate modal parameters.

This method can also be used for laboratory test analysis, or for the nonlinear stage (Kaya and McNiven, 1978).

Part Four

Seismic Considerations in Structural Engineering

12 Seismic hazard and risk analysis and zonation

12.1 Seismic hazard and risk

Engineering considerations of earthquakes are the earthquake load and the safety of engineering works against it. Before the occurrence of an earthquake, engineers need to know the possibility of a certain strength and repetition rate of ground motion occurring at a site and the dynamic properties and strength of the engineering works in order to design future works safely and economically and to know the possible behavior of existing works if such an earthquake occurs and the necessary measures that should be taken. The social considerations of earthquakes are similar but more on the side of safety, economic risk and political or public measures of disaster reduction.

The process of evaluating the design parameters of earthquake ground motion at the engineering site is called seismic hazard assessment (SHA). The safety of an engineering work is controlled by both the strength of earthquake motion and the seismic capacity of the engineering work. Evaluation of the possible loss from the earthquake is referred to as seismic risk assessment (SRA). Put simply, SHA considers the action and SRA the result.

During their development in the last few decades, SHA and SRA have experienced two stages. In the first stage, SR (seismic risk) was used to mean both SH (seismic hazard) and SR, and was treated as a deterministic quantity; for example, it was required to specify the maximum possible design values of earthquake motion, a peak or effective peak acceleration with a standard response design spectrum; and the structure was then designed for it, both structure and ground motion being treated as deterministic and the uncertainty taken care of by an empirical safety factor. In the present stage, firstly, SH and SR are distinguished; and secondly, uncertainties in both earthquake ground motion assessment and seismic structural strength evaluation are better understood and explicitly treated in practice. For example, the static design of structures has been shifted from a single empirical factor of safety to a probabilistic design format based on reliability theory almost worldwide; and the probabilistic SHA method has been widely used as the major method for earthquake zonation mapping, from which design parameters of seismic design are specified. The foundation of this development is the realization of the large uncertainties involved, the accumulation of strong earthquake ground motion data and the related research. These uncertainties involve at least:

- the prediction of the time, location and strength of the future earthquake and ground motion;
- the quality, strength and dynamic properties of structures;

- the relation of damage level with a few ground motion parameters;
- the consequence of the damage level on casualty numbers and economic effects.

It is common to express the relation between seismic hazard and seismic risk in the following form:

$$\mathrm{SR} = \mathrm{SH} * \sum (\mathrm{VUL} * \mathrm{LS} * \mathrm{SOC})_i \tag{12.1}$$

where VUL represents the vulnerability or capacity of the structure, i.e. the probability $P[S|Y]$ of damage states S of the structure for various possible earthquake excitations Y; the convolution SH $*$ *VUL* is the expected damage of the structure under the possible earthquake hazard; LS is the possible loss, in terms of dual indices of casualties and economic loss, from the possible damage of the structure; SOC is the society, capital or engineering work, exposed to hazard, including average number of persons under the influence of the structure, and the total number of that type of structure or other item considered in the society. SH, SR and LS are vectors but VUL and SOC are matrices. The subscript i refers to the ith type of structure or system, and the summation is over the number of types of structure to be considered in an area.

Take the following assumed simple case as an example. Suppose there are two kinds of buildings, reinforced concrete (RC) and brick, in a small town and there are N_1 RC buildings and N_2 brick buildings. If SH is given in terms of intensity, and the SH of the town studied is given in terms of the probability P of occurrence of intensities I as follows: $P=0$, 0.40, 0.60, 0 for $I=\mathrm{V}$ or smaller, VI, VII and VIII or more respectively, or in vector form,

$$\mathrm{SH} = [0.4\ 0.6\ 0] \text{ or } [0.4\ 0.6]$$

Suppose only three damage levels (no damage, moderate and collapse) are considered and the following vulnerability matrices in terms of probability of occurrence are obtained from past earthquakes:

$$(\mathrm{VUL})_1 = \begin{bmatrix} 0.9 & 0.1 & 0 \\ 0.4 & 0.6 & 0 \end{bmatrix}$$

for a certain type of RC building, and

$$(\mathrm{VUL})_2 = \begin{bmatrix} 0.5 & 0.5 & 0 \\ 0.2 & 0.7 & 0.1 \end{bmatrix}$$

for a certain type of brick building, where the first and second rows refer to intensity VI and VII respectively, and the first to third columns to damage levels ‘no damage’, ‘moderate damage’ and ‘collapse’ respectively. For example, the probability of moderate damage at intensity VII is 0.6 for RC buildings and 0.7 for brick buildings. Suppose the loss matrix is

$$\mathrm{LS} = \begin{bmatrix} 0 & 0 \\ 0.02 & 0.50 \\ 0.20 & 1.00 \end{bmatrix}$$

where the three rows are respectively loss for no damage, moderate damage and collapse. The first column refers to death ratio of the persons in the building and the second column the loss ratio of the cost of the building. Let L be the average number of persons in the building and R_1 and R_2 be the cost of the RC and brick buildings.

The expected risk for this town is then

$$\mathrm{SR} = \mathrm{SH} * \sum (\mathrm{VUL} * \mathrm{LS} * \mathrm{SOC})_i$$

$$= [0.4\ 0.6] * \begin{bmatrix} 0.9 & 1.0 & 0 \\ 0.4 & 0.6 & 0 \end{bmatrix} * \begin{bmatrix} 0 & 0 \\ 0.02 & 0.50 \\ 0.20 & 1.00 \end{bmatrix} * \begin{bmatrix} L \\ N_1 R_1 \end{bmatrix}$$

$$+ [0.4\ 0.6] * \begin{bmatrix} 0.5 & 0.5 & 0 \\ 0.2 & 0.7 & 0.1 \end{bmatrix} * \begin{bmatrix} 0 & 0 \\ 0.02 & 0.50 \\ 0.20 & 1.00 \end{bmatrix} * \begin{bmatrix} L \\ N_2 R_2 \end{bmatrix}$$

$$= 0.0324L + (0.20N_1 R_1 + 0.37N_2 R_2)$$

where the first term represents casualties and the second term economic loss, and they are not additive.

In addition to buildings, the current status of SRA may include the damage effect of other engineering and nonengineering works, such as society functioning. All of these items should have their own VUL, LS and SOC descriptions and their own classification of damage levels; if the dual parameter definition of loss is used in LS, i.e. casualties and money, LS will be $n \times 2$ where n is the number of damage levels and SOC a vector of two elements, one being the number of persons exposed to the danger and the other a product of the number of individuals and the average unit cost of that item.

SRA can be carried out in a deterministic sense, which is simple but not reasonable because of the large uncertainties involved in almost every stage of the assessment. In the case of the deterministic approach, SH = 1 for a given intensity and VUL is a scalar quantity.

The main task of SHA is the evaluation of seismicity, which is the most important, the most difficult and the most uncertain job.

12.2 Seismicity analysis

In section 2.4, a general introduction has been given to seismicity, which is the rule for or knowledge on the time, space, strength and probability of occurrence of earthquake. Seismicity of a region is controlled by two factors: the stress and strain field from plate-tectonic movement and the strength and deformation capacity of the crust medium where the earthquake occurs; seismicity thus has very strong regional characteristics. Interplate seismicity is quite different from intraplate; regions in one plate may also vary over a large range because the crust formation varies. It is important then to study the seismicity of a region.

12.2.1 Data for a seismicity study

(a) Historical earthquake records

Earthquake history records are the direct and dependable data for a seismicity study. Written historical records may be traced back to a few thousand years ago for some parts of the world and with systematic and reliable data of about two thousand years, such as for some regions of China. For example, Table 12.1 lists the historic earthquakes, of magnitude not less than 6, recorded in the North China (32°–42°N; 105°–125°E) since 780 BC, which may be the richest in a region in the world. Analysis showed, however, that only those after AD 1400 are reliable and complete. For earthquakes before 1900, the magnitudes are converted from intensity; for those after 1900, they are instrumental. The focal depth of these earthquakes was roughly 15–40 km.

(b) Seismo-geological data

Complete historical earthquake data have only existed for a few hundred years which is too short compared to the average return period of a few thousand years of very strong earthquakes and are therefore not sufficient to estimate the seismicity of very strong earthquakes. Tectono-geological data are nature's record of large earthquake events over a long period and of millions of years, and are thus indispensable to compensate for the shortcomings of the historical earthquake data. The visible traces of faults or other lineament structures and the instrument-detectable deep structures are all the results of crustal movement, of which an earthquake is an important one of abrupt nature. The difficulty is twofold: to find the deep structure and to identify those due to earthquakes from other tectonic movement. Important advances have been made in the last two decades.

There are many tectono-geological traces that very strong earthquakes may leave in the Earth's crust. The most important are faults and offsets of faults. Earthquake cliffs are traces of vertical differential movement of the two sides of the fault, and offsets of lineament structures such as streams and hill ridges are those of horizontal differential displacement on the fault. Phenomena from sand liquefaction are also evidence of strong earthquakes.

Trenching across a known fault is the popular method to discover paleo-earthquake evidence; displacement of offsets is used to estimate the magnitude and radioactivity of the material related to the fault to estimate the time of the earthquake. As an example, scientists in the United States have used this method to identify strong paleo-earthquakes which occurred on a segment of the San Andreas Fault, in addition to the 1857 earthquake, in AD 260, 350, 590, 735, 845, 935, 1015, 1083, 1350, 1550 and 1720. Similar work has been done in China and other countries.

(c) Instrumental earthquake data

Instrumental earthquake data are much better than historical data in the following respects: accurate location of the focus, better determination of magnitude, recording of very small earthquakes and possible determination of the mechanism of the earthquake which will in turn provide better

Table 12.1 Earthquakes in northern China of $M \geqslant 6$

No.	Date	Lat. (°N)	Long. (198E)	*M*
1	780[a]	34.5	107.8	6–7
2	231[a]	39.8	114.4	6½
3	22 February 186[a]	33.8	105.6	6–7
4	01 June 70[a]	36.3	119.0	7
5	23 October 46	33.0	112.5	6½
6	23 February 128	34.7	105.4	6½
7	October 143	34.7	105.3	7
8	December 344	35.4	114.3	6
9	16 August 462	35.0	116.8	6
10	21 May 512	38.9	112.8	7½
11	13 December 600	34.5	106.5	≥ 6
12	12 September 649	36.1	111.5	6½
13	16 August 701	33.0	121.0	6
14	19 March 734	34.6	105.6	7
15	777	37.8	115.2	6
16	08 March 788	32.4	109.9	6½
17	27 May 793	34.9	110.4	6
18	20 October 849	40.8	109.8	6¾
19	14 July 876	37.8	105.9	6½
20	April 1022	39.8	113.1	6½
21	09 January 1038	38.4	112.9	7¼
22	1057	39.7	116.3	6¾
23	14 August 1068	38.5	116.1	6½
24	15 January 1102	37.7	112.4	6½
25	March 1143	38.5	106.3	6½
26	March 1181	35.8	111.5	6¼
27	04 December 1209	36.0	111.8	6½
28	02 June 1219	36.0	106.2	6½
29	27 September 1290	41.6	119.3	6¾
30	25 August 1291	36.1	111.5	6½
31	17 September 1303	36.3	111.7	8
32	03 May 1305	39.8	113.1	6½
33	12 September 1306	35.9	106.1	6½
34	05 October 1314	36.6	113.8	6
35	08 September 1337	40.4	115.7	6½
36	18 April 1352	35.6	105.3	7
37	07 August 1366	38.3	112.5	6
38	13 May 1477	38.5	106.3	6½
39	09 March 1481	33.5	116.2	6
40	29 January 1484	40.5	116.5	6¾
41	10 August 1487	34.4	108.9	6¼
42	10 April 1495	37.6	105.6	6¼
43	19 January 1501	34.8	110.1	7
44	17 October 1502	35.7	115.3	6½
45	09 October 1505	32.8	122.7	6¾
46	04 February 1524	34.0	114.1	6
47	22 October 1536	39.8	116.8	6
48	12 September 1548	38.0	121.0	7
49	23 January 1556	34.5	109.7	8
50	25 July 1561	37.5	106.2	7¼
51	01 April 1568	34.4	109.0	6
52	25 April 1568	39.0	119.0	6
53	15 May 1568	34.4	109.0	6¾
54	10 April 1587	35.6	114.0	6
55	06 October 1597	38.5	120.0	7
56	25 October 1604	34.2	105.0	6
57	23 October 1614	37.2	112.5	6½

Table 12.1 (*contd.*)

58	20 May 1618	37.0	111.9	6
59	16 November 1618	39.8	114.5	6½
60	18 March 1622	35.5	116.0	6
61	25 October 1622	36.5	106.3	7
62	10 February 1624	32.4	119.4	6
63	17 April 1624	39.8	118.8	6¼
64	04 July 1624	35.4	105.9	6
65	28 June 1626	39.4	114.2	7
66	07 October 1628	40.6	114.2	6¼
67	14 January 1634	34.0	105.2	6
68	21 July 1654	34.3	105.5	8
69	03 February 1658	39.4	115.7	6
70	16 April 1665	39.9	116.6	6½
71	25 July 1668	34.8	118.5	8½
72	18 October 1673	40.5	113.5	6½
73	02 September 1679	40.0	117.0	8
74	22 November 1683	38.7	112.7	7
75	18 May 1695	36.0	111.5	7½
76	28 September 1704	34.9	107.0	6
77	14 October 1709	37.4	105.3	7½
78	19 June 1718	35.0	105.2	7½
79	12 July 1720	40.4	115.2	6¾
80	30 September 1730	40.0	116.2	6½
81	03 January 1739	38.8	106.5	8
82	27 June 1764	33.0	121.0	6
83	02 September 1765	34.8	105.0	6½
84	23 October 1815	34.8	111.2	6¾
85	03 August 1820	34.1	113.9	6
86	19 November 1829	36.6	118.5	6¼
87	12 June 1830	36.4	114.3	7½
88	28 September 1831	32.8	116.8	6¼
89	04 August 1846	33.5	122.0	7
90	26 May 1852	37.5	105.2	6
91	16 December 1852	33.5	122.0	6¾
92	14 April 1853	33.5	111.5	6¾
93	15 April 1853	33.5	121.5	6¼
94	19 July 1861	39.4	122.0	6
95	02 December 1882	38.1	115.5	6
96	14 January 1885	34.5	105.7	6
97	13 June 1888	38.5	119.0	7½
98	08 January 1910	35.0	122.0	6¾
99	16 December 1920	36.7	104.9	8.5
100	25 December 1920	36.6	105.2	7.0
101	28 December 1920	37.1	105.3	6½
102	07 January 1921	37.0	105.0	6
103	12 April 1921	35.8	106.2	6½
104	01 December 1921	33.7	122.0	6½
105	20 September 1922	39.2	120.5	6½
106	03 February 1927	33.5	121.0	6½
107	14 January 1929	40.7	111.3	6
108	22 August 1932	36.1	121.6	6¼
109	21 January 1934	41.1	108.3	6¼
110	01 August 1936	34.2	105.7	6.0
111	01 August 1937	35.4	115.1	7
112	01 August 1937	35.3	115.2	6¾
113	19 January 1940	42.7	121.3	6
114	19 December 1944	39.7	124.3	6¾
115	23 September 1945	39.5	119.8	6¼
116	23 May 1948	37.7	121.8	6

117	08 March 1966	37.4	114.8	6.8
118	22 March 1966	37.4	114.8	6.7
119	22 March 1966	37.5	115.1	7.2
120	26 March 1966	37.7	115.3	6.2
121	29 March 1966	37.4	115.8	6.8
122	27 March 1967	38.5	116.5	6.3
123	18 July 1769	38.2	119.4	7.4
124	04 February 1975	40.7	122.8	7.3
125	06 April 1976	40.2	112.1	6.2
126	28 July 1976	39.4	117.8	7.8
127	28 July 1976	39.2	117.8	6.2
128	28 July 1976	39.7	118.5	7.1
129	23 September 1976	39.9	106.4	6.2
130	15 November 1976	39.4	117.7	6.9
131	12 May 1977	39.2	117.7	6.2
132	25 August 1979	41.2	108.1	6.0
133	21 May 1984	32.5	121.6	6.2
134	28 March 1991	40.0	113.5	6.1

[a]Years BC.

understanding of seismicity of a region. However, they span a much shorter period.

12.2.2 Law of seismicity

Earthquake occurrence is a physical process which is directly connected to the mechanical properties of, and stress distribution in, the Earth's crust. The Earth's crust is so complicated and unknown in detail that an earthquake is generally considered as a random phenomenon, the average properties being predictable but not the individual event. There are two ways to describe the law of seismicity in a region: macroscopic and statistical. The macroscopic approach will be discussed in this section and the statistical one in the following.

Two principles or hypotheses were commonly used in the 1950s to describe the law of seismicity: the first is the repetition principle, i.e. an earthquake of similar strength may occur again in the province; and the second is the analog principle, i.e. seismicity is similar for seismic provinces of similar tectono-geology. There are two assumptions made in this description:

1. various possible phases or stages have been included in the past seismicity data of the region;
2. earthquakes of magnitudes possible in the region may occur at any time and with no connection with the previous one.

There are three defects in this hypothesis:

1. no possibility is considered for an earthquake of magnitude greater than the maximum that has occurred in the region;
2. no probability sense is given to the occurrence of an earthquake;
3. no consideration is given to the possible temporal nonstationarity and spatial inhomogeneity of seismicity in the province.

On the basis of the Chinese strong earthquake data, Chinese seismologists believe that there is a strong correlation between the tectono-geological background and earthquakes of very large magnitude and that this correlation varies for different regions. For example, in North China, an earthquake of magnitude greater than 7 3/4 occurs only in areas where the following three conditions exist (Shi, 1987):

1. recent structural horizontal stress field;
2. causative belt controlled by deep and large fault;
3. segment of maximum horizontal slip in an echelon strike-slip fault in a NNE direction, often intersecting another fault in a NW direction.

12.2.3 Statistical analysis

An earthquake is universally considered as a random phenomenon, i.e. a phenomenon where average properties may be determined but where the individual events are unpredictable. The most popular mathematical model for earthquakes is the Poisson distribution.

(a) Poisson distribution

Earthquake occurrence is random in space and time. Suppose that the magnitude of the earthquake is not considered and that the space or time t is divided into many intervals of small area or time interval Δt. Consider the random event where an earthquake occurs inside the interval Δt; if this event satisfies the conditions of independence, stationarity and nonrepetition, it is a Poisson event. It can be proved (e.g. Ang and Tang, 1975, p. 115) that the probability of n events occurring in a period of time t is

$$P(n)=\frac{(\nu t)^n \exp(-\nu t)}{n!} \quad (n=0, 1, 2, \ldots) \tag{12.2a}$$

which is Poisson distribution. Its expectation is $E[n]=\nu t$, and its standard deviation is also νt. In seismicity studies, the unit of t is years and ν is the average annual rate of occurrence.

Let P_1 be the probability of ground motion $Y>y$ at a certain site when one earthquake occurs, the probability of $Y_T>y$ for n times in T years at the site is

$$P(n)=\frac{\exp(-P_1\nu T)(P_1\nu T)^n}{n!}$$

The probabilities of $Y_T\leqslant y$ and $Y_T>y$ at the site are respectively

$$P[Y_T\leqslant y]=P(n=0)=\exp(-P_1\nu T) \tag{12.2b}$$

$$P[Y_T>y]=1-\exp(-P_1\nu T) \tag{12.2c}$$

Let the return period of $Y>y$ be T_Y, then

$$T_Y=\frac{1}{P_1\nu}=-\frac{T}{\ln(1-P[y_T>y])} \tag{12.2d}$$

For example, if $P[y_T > y]$ is 10% in $T = 50$ years, then its return period is

$$T_Y = \frac{-50}{\ln(1-0.1)} = 475\,\text{years}$$

If $P_1 \nu T = T/T_Y = 1$, then the probability is

$$P[Y_T > y] = 1 - \frac{1}{e} = 0.63$$

which means that the probability of maximum ground motion within a return period is 63%.

(b) Exponential distribution

Let t be the time interval between two consecutive events. For a Poisson event, the probability is

$$P(0) = P(t) = 1 - F(t) = \frac{(\nu t)^0 \exp(-\nu t)}{0!} = \exp(-\nu t) \tag{12.3a}$$

where

$$F(t) = 1 - \exp(-\nu t) \tag{12.3b}$$

is the probability of occurrence of at least one event in t, and the probability density function is

$$f(t) = \frac{\mathrm{d}F}{\mathrm{d}t} = \nu \exp(-\nu t) \tag{12.4}$$

which is the exponential distribution. The expectation and variance of this distribution are respectively $1/\nu$ and $1/\nu^2$; therefore, the average recurrence period of a Poisson event is $1/\nu$. From equation 12.3, for $t = 1$ years and very small ν,

$$F(t=1) = 1 - \exp(-\nu t) \cong 1 - \left(1 - \nu + \frac{1}{2}\nu^2 - \ldots\right) \cong \nu$$

which means that the annual probability of occurrence is

$$P_{1\text{yr}} = F(t=1) = \nu$$

(c) Lack of memory

The poisson distribution has no memory; the occurrence of one event is not affected by the events which occurred before it.

12.2.4 Magnitude–frequency relationship

In section 2.4, we explained the Gutenberg–Richter magnitude–frequency relationship and gave some examples of seismic provinces in China. Seismicity anywhere in the world follows this relation, linear in $\log N$ and M coordinates up to a certain upper limit M_u. The most commonly used form is the probability $F(M)$ of occurrence of earthquake of magnitude not more than M

$$F(M) = k_1\{1 - \exp[-\beta(M - M_0)]\} \tag{12.5}$$

$$\frac{1}{k_1} = 1 - \exp[-\beta(M_u - M_0)]$$

and the total number of earthquakes of magnitude in the range M_0–M_u

$$N_T = \exp(\alpha)[\exp(-\beta M_0) - \exp(-\beta M_u)] \tag{12.6}$$

Under the assumption of Poisson distribution, the seismicity can be described by the foregoing two equations with four parameters: the annual rate v of earthquake occurrence; the lower and upper limits M_0 and M_u; and the relative distribution of small and large earthquakes, or the slope β of the straight line given by equation 12.6 in $\ln N$–M coordinates, where the annual rate ν is equal to N_T/T and T is the number of years spanned by the earthquake data.

The lower limit magnitude M_0 is taken as the minimum magnitude for which completeness of data is certain; in China it is usually taken as 4¾ in north China and 5 or more in less populated regions. The annual rate $\nu = N_T/T$ is estimated from the total number of earthquakes.

The most difficult problem is the estimation of the upper limit magnitude M_u, and its estimation is very sensitive to SHA. It is estimated usually by a combination of the following considerations:

1. the maximum historical earthquake;
2. seismicity;
3. statistical analysis;
4. tectono-geological analysis.

12.2.5 Characteristic earthquake

Although there are many cases where the Gutenberg–Richter magnitude frequency relation is a good description of seismicity of a province, there are some cases where it is not so, such as the province or seismic belt where earthquakes with characteristic magnitude occur very often, for example, on the Xianshuihe Fault in southwestern China (Allen *et al.*, 1987). If the data are plotted in $\log N$–M coordinates, the curve is not a straight line but peaks at a characteristic magnitude, 7 1/2 in the case of the Xianshuihe fault. In this case, an ordinary Gutenberg–Richter relation plus a characteristic magnitude will be a better representation of the seismicity (Cornell and Winterstein, 1988).

12.2.6 Sectional model and potential source

At the end of the 1980s, the Chinese suggested a sectional model of seismicity analysis, which considers the sectional divisions of time, space and magnitude of an earthquake (Shi, 1987; Hu, 1991). A Poisson distribution is assumed inside the three-dimensional sections of time, space and magnitude. The time span is divided into quiescent and active stages, with

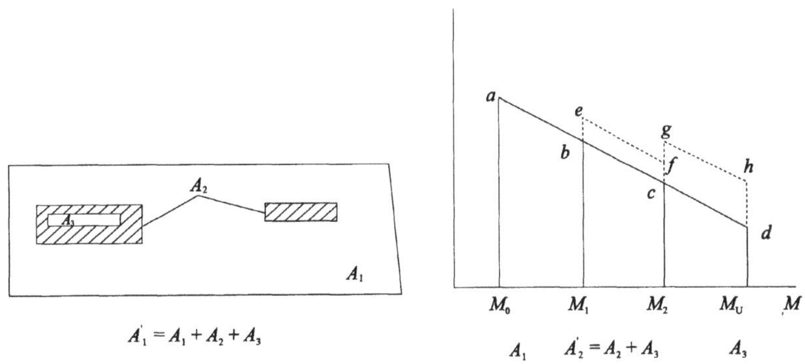

Figure 12.1 Sectional Poisson model and magnitude–frequency relation.

different annual rates and upper limit magnitudes, on the basis of historical earthquake data. The upper limit magnitude of the source cannot be greater than that of the province and the seismicity of all the sources considered together should be the same as that of the province. The space of a seismic province is divided into potential seismic sources with different seismicity according to the seismo-tectonic backgrounds and past seismicity of the sources. A special feature of the method is that the magnitude is also divided into several sections, from 6 to 8.5 with an increment of 0.5.

The ordinary Poisson model assumes that an earthquake of magnitude in the whole range M_0-M_u occurs with equal probability anywhere in the province. The sectional model assumes that an earthquake of magnitude in one magnitude section, say, M_i-M_{i+1}, occurs with equal probability in those sources with upper limit magnitude not less than M_{i+1} in one temporal stage of seismicity. As a result of this assumption, the seismicity in a province follows a Poisson distribution, and the Gutenberg–Richter magnitude-frequency relation (equation 12.5) holds in the province; but in one potential source, a Poisson assumption does not hold for the whole magnitude range but does hold for one magnitude section. The case shown in Figure 12.1 serves as an example, where there are two sources 2 of total area A_2 of upper limit magnitude M_2, and one source 1 of area A_3 of upper limit magnitude $M_3=M_u$, and the rest of the province of area A_1 is the source 3 of an upper limit magnitude M_1. The area of the province is then $A=A_1+A_2+A_3$ with an annual rate ν estimated by the ordinary method from data for the whole province. According to the assumption of this method, the annual rate of each source in each magnitude section will be those shown in Table 12.2 with values given by the following equation:

Table 12.2 Seismicity of each section

Source	Total area	Upper limit magnitude	M_0–M_1	annual rate M_1–M_2	M_2–M_u	Segment in Figure 12.1(b)
1	A_1	M_1	$\nu_1^{(1)}$	–	–	ab
2	A_2	M_2	$\nu_1^{(2)}$	$\nu_2^{(2)}$	–	ab, ef
3	A_3	$M_3=M_u$	$\nu_1^{(3)}$	$\nu_2^{(3)}$	$\nu_3^{(3)}$	ab, ef, gh

$$\nu_j^{(k)} = \frac{\nu_j A' A_k}{A_j'^2} \quad (j=1, 2, 3; \quad k=j, \quad j+1, \ldots 3) \tag{12.7}$$

where $A_j' = A_j + \ldots + A_3$; ν_j' is the annual rate of the whole province in the magnitude range M_{j-1}–M_j; $\nu = \nu_1 + \nu_2 + \nu_3$.

12.3 Seismic hazard assessment

There are two kinds of seismic hazard assessment (SHA). The early SHA was usually deterministic; the probabilistic method was introduced at the end of the 1960s by Cornell (1968).

No matter what kind of method is used for SHA, it usually consists of the following steps, as shown in Figure 12.5 (section 12.5.1).

1. Determine the seismic provinces or belts where future earthquakes may occur according to the same law of seismicity.
2. Determine the parameters of seismicity for each province and source (Section 12.2).
3. Select the parameters for seismic hazard, peak acceleration, response spectrum and/or intensity, and determine their attenuation laws suitable for the region.
4. Determine the seismic hazard at the sites.

12.3.1 Parameters for seismic hazard

The most widely used parameters for hazard are intensity and peak acceleration; peak velocity and response spectrum are also common. They are selected according to the job on hand.

12.3.2 Deterministic SHA

Although an earthquake is universally considered to be a random event, for simplicity reasons, the uncertainty was not considered explicitly and a fuzzy term 'probable maximum' was used as a subjective deterministic quantity. According to this idea, future probable maximum earthquakes were estimated on the basis of past earthquakes and seismotectonic background, and this is the step which makes a random or uncertain event a deterministic one. This was a reasonable approach 30 years ago when structural design was based on a similar idea of factor of safety. Many seismologists still follow this approach. In this approach, seismic hazard is specified by a given intensity or peak value of acceleration without specifying the possibility or probability of the assessment.

12.3.3 Probabilistic SHA procedure

Probabilistic SHA consists of four main steps.

1. Define the possible seismic provinces around a site, where an earthquake will be a significant hazard to the site.
2. Determine the seismicity of each province, assuming the probability of having an earthquake at any location is uniform in the province.

3. Find the seismic hazard to the site from the sum of the contributions of all provinces with known attenuation laws of ground motion. In this approach, seismic hazard is given by ground motion Y with given probability of exceedance

$$P[Y \geqslant y] = \sum p[Y \geqslant y|E_j] \cdot P[E_j] \tag{12.8a}$$

$$P[Y \geqslant y|E_j] = \int \ldots \int P_j[Y \geqslant y|x_1, x_2, x_3, \ldots] \cdot f_j(x_1)$$

$$\cdot f_j(x_2|x_1) \cdot f_j(x_3|x_1, x_2) \ldots \mathrm{d}x_3 \mathrm{d}x_2 \mathrm{d}x_1 \tag{12.8b}$$

where $P_j[Y \geqslant y|E_j]$ is the probability of $Y \geqslant y$ when an earthquake occurs in province j; x_i $(i = 1, 2, \ldots)$ is the uncertain factor considered, for example the earthquake magnitude and position variables; and $f(\cdot)$ is the probability or conditional probability density function. The probability of exceedance may be the annual probability $P_{1\text{yr}}$, or probability P_T in T years, or the return period $T_r = 1/P_{1\text{yr}}$.
4. Modify the result to consider the uncertainties in the process.

When spectrum $S(T)$ is required, the procedure is to find the ordinates $Y = S(T_i)$ of the spectrum one by one for $i = 1, 2, \ldots$ and all the ordinates of the same probability of exceedance are often used for the spectrum of the site which is usually referred to as response spectrum of uniform probability.

12.3.4 Probabilistic SHA for a point source

Cornell (1968) was the first to formulate the probabilistic approach of SHA, modified later by others (Der Kiureghian and Ang, 1977); it is the most widely used approach. For simplicity, he assumed a point source for each earthquake, but the seismic province can be a point, a line, an area or a volume.

Under the assumption of a point earthquake source and neglecting some uncertainties, equation 12.8b becomes

$$P[Y \geqslant y|E_j] = \int\int P_j[Y \geqslant y|M, R] f_j(M) f_j(R|M) \,\mathrm{d}M \mathrm{d}R \tag{12.9}$$

Suppose there are n provinces around the site and let ν_j be the average annual rate of earthquake of magnitude $M \geqslant M_0$ occurring in the jth province; the total annual rate is then $\nu = \sum \nu_j$. The probability of one earthquake of $M \geqslant M_0$ occurring in province j is then

$$P(E_j] = \frac{\nu_j}{\nu} \tag{12.10}$$

From equation 12.8a, it is obvious that

$$P[Y \geqslant y] = \frac{1}{\nu} \sum \nu_j P[Y \geqslant y|E_j] \tag{12.11}$$

The occurrence of a future large earthquake in a province is usually considered a Poisson event, i.e. independent in time and space; the ground motion $Y \geqslant y$ at a site due to such as earthquake is also a Poisson event with average rate $\nu P[Y \geqslant y] = \sum \nu_j P[Y \geqslant y|E_j]$. Referring to equation 12.3b, the annual probability of $Y \geqslant y$ is

$$P_{1\text{yr}}[Y \geqslant y] = 1 - \exp\left\{-\sum \nu_j P[Y \geqslant y|E_j]\right\} \tag{12.12}$$

When the annual probability of exceedance is very small, say $\leqslant 0.05$,

$$P_{1\text{yr}}[Y \geqslant y] \cong \sum \nu_j P[Y \geqslant y|E_j] \tag{12.13}$$

and the return period is

$$T_r = \frac{1}{\sum \nu_j P[Y \geqslant y|E_j]} \tag{12.14}$$

It is then easy to find the probability of having ground motion over y at least once in T years. From equation 12.13, the accumulated probability of $Y<y$ in one year is

$$F_{1\text{yr}}(y) = 1 - P_{1\text{yr}}(Y>y) = 1 - \sum P[Y>y|E_j]\nu_j \tag{12.15}$$

Consider now the time variable as a discrete interval of one year and assume that the maximum of y in one year is statistically independent and has the same probability of exceedance. Following the Bernoulli independent trial sequence, the probability of exceeding a given Y in T years is

$$P_T(Y>y) = 1 - F_{1\text{yr}}^T(y) = 1 - \{1 - \sum \nu_j P[Y \geqslant y|E_j]\}^T \tag{12.16}$$

(a) Calculation of $P[Y \geqslant y|E_j]$

First express equation 12.11 as follows:

$$P[Y \geqslant y|E_j] = 1 - F(y) = \int_d^{r_0} P[Y \geqslant y|E_{j,r}] f_j(r)\,dr \tag{12.17}$$

where $E_{j,r}$ is an event occurring in province j at distance r from the site, $f_j(r)$ is the probability density function and the definitions of d and r_0 are given in Figure 12.2.

From the attenuation law of intensity $y=I$,

$$y = c_1 + c_2 M - c_3 \ln R \tag{12.18}$$

it is found that when M and R are independent,

$$\begin{aligned} P[Y \geqslant y|R=r] &= P[c_1 + c_2 M - c_3 \ln r \geqslant y|R=r] \\ &= \exp\left[-\beta\left(\frac{y + c_3 \ln r - c_1}{c_2} - M_0\right)\right] \end{aligned} \tag{12.19}$$

where

$$\frac{y + c_3 \ln r - c_1}{c_2} \geqslant M_0 \quad \text{or} \quad y \geqslant c_1 + c_2 M_0 - c_3 \ln r$$

If y is smaller than the right-hand side of this expression, the probability of $Y \geqslant y$ is 1 (for $M > M_0$ and $R = r$).

(b) Integration of equation 12.17

Integrate for all possible r. If the probability density of having an earthquake of $M \geqslant M_0$ at r is $f(r)$, the final result through equation 12.17 will be

$$P[Y \geqslant y] = 1 - F(y) = \int_d^{r_0} P[Y \geqslant y | R = r]\, f(r)\, \mathrm{d}r \qquad (12.20)$$

Take the line province as an example. The above-mentioned procedure will be illustrated for the case of line province as shown in Figure 12.2. The dashed line of length L at depth h represents the fault or the linear province, on which an earthquake may occur at any point, say point O, with uniform probability. Suppose site D is in a vertical plane through the center C of the fault. The probability density $f(r)$ can easily be derived as follows. The probability function of earthquake occurrence at distance $R \leqslant r$ is

$$F(r) = P[R \leqslant r] = P[x^2 + d^2 \leqslant r^2]$$
$$= P[|x|] \leqslant (r^2 - d^2)^{1/2}] = \frac{(r^2 - d^2)^{1/2}}{L/2} \quad (d \leqslant r \leqslant r_0) \qquad (12.21)$$

which indicates that the probability of occurrence of an earthquake at any place on the line province ($|x| \leqslant L/2$) is uniformly distributed, because the accumulated probability increases linearly from 0 to 1 when $|x|$ increases from 0 to $L/2$. The probability density function $f(r)$ is then

$$f(r) = \frac{\mathrm{d}}{\mathrm{d}r}\left(\frac{2(r^2 - d^2)^{1/2}}{L}\right) = \frac{2r}{L(r^2 - d^2)^{1/2}} \quad (d \leqslant r \leqslant r_0) \qquad (12.22)$$

With equations 12.19 and 12.22, equation 12.17 can be integrated but the result is complicated; an approximate solution can be obtained for large intensities.

(c) Numerical integration of equation 12.17

This equation is usually integrated numerically by dividing each province or source into elements of equal distance R under the assumption that the

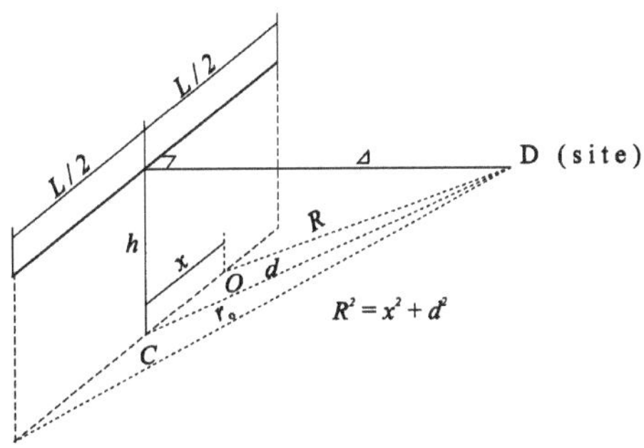

Figure 12.2 Line province and site.

earthquake occurrence is uniformly distributed within one province or source. The element is so small that R can be taken as constant in the element and equation 12.17 can be integrated by summation of the probabilities of all the elements in that province or source. Several types of computer program are widely available (e.g. McGuire, 1976).

12.3.5 Probabilistic SHA: fault-rupture model

This model was introduced by Der Kiureghian and Ang (1977), who considered the finite dimension of rupture length as a function of the earthquake magnitude. The effect of finite rupture length is large for sites very close to a fault. Cornell's point source method is widely popular, but Der Kiureghian and Ang's rupture model is popular in China.

12.3.6 Modification to consider uncertainty

Der Kiureghian and Ang (1977) considered the uncertainties from ground motion evaluation and the fault length–magnitude relation. They assumed that the random ground motion Y_a may be expressed in the following manner by a random factor ε and an average Y of Y_a

$$Y_a = \varepsilon Y$$

where Y_a and thus ε, taking only positive values, follow log-normal distribution and the mean and coefficient of variation of $\ln \varepsilon$ are respectively 0 and Ω. When considering the randomness of both ground motion (ε_1) and fault rupture length (ε_2), they assume that the resulting randomness is

$$\varepsilon = \varepsilon_1 \varepsilon_2$$

whose mean and coefficient of variance are respectively

$$E[\varepsilon] \approx 1.0, \quad \Omega \cong (\Omega_1^2 + \Omega_2^2)^{1/2} \tag{12.23}$$

Past results suggest that $\Omega_1 = 0.54$–0.84, $\Omega^2 = 0.17$–0.30.

The probability of exceedance in equations 12.8, with modification for uncertainty, becomes

$$P[Y_a \geqslant y] = P[\varepsilon Y \geqslant y] = \int_0^\infty P[\varepsilon Y \geqslant y | \varepsilon = \mu] f(\mu)\, d\mu$$

$$= \int_0^\infty P[Y \geqslant y/\mu] f(\mu) d\mu \tag{12.24}$$

where $f(\mu)$ is the probability density function of μ. Because ε_1 is much larger than ε_2, the probability density function $f(\varepsilon)$ of ε may be taken the same as ε_1, i.e. log-normal.

In integrating Equation 12.24, two points may be mentioned here. First, the integration is carried out numerically, and the limits of integration may be taken as $\mu/\sigma = -k$, k and $k \geqslant 5$, where σ is the standard deviation of ε. Second, if modification for uncertainty is calculated for small y values,

$P[Y \geq y/\mu]$ near the lower limit of integration is nearly 1, which should be included, although it is not of design interest.

12.3.8 Probabilistic SHA: another method

There is another probabilistic method of SHA, which relies completely on historical earthquake data with no help from tectono-geology. It assumes that the past earthquake records can be used exclusively to estimate future seismicity. For regions with long and sufficiently detailed earthquake records, this assumption is reasonable. This method was used before the probabilistic SHA methods described above, firstly in Japan and is still commonly used there (Okamoto, 1984), perhaps because of Japan's special background of seismicity.

There are two possible approaches for this method, both starting from historical earthquake data. One uses the data recorded in a small area directly, such as the intensity data I for a city or a county; and the other uses the epicenter and magnitude of each earthquake and computes the effect, either intensity or ground motion, at a city or a site from attenuation relations. The second approach may also be used as a supplement to the first to find the missing intensity data for the city for some earthquakes. If ground motion, such as peak acceleration or response spectrum, is required, the second approach is used.

Let $Y_i (i=1, 2, \ldots, n)$ be a set of intensity or ground motion data for a city. A statistical distribution $p(Y)$ is assumed theoretically or empirically from the data set, such as an exponential or extreme value distribution. If the exponential is assumed,

$$p(Y) = \frac{1}{b}\exp\left(-\frac{Y}{b}\right) \tag{12.25}$$

the probability of Y exceeding a given value y is then

$$P[Y \geq y] = \int_a^\infty p(A)\,\mathrm{d}A = \exp\left(-\frac{a}{b}\right) \tag{12.26}$$

which is the probability P_1 of $Y \geq y$ for one earthquake and it can be used in equations 12.14–12.16 to find the probability of $Y \geq y$ in T years or the return period T_r of ground motion $Y \geq y$.

12.4 Structural vulnerability and seismic risk assessment

As explained at the beginning of this chapter, structural vulnerability is a measure of the seismic strength or capacity of a structure. Firstly, various limit or damage states corresponding to different levels of damage of the structure are defined according to the functional requirements of the structure. These limit states may be defined in deterministic terms as physical quantities or in fuzzy terms as subjective gradation; the latter is more popular now because of its simplicity and the availability of data.

Secondly, considering the randomness of the structural strength, the strength or capacity of the structure corresponding to any limit or damage state should be given in probabilistic terms and the vulnerability of a structure is then defined by a probability matrix VUL. The evaluation of vulnerability matrix VUL for various types of structures and systems is the main task of seismic risk assessment (SRA), as shown in equation 12.1.

12.4.1 Damage states and vulnerability

In making engineering decisions, it is necessary to know the possible risk that could occur in the structure's lifetime for various design choices in order to reach an optimum design between economy and safety. Seismic risk is controlled by the following two factors: seismic structural response Q and the strength R or the allowable value of the response. The damage state is then defined by Q and R. For seismic design of ordinary structures, the controlling quantity is usually story shear or story drift; the maximum value of story shear or drift is then the considered response Q, which is determined by the ground motion or the seismic hazard and the dynamic structural property; the allowable story shear or drift is the capacity R, which is determined by the capacity of the members of that story.

(a) Damage state

The damage state of a structure may be defined by its capacity R, which may be the allowable value of Q, either the story shear or drift Δ, corresponding to a specific damage level. Let the subscripts y and max refer to the elastic limit state and ultimate limit state. The allowable ductility factor is $\mu_{max} = \Delta_{max}/\Delta_y$. The damage state is defined by the relation between Q and R; and the vulnerability by the probability of the occurrence of these damage states. Special cases of two-state (S_0,–S_1) and five-state (S_0–S_4) damage criteria are given below as an illustration, where R_i defines the boundary of state S_i.

1. Two-state damage criterion:

$$S_0\text{, intact } (Q < R_0);$$

$$S_1\text{, damage } (Q > R_0);$$

2. Five-state damage criterion:

$$S_0\text{, intact } (Q < R_0);$$

$$S_1\text{, slight } (R_0 < Q < R_1);$$

$$S_2\text{, moderate } (R_1 < Q < R_2);$$

$$S_3\text{, serious } (R_2 < Q < R_3);$$

$$S_4\text{, complete } (R_3 < Q).$$

Figure 12.3 shows the three-state case, where $Q < R_y$ or $\Delta < \Delta_y$ is the intact state S_0, $\Delta_y < \Delta < \Delta_{max}$ the moderate damage state S_1 and $\Delta > \Delta_{max}$ the collapse or complete damage state S_2.

Figure 12.3 Structural response and capacity.

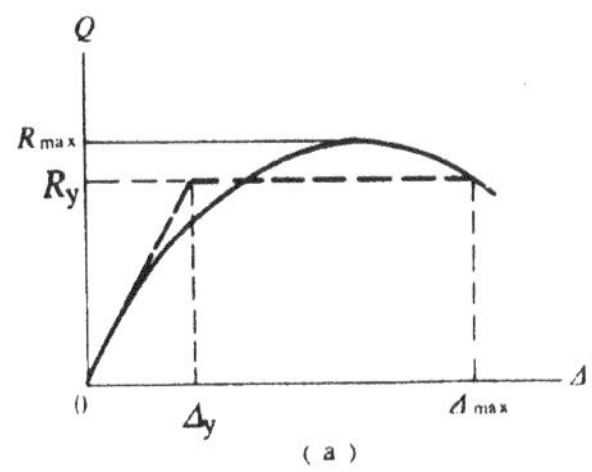

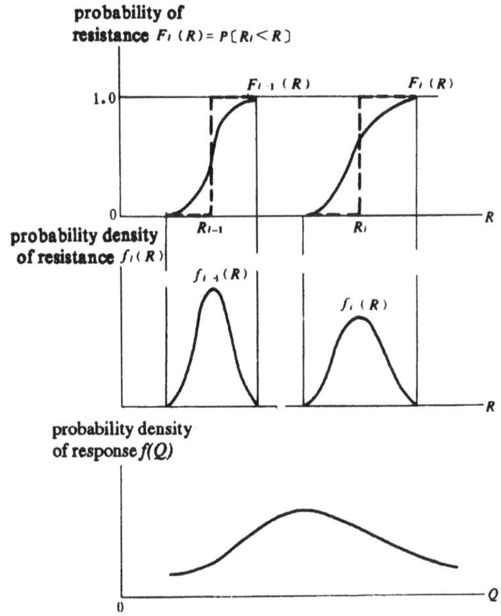

Some other classifications of damage states were suggested but are not widely used. One six-state case is as follows: $R_0 = 1\%$ of the constructional cost of the structure, $R_1 = 4\%$, and $R_2 = 10\%$, $R_3 = 30\%$ and $R_4 = 50\%$.

(c) Vulnerability information from 1976 Tangshan earthquake

Damage information accumulated from past earthquakes may be used to obtain the vulnerability of ordinary structures, and data from the 1976 Tangshan earthquake are given in Tables 12.3–12.8 as examples (Liu, C1985–86, Vol. 2). They are usually given in terms of intensity.

Table 12.3 lists the statistics for multistory brick buildings in the Tangshan area, where S_0 is intact, S_1 slight, S_2 moderate, S_3 serious, S_4 partial collapse and S_5 complete collapse. Results for the percentage of buildings for each damage state are also given in the table, where intensities VIII and VIII–IX are combined into VIII. The data for Beijing are included, where intensity was rated as VI, from the same source. The results are shown in the table in parentheses, and there is almost no difference from the data in Tangshan for intensity VI+. The numbers of buildings for these two cases are relatively large, but the difference may reach 2%, which is perhaps the order of accuracy. In the rest of this section, the percentages of buildings damaged will be rounded to integer values.

The statistics for other types of structure in Tables 12.4–12.8 are given on a five-state scale where S_0 is intact, S_1 slight, S_2 moderate, S_3 serious and S_4 complete damage.

Table 12.3 Damage statistics for multistory brick buildings

Intensity I	Total	S_0	S_1	S_2	S_3	S_4	S_5
		Number damaged					
VI+	574	265	231	70	8	0	0
VII	23	10	5	5	3	0	0
VIII	17	2	6	5	4	0	0
VIII–IX	17	2	4	5	4	2	0
VIII, VIII–IX	34	4	10	10	8	2	0
IX	382	5	26	131	124	38	58
X	337	3	17	22	67	64	165
XI	952	3	14	45	111	115	664
		Percentage damaged					
VI	3564	45.5	38.0	11.9	4.6	0	0
VI+	574	46.2	40.2	12.2	1.4	0	0
VII	23	43.5	21.7	21.7	13.1	0	0
VIII	34	11.8	29.4	29.4	23.5	5.9	0
IX	382	1.3	6.8	34.3	32.5	9.9	15.2
X	337	0.8	5.0	6.5	19.8	19.0	48.9
XI	952	0.3	1.5	4.7	11.7	12.1	69.7

Table 12.4 Damage statistics for single-story industrial factories[a]

Intensity I	Total no.	Percentage				
		S_0	S_1	S_2	S_3	S_4
VI	43	84	16	0	0	0
VII	19	21	37	26	16	0
VIII	207	16	17	36	27	4
IX	41	12	2	15	46	25
X	100	2	10	18	28	42
XI	57	4	4	7	12	73

[a] This type of building includes reinforced concrete and brick columns. The numbers of each kind may not be sufficient to distinguish between them.

Table 12.5 Damage statistics for tanks on masonry supports

Intensity I	Total no.	Percentage				
		S_0	S_1	S_2	S_3	S_4
VI	38	73	24	3	0	0
VII	54	24	33	30	6	7
VIII	107	10	28	37	10	15
IX	53	8	8	8	6	70
X	16	0	19	25	31	25
XI	23	0	0	13	4	83

12.4.2 SRA procedure

In many cases, the seismic risk for one type of engineering work is studied, and only the first part of the problem, i.e. $\mathrm{SR} = SH * \mathrm{VUL}$, which is the probability of occurrence of damage state of structure or other system

Table 12.6 Damage statistics for Brick chimneys

Intensity I	Total no.	Percentage				
		S_0	S_1	S_2	S_3	S_4
VI	6000	93.1	4.8	4.8	1.8	0.3
VII	101	43	4	10	25	18
VIII	45	0	18	11	18	53
IX	33	0	3	3	6	88
X	137	0	0	5	12	83
XI	18	0	0	0	0	100

Table 12.7 Damage statistics for highway beam bridges

Intensity I	Total no.	Percentage				
		S_0	S_1	S_2	S_3	S_4
VII	54	46	26	24	4	0
VIII	39	8	15	41	23	13
IX	14	7	7	7	15	64
X–XI	23	17	13	22	30	17

Table 12.8 Damage statistics for earth dams, water gates and pump stations

I	Earth Dams					Water gates					Pump stations				
	Total no.	Percentage				Total no.	Percentage				Total no.	Percentage			
		S_0	S_1	S_2	S_3		S_0	S_1	S_2	S_3		S_0	S_1	S_2	S_3
VI	7	12	59	29	0	5	20	80	0	0	0	0	0	0	0
VII	37	21	21	33	25	19	16	42	21	21	11	64	36	0	0
VIII	10	10	20	30	40	12	17	25	25	33	19	68	11	21	0
IX	1	0	0	0	100	10	10	40	50	0	12	0	25	17	58
X	0	0	0	0	0	1	0	0	0	100	3	0	0	0	100
XI	0	0	0	0	0	1	0	0	0	100	0	0	0	0	0

under given seismic hazard is considered. If both the damage state and seismic hazard are given in discrete quantities, the probability of occurrence of the ith damage state S_i is

$$P[S_i] = \sum P[S_i | Y_j] P[Y_j] = \sum (P[< R_i | Y_j] - P[< R_{i-1} | Y_j] P[Y_j] \quad (12.27a)$$

$$= \sum P[S_i | Y_j](P[Y_j] - P[Y_{j+1}]) \quad (12.27b)$$

Equation 12.27a applies for a discrete ground motion quantity Y_j, such as intensity, while equation 12.27b for discrete divisions of a continuous ground motion quantity Y_j, such as a range of acceleration A_j–A_{j+1}. In the equations, $P[< R_i | Y_j]$ is the conditional probability of response Q less than strength R_i or $Q < R_i$ for given condition $Y = Y_j$; and $P[S_i | Y_j]$ is the probability of occurrence of the damage state S_i, giving $Y = Y_j$, which is, by definition,

$$P[S_i | Y_j] = P[< R_i | Y_j] - P[< R_{i-1} | Y_j] = P[> R_{i-1} | Y_j] - P[> R_i | Y_j] \quad (12.27)$$

For the two-state case, because there is only one boundary of strength R_0, equations 12.27a and b become

$$P[S_i]=\sum P[S_i|\,Y_j]\,P[Y_j]=\sum P[>R_0|\,Y_j]\,P\,Y_j] \qquad (12.28)$$

If the ground motion Y is given as a continuous quantity, such as acceleration, the result will be

$$P[S_i]=\int (P[<R_i|\,Y]-P[<R_{i-1}|\,Y])\,f(Y)\,\mathrm{d}Y \qquad (12.29)$$

where $f(Y)$ is the probability density function of Y and its accumulated probability is

$$F(y)=P[<Y]=\int_0^Y f(y)\,\mathrm{d}y$$

If both ground motion and structural response are given as continuous quantities, it will be

$$P[S_i]=(P[<R_i]-P[<R_{i-1}])=\int_{R_{i-1}}^{R_i}\frac{\mathrm{d}P[<R]}{\mathrm{d}Y}\,\mathrm{d}Y$$

$$\int_{R_{i-1}}^{R_i} F(R|\,Y)\,f(Y)\,\mathrm{d}Y \qquad (12.30)$$

where $F(R|\,Y)=P[<R|\,Y]$ is the conditional probability of $Q<R$, giving Y; and $\mathrm{d}P[<R]/\mathrm{d}Y=F(R|\,Y)\,f(Y)$ is the probability density.

Relations given above are for deterministic structural strength R_i. Because of uncertainties in geometric dimension, strength and rigidity of structures, R_i is usually uncertain or random, and is described by its probability distribution $f_i(R)$. As shown in Figure 12.3, the dashed line represents the deterministic strength, and the solid line the random, where $F_i(R)=P_i[<R]$ is the accumulated probability function of strength R_i and $f_i(R)=\mathrm{d}F_i/\mathrm{d}R$ is the density; the assembly of all $f_i(R)$ is the vulnerability of structure.

To find the seismic risk for this case, it is necessary to compute first the probability density $f(Q)$ of the response Q for all possible Y values,

$$f(Q)=\int_0^\infty f(Q|\,Y)\,f(Y)\,\mathrm{d}Y \qquad (12.31)$$

where $f(Q|\,Y)=\mathrm{d}F(Q|\,Y)/\,\mathrm{d}Q$ is the probability density of Q for given Y. For random strength R_i, the probability of occurrence of the damage state $S_i(R_{i-1}<Q<R_i)$ is

$$P[S_i]=P[>R_{i-1}]-P[>R_i]=\int_0^\infty F_{i-1}(Q)f(Q)\,\mathrm{d}Q-\int_0^\infty F_i(Q)f(Q)\,\mathrm{d}Q$$

$$=\int_0^\infty [F_{i-1}(Q)-F_i(Q)]f(Q)\,\mathrm{d}Q \tag{12.32}$$

It is clear then that SRA deals with three factors: seismicity or seismic hazard $f(Y)$, structural response analysis $F(Q|Y)$ or $f(Q|Y)$, and structural strength R. The first one is the concern of the seismologist while the other two are that of the engineer.

12.4.3 SRA

(a) Random ground motion: two-state case

Two-state damage has been defined in the foregoing section and the calculation of its risk is given by equation 12.28; in this section, only the evaluation of conditional probability $P[S_1|Y_j]$ and probability $P[Y_j]$ of ground motion exceeding Y_j is discussed, with some examples of SRA.

Probability of damage $P[S_1]$. It is possible to express the earthquake response Q of a structure, such as maximum internal force or deformation, in terms of ground motion Y, where Y may be intensity, peak acceleration or velocity, through the input–output relation

$$Q=Q(Y) \quad \text{or} \quad Y=Y(Q) \tag{12.33}$$

This relation is linear in linear structures, but may be very complicated in nonlinear cases. It is common to assume that the probability $P=P[>R_0]$ of response Q exceeding the structural capacity or allowable strength R_0 of that damage state is equal to the probability p of ground acceleration A exceeding $A=A(R_0)$, where $A=A(Q)$ is evaluated from equation 12.33. For simplicity, let it be

$$p=cA^{-k} \tag{12.34}$$

where k may have values in the range 1.5–3.0. In deriving this relation, the input–output relation (equation 12.33) and structural resistance R_0 are considered; the coefficients c and k are then functions of the dynamic and strength properties of the structure as well as the seismicity and ground motion attenuation. The only factor not considered explicitly is the probability of occurrence of A at the site. Let ν be the annual rate of earthquake; the annual rate of ground motion exceeding A at the site will be

$$\nu_a=\nu p=\nu cA^{-k} \tag{12.35a}$$

If intensity is used in place of ground motion, it is

$$\nu_i=\nu\exp(-c_i I) \tag{12.35b}$$

where c_i may have values in the range of 1.0–2.0.

Let S_1^n be the event where the damage state S_1 occurred n times in the lifetime T years of the structure. On the basis of a Poisson assumption,

$$P[S_1^n] = (\nu_a T)^n \frac{\exp(-\nu_a T)}{n!} \tag{12.36}$$

The probability of no damage of the structure in T years is then

$$P[S_0] = P[S_1^0] = \exp(-\nu_a T) = \exp(-\nu c T A^{-k}) \tag{12.37}$$

which is an extreme value distribution of type II. The probability of occurrence of damage at least once in T years is finally

$$P[S_1] = 1 - P[S_0] = 1 - \exp(-\nu c T A^{-k}) \tag{12.38}$$

Compared with equation (12.28), this equation is derived by using equations 12.33 and 12.34 for conditional probability $P[S_1|A_j]$ and $P[A_j]$, and by computation of the Poisson distribution for the summation.

Example. Suppose the average annual rate of ground motion exceeding A is

$$\nu_a = 0.003A^{-1.5}$$

where A is the ratio of acceleration to g. The return period for $A \geqslant 0.1g$, 0.29 and 0.3g is $T = 1/\nu_a = 1/0.0095 = 105$, 300, and 550 years respectively. For a building designed for no damage at 0.2g,

$$\nu_a = 0.0003(0.2)^{-1.5} = 0.0035$$

$$P[S_1] = 1 - \exp[-0.00335(50)] = 0.15$$

Similarly, the risk or the probability of occurrence in 50 years of S_1 is 0.09 and 0.06 respectively for a building designed for 0.3g and 0.4g.

(b) Ground motion in design

SHA may be used to define ground motion in design. For example, as a design value A, one can select empirically the ground acceleration corresponding to a probability of exceedance of 10% in the lifetime of a structure. From equation 12.37,

$$P[S_0] = 1 - 0.1 = 0.9 = \exp(-\nu c T A_{0.1}^{-k})$$

$$\nu c T A_{0.1}^{-k} = 0.1054$$

$$A_{0.1} = \left(\frac{\nu c T}{0.1054}\right)^{1/k}$$

If intensity I, equation 12.35b, is used, the risk is then

$$P[S_1] = 1 - \exp(-\nu_i T) = 1 - \exp[\nu T \exp(-c_1 I)] \tag{12.39}$$

(c) Random ground motion and structural strength

When both ground motion and structural strength are random, the damage probability matrix, or the probability $P[S_j|Y]$ of damage state S_j for given ground motion Y, is

$$P[S_j|Y]=\int_0^{\infty} f(R)\, P[S_j|R,\ Y]\,\mathrm{d}R \tag{12.40}$$

where $f(R)$ is the probability density function of the structural strength R, which is usually assumed to be log-normal.

Yin (1991) performed SRA for several kinds of typical structures, for which there are enough damage data to deduce their damage probability matrices. For example, for multistory (two to six) unreinforced brick buildings, he calculated the structural strength by the following equation:

$$R=\frac{\sum_{l}^{n} i^2}{\sum_{l}^{n} i \sum_{j}^{n} i}\,\frac{\sum_{j}^{m} F_i}{2A_j}\,\tau(1+\sum C_k) \tag{12.41}$$

where F_i is the cross-sectional area of the ith wall at the jth story; A_j is the floor area of the jth story; τ is the design shear strength of the wall; and C_k is the modification factor of the kth item important to the quality of the building, as given in Table 12.9. The first factor in equation 12.41 considers the variation of story shear in the building for a reversed triangular distribution of inertial force, where i is the floor number, j the floor considered, n the total number of stories and m the number of walls of that floor. R is in kg/m^2.

On the basis of analysis of damage data, Yin found the relation between the mean calculated strength and the damage state as shown in Figure 12.4. Under the assumption of normal distribution of the strength of the buildings of each damage state, the probability of a building with strength R in the interval a to b at the jth damage state S_j for given earthquake intensity I is

$$P[S_j|R,\ I]=P[S_j|a\leqslant R\leqslant b,\ I]=\Phi\left(\frac{b-R_j}{\sigma}\right)-\Phi\left(\frac{a-R_j}{\sigma}\right) \tag{12.42}$$

where R_j and σ are respectively the mean strength and standard deviation of the jth damage state. Tables 12.10 and 12.11 give examples for unreinforced brick multistory buildings.

Table 12.9 Modification factor C_k

Item	Yes	No
Wall spacings satisfy the code requirements	0	− 0.1
Rigid floor	0	− 0.1
Obvious quality defects	− 0.2	0
Regular plan and elevation layout	0	− 0.1
Code TJ 11–78	+ 0.3	0
Code TJ 11–74	+ 0.2	0

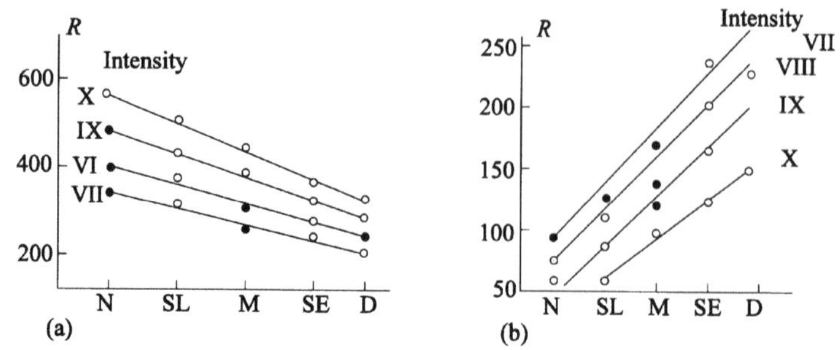

Figure 12.4 Relation between damage state S and building strength R: (a) brick building; (b) factory building. N = no damage; SL = slight; M = moderate; SE = severe; D = destructive.

Table 12.10 Damage probability matrix for unreinforced brick multistory buildings, I = VIII

R(kg/m^2)	$P[S_j\|a \leq R \leq b, I = \text{VIII}]$ (%)				
	S_0	S_1	S_2	S_3	S_4
≤ 100	–	–	–	–	100
101–150	–	–	–	13	87
151–200	–	–	4	27	69
201–250	–	–	17	47	36
251–300	–	10	40	30	12
301–350	16	32	39	13	–
351–400	30	44	16	2	–
401–450	61	35	4	–	–
451–500	79	21	–	–	–
> 500	100	–	–	–	–

Table 12.11 Damage probability matrix for unreinforced brick multistory building, I = IX

R(kg/m^2)	$P[S_j]\|a \leq R \leq b, I = \text{IX}]$ (%)				
	S_0	S_1	S_2	S_3	S_4
≤ 150	–	–	–	–	100
151–200	–	–	–	13	87
201–250	–	–	–	31	69
251–300	–	–	14	54	32
301–350	–	6	41	44	9
351–400	6	25	54	15	–
401–450	26	49	25	–	–
451–500	52	42	6	–	–
501–550	74	26	–	–	–
551–600	90	10	–	–	–
> 600	100	–	–	–	–

12.5 Earthquake zonation

Earthquake zonation started about 60 years ago, and up to the 1980s, all major seismic countries had their own zoning map. Although the purpose of zonation remains nearly the same, the content and methodology have changed gradually.

12.5.1 Purpose and content of earthquake zonation

The purpose of earthquake zonation is to show the seismic hazard distribution in a large region. There are roughly three kinds of earthquake zonation for different purposes and with different indexes. The first is regionalization of seismicity, which depicts a large region divided into provinces of different seismicity, including the rate or frequency, magnitude and energy of earthquakes, with the purpose of showing the geographical distribution of seismicity for the study of seismic hazard and tectonic structure distribution. From this geologists may determine crust structure; seismologists may understand and predict earthquakes; and engineers may evaluate seismic hazard.

The second is damage zonation, which depicts a large region in terms of damage caused by earthquakes, including casualties, engineering damage (both structural and ground) and economic loss, with the purpose of providing data for government and society, in order to plan the best use of land, and for disaster mitigation and relief. This kind of zonation should provide maps to show the geographical distribution of various kinds and levels of damage. Government and society are concerned with disaster reduction; seismologists with earthquake prediction accuracy; and engineers with reduction of engineering damage.

The third is ground motion zonation in terms of amplitude, spectrum and duration of ground motion, together with the probability of their occurrence. It is the zonation of seismic hazard, used primarily for engineering purposes in the design of new structures and the evaluation and retrofitting of existing structures. Engineering requirements have changed greatly from a very simple intensity with no probability meaning in the 1950s to multiple parameters such as simultaneous statement of peak values of acceleration, velocity and displacement or spectrum, with a statement of probability of occurrence such as 10% probability of exceedance in 20, 50, 100 and 500 years to satisfy the requirements of multiple level design. For engineers, seismicity zonation is a prerequisite, damage zonation is a rough prediction of the result, but ground motion or seismic hazard zonation forms the basis of their design.

The advantage of making a zoning map is to have a group of specialists to concentrate on the problem and to provide for related institutions or government a general view of what is known on the hazard of earthquake distribution of the whole region so that land-use planning, city planning, site selection and seismic designing and retrofitting of engineering works can be based upon it. The mapping of a very large region has to be rough or averaged with some important factors not considered, for example the site conditions; for very critical engineering works, such as nuclear power plants, an individual SHA has to be made.

Figure 12.5 shows schematically the prerequisites of seismic zonation, the zonation and its application.

12.5.2 Seismicity zonation

Here seismicity is used to mean some parameter related to the earthquake itself, to differentiate it from the other two forms of zonation in terms of

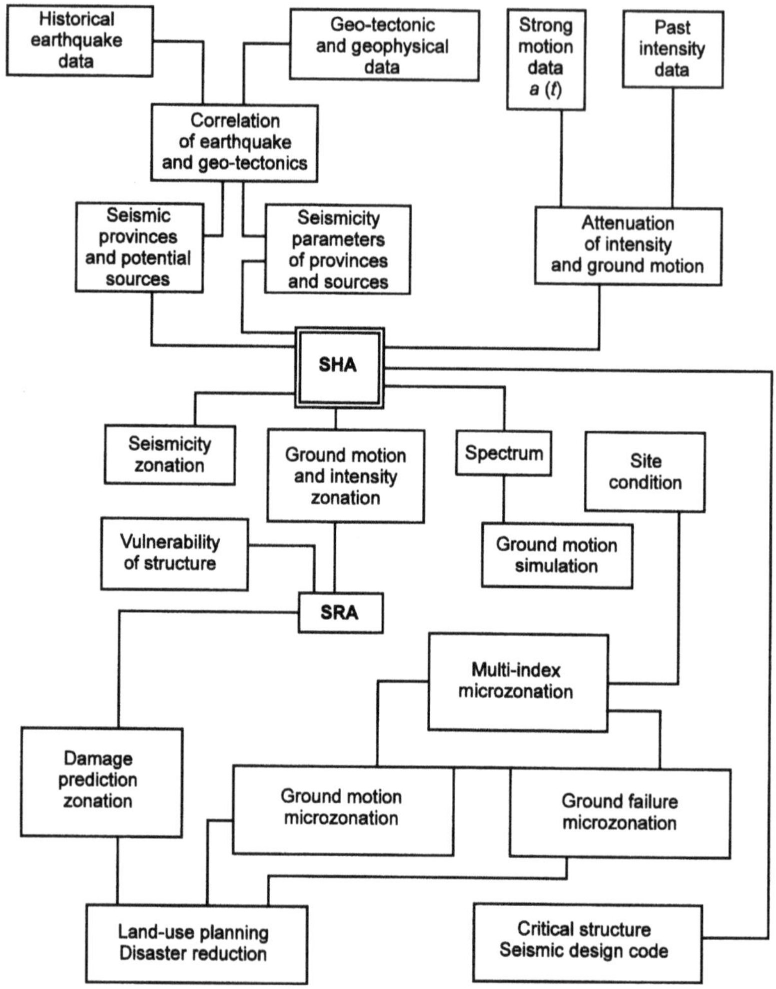

Figure 12.5 Seismic zonation and its relation to other work.

the effect of the earthquake. For example, Figure 12.6 is a map giving the possible maximum magnitude of earthquakes in the period 1973–2073 in North China, which is an intermediate result of the current intensity zoning map of China and obtained from a deterministic approach given in section 12.2.2. Figure 12.7 shows potential province or source zoning used in the Western United States. The potential province or source zonation in probabilistic SHA, in which the seismicity is assumed uniformly distributed, may be considered as seismicity zonation.

Lomnitz (1974, Chapter 12) suggested a map of historical seismicity, which is a contour line map of the sum of the released energy within a given period of time in a certain area. The energy is converted from magnitude M by Richter's relation

Figure 12.6 Seismicity map of northern China (1977).

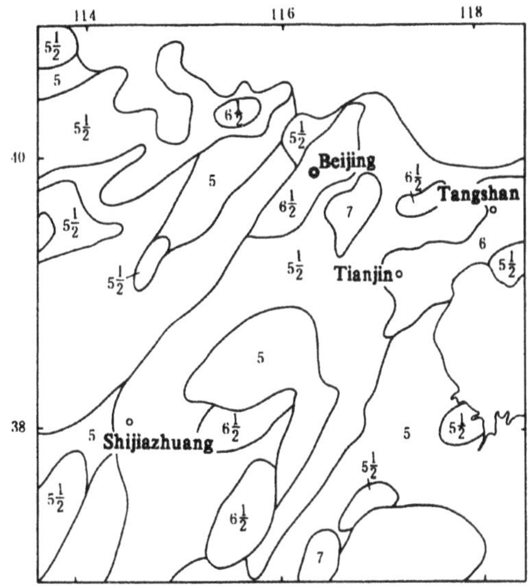

$$\log E = 1.5M + 11.4$$

This kind of map has not been used in the engineering field.

12.5.3 Intensity zonation

Intensity zonation was the form of seismic zonation to be used for engineering purposes, intensity being the oldest popular way to represent ground motion. Seismic zonation of this kind was first studied in the former Soviet Union and became systematic in 1947. The 1949 seismic zonation map completed by Gorshkov specified five zones respectively for intensities V–IX and was approved as a legal map for engineering design. In 1948, Ulrich of the United States Coast and Geodetic Survey (USCGS, the forerunner of the US Geological Survey, USGS) suggested a seismic zonation map of the USA, dividing the nation into four zones, 0, 1, 2 and 3, corresponding to no damage, slight, moderate and heavy damage, or intensity <V, V–VI, VII and ≥VIII respectively. The zonation by number or letter, called engineering zonation, is common worldwide. In 1950, the USCGS published this map, but withdrew it in 1952. However, it was adopted in the Uniform Building Code (UBC) in 1955 and its later editions. Figure 12.8 is the 1969 edition. In 1959, Richter followed the Soviet Union's approach and suggested an intensity map of the United States with zones of intensities V, VI, VII, VIII, IX (occasional) and IX (often), but it was not used in practice. The zoning map currently used in the USA is that in Figure 12.8.

The earthquake zonation map of Russia has been an intensity map since 1937, with seismic zones VI, VII, VIII and ≥IX. The 1937 edition was prepared by the Institute of Geophysics, USSR Academy of Sciences, and

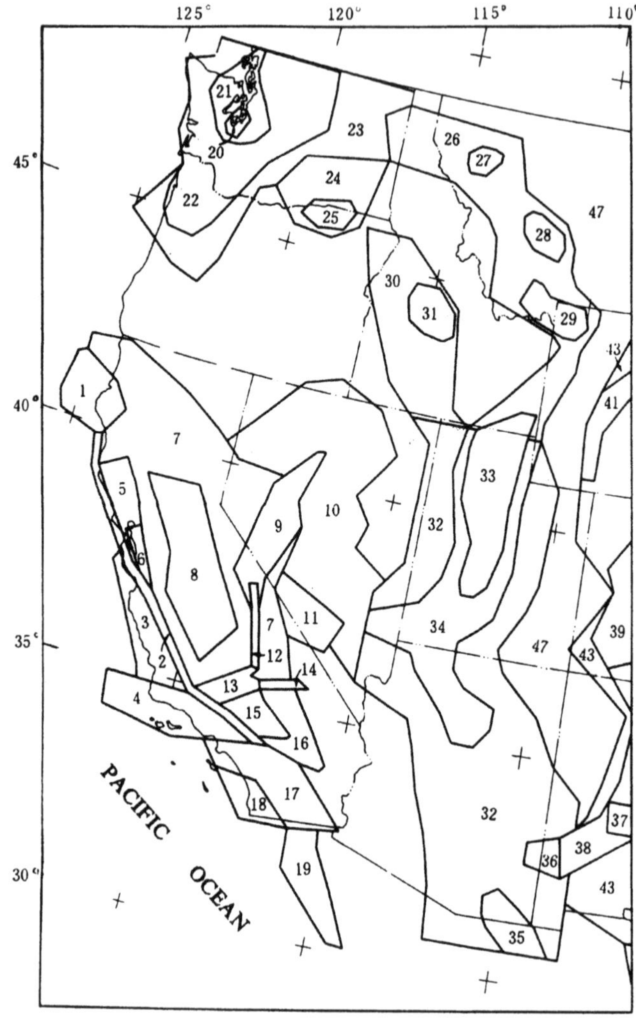

Figure 12.7 Seismic provinces of western USA.

adopted in the official seismic design code. Their SHA is deterministic, and the zoned intensity has no explicit probability indication.

Intensity zonation may be considered as a form of damage zonation because intensity is defined primarily by damage or the effects of an earthquake. But on the other hand, it may be considered as a ground motion zonation because it has been used widely as a design load specification. For example, intensity VII means $0.13g$ and intensity IX means $0.5g$ in some seismic design codes. The former is reasonable and sound, but the latter is also used. The reason for the existence of intensity zonation is the lack of acceleration data in many regions in the world and also the persistence of tradition.

Figure 12.8 Earthquake zoning map of USA, in use since 1955.

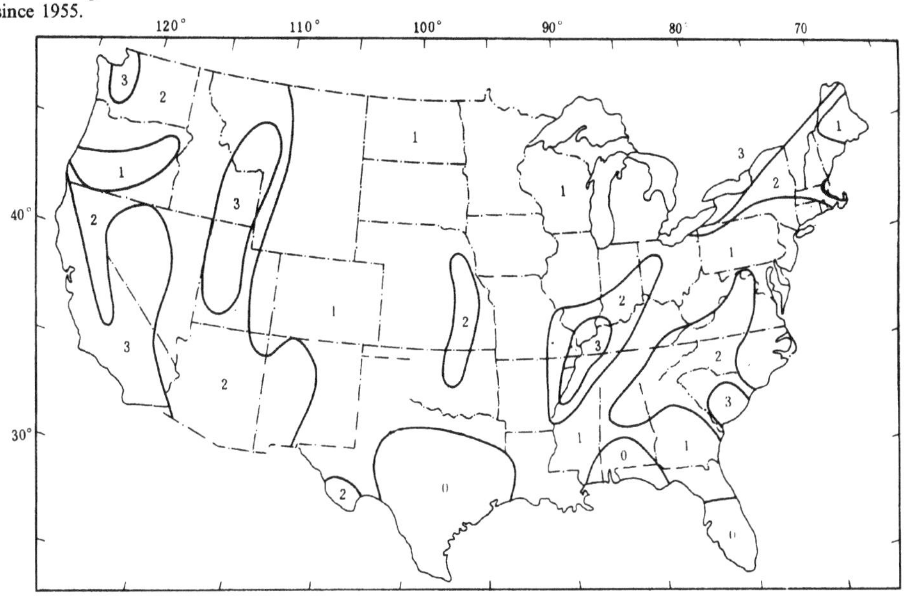

12.5.4 *Ground motion zonation*

(a) Indirect ground motion zonation map

The early editions of ground motion zonation maps were obtained from ground motion attenuation converted directly from intensity through a relation between intensity I and acceleration A,

$$I=f_1(A)$$

This was then substituted into an intensity attenuation law

$$I=f_2(M, R)$$

where M and R are respectively earthquake magnitude and distance, to obtain the following converted ground motion attenuation:

$$A=f(M, R)$$

The result is almost the same as that obtained by converting the intensity from the intensity map into ground motion.

One convenience of this map is that the final zonation is given as a physical quantity which can vary continuously in any fashion to satisfy engineering requirements, especially necessary for regions of very strong earthquakes.

An indirect ground motion map can be either an acceleration or a velocity map, depending on how intensity is related to ground motion; it is not possible to have a map for a second parameter, such as response spectrum ordinates.

(b) Ground motion zonation map

This zonation map is obtained directly from ground motion attenuation based on strong motion observation data, or on those from a reference region (section 5.5.3). This kind of map was first completed in the United States, and later in Canada, Japan and a few other countries, including China recently.

Before 1976, zonation maps in the United States were based on intensity, as shown clearly in Figure 12.8. In 1976, the USGS published its ground motion zonation map (Algermissen and Perkins, 1976), which was in terms of bedrock acceleration contours of 0.04, 0.1, 0.2, 0.4 and 0.6 g with a probability of exceedance of 10% in 50 years, and an acceleration inside the highest contour of 0.8g. In 1978, the model code ATC-03 (ATC, 1978) published ground motion zonation maps of effective peak acceleration (EPA) and velocity (EPV). Ground motion zonation with two parameters EPA and EPV with a set of probabilities of exceedance is perhaps one of the best possible ways to satisfy different requirements of engineering problems (Basham *et al.*, 1982; FEMA, 1988).

The idea of multiple parameters in place of a response spectrum led to a set of ground motion zoning maps, such as the ATC, EPA and EPV maps. This kind of zonation map first started in Japan; Kanai and Suzuki (1968) completed a bedrock velocity (v_0) map of Japan with return periods of 75, 100 and 200 years with contours of $v_0 T_g^{1/2}/5 = 1$, 3, 4, 5, 6, 8 and 10, where the predominant period T_g is in seconds and bedrock velocity v_0 in centimeters per second. Ohsaki *et al.* (1980) suggested a bedrock velocity map of Japan with return periods of 50, 100 and 200 years, which is considerably greater than that of Kanai's, due to differences in attenuation laws: Ohsaki used attenuation from strong motion records obtained in Japan while Kanai used the earlier weak motion records in mines. Hattori (1980a,b) completed several sets of ground motion zonation maps, not only for Japan but for many parts of the world; he used parameters of bedrock velocity (0.5, 1, 2, 5, 10, 15, 20, 30, 40 cm/s) and soil acceleration (0.02, 0.05, 0.1, 0.15, 0.2, 0.3 and 0.4g).

Oliveira (1980) completed ground motion zonation maps of Portugal with parameters of acceleration, velocity and displacement of return period 1000 years, because of the weak seismicity. Durate (1980) also completed a set of ground zonation maps for Portugal; in addition to acceleration a, velocity v and displacement d, he considered also the strong motion duration T_d together with the effect of different combinations of magnitude M and distance R on the combination of a, v, d and T_d; this was a new type of zonation map.

Katayama (1980) is thought to have been the first to complete a set of zonation maps of Japan with a series of response ordinates for periods from 0.2 to 4.0s with return period of 75 years. In a workshop on Ground Motion Parameters for Seismic Hazard Mapping (Whitman, 1989), it was recommended to have zonation maps of the United States with many ordinates (perhaps at $T=0.1$, 0.2, 0.5, 1, 2 and 5 s) of the response spectrum and several probability levels (perhaps 0.02, 0.01, 0.002, 0.001, 0.0005 and 0.0002 per year); it was also recommended to adopt them gradually first in the National Earthquake Hazard Reduction Program (NEHRP) Provisions and then in UBC. This was only the recommendation of a group of specialists, but it may be an indication of the trend in zonation. The 1981 Japanese Building Code, the 1982 API code and the 1989 Chinese Building Code have adopted the two-level design principle and the ATC-03 model code, 1985 Canadian code and others have adopted the two-parameter motion zonation method for the specification of response spectrum. This seems an initial step towards ground motion zonation using multiple parameters with multiple probability levels.

12.6 Microzonation

Microzonation is sometimes used as a brief term for earthquake or seismic microzonation. For example, the International Conference on Earthquake Microzonation was called the International Conference on Microzonation before 1982 for the first two conferences. After the third in 1982, this conference was not held again until 1991.

Seismic microzonation is one of the most active branches of earthquake engineering, and its features are described below.

12.6.1 Purpose and content of microzonation

For important engineering works such as nuclear power plants, high dams, critical industrial facilities and major city structures, it is wise to carry out microzonation, which includes not only the evaluation of the effects of local site conditions but also a more detailed evaluation of SHA because of the critical consequences of damage. The results of microzonation will be used for structural design in place of the more general, rough SHA given by zonation.

In comparison with seismic zonation, microzonation deals with a small area, such as a city. Microzonation considers the local variation of site conditions in a background environment of given seismicity, while zonation considers only vague average site conditions; microzonation usually uses multiple indices, including acceleration, velocity or response spectrum, as well as different kinds of ground failure (sand liquefaction, slope slide and settlement) to serve engineering purposes directly, while zonation usually uses only intensity or acceleration, or two parameters, acceleration and velocity; microzonation serves the engineering design of specific structures while zonation serves only ordinary structures.

12.6.2 Intensity microzonation

Medvedev (1962), a Russian engineering seismologist, suggested the principle and systematic method of seismic intensity microzonation in 1952 (section 4.5). From his experience of earthquakes, he suggested that intensity on solid sites is lower by one grade and intensity on weak sites is higher by one grade than that on average sites for which the intensity is specified by seismic zonation. Microzonation therefore classifies a city into three site conditions: solid, average or moderate and weak sites; and the result is respectively $I-I$, I and $I+1$, where I is the intensity from the zonation map. The modified intensity is called the site intensity. Many cities in Russia and some in other countries (such as Beijing in China in the 1950s and Skopje in former Yugoslavia) have been microzoned this way.

This approach is simple and comparable with damage levels experienced. Its disadvantage is that it is too simple to suit the requirements of modern development. The velocity and displacement of ground motion are indeed highest on weak ground and lowest on solid ground, but not the acceleration, response spectrum or duration of shaking; on the other hand, the ground failure, rather limited to weak sites, should not be treated by increasing the design value of intensity, which results primarily in strengthening the superstructure and is not a good measure to prevent foundation failure.

12.6.3 Site microzonation

Site microzonation of a city is performed in terms of site classification. Based on dynamic strength and properties, engineering sites are usually classified into a few classes, say I, II, III and IV as in China Code GBJ 11–89 and the ATC-03 model code.

Japan and China were among the earliest group of countries to complete site microzonation maps for city or industrial areas. Early in the 1940s, the Japanese realized the important influence of site conditions on damage to structures (section 4.5). In 1951, their seismic building design code specified different seismic forces for different buildings on different sites (section 14.2.1). In the 1960s, Kanai suggested a method of site classification according to amplitude and frequency content of ambient vibration at the site. This method of microzonation has had international influence and has been applied to some cities in Chile and former Yugoslavia; and the 1972–74 Program of the United Nations has recommended it together with Medvedev's method.

The first method of microzonation in China started in 1964 when a seismic design code (draft) was published (section 14.2.3). In that code draft, the effect of site conditions was considered by modifying the design spectrum instead of intensity or acceleration. This method has been applied to several cities and industrial areas, such as Beijing and Tianjin, and to many other projects.

12.6.4 Earthquake-related microzonation

The two kinds of microzonation mentioned above are actually site microzonation, i.e. the effect of site conditions is specified separately from, and can be added to, the result of zonation which is the effect of seismicity. However, it is now known that the site effects and seismicity effects are correlated. For example, the shape of the response spectrum was assumed constant in the early zonation maps and site microzonation assumed a constant spectrum shape for each site condition, which means they are independent; but the new trend of zonation considers that the spectrum is a coupled function of both site conditions and earthquake environment. If this is considered, there are two ways to microzone a city as follows.

A simple way is to express the site effects as a site spectral transfer function $H_s(\omega) = G_s(\omega)/G_0(\omega)$ or the ratio of site spectrum $G_s(\omega)$ and bedrock spectrum $G_0(\omega)$. If the zonation can provide the bedrock spectrum, and the site spectral transfer function can be given for each site type, the final result is then $G_s(\omega) = G_0(\omega) \cdot \mathrm{H}_s(\omega)$. In finding the transfer function $H_s(\omega)$ linearity of site response is assumed.

A complicated way is to find the site spectrum $G_s(\omega)$ with consideration of nonlinear behavior of the site under strong earthquake motion as follows.

1. Find the response spectrum $G_0(\omega)$ and the time envelope function of bedrock ground motion from a comprehensive SHA for the city, considering the seismicity of the environment.
2. Generate a time history $a_0(t)$ of bedrock motion complying with the given ground motion parameters.
3. Find the site responses ($G_s(\omega)$ and others) of the sites, considering the specific site conditions together with the nonlinear dynamic properties of the site materials.
4. Microzone the city into subsections of similar response spectrum shape and peak acceleration or with several microzonation maps of ordinates of different periods of the spectrum.

This kind of microzonation has been performed in the United States, Japan and China. Hays from the USGS (Hays *et al.*, 1978) completed such a microzonation map for Salt Lake City, and provided microzonation maps of the site spectral transfer function $H_s(\omega)$ for one- and two-story buildings of period $T=0.1$–0.2 s and for three- to seven-story buildings of $T=0.2$–0.7 s. Kobayashi and Kagami (1972) published the results of microzonation of Kawasaki City, which lies between Tokyo and Yokohama, in terms of the acceleration spectrum $S_a(T)$ for $T=0.3$, 0.5 and 1.0 s; and the required ductility factor at various limit strengths ($Q_y/\mathrm{m} = S_a/\mu = 0.3$) and the design seismic coefficient $k = a_{max}/g$ (with $\mu = 3$ and 5) of a structure with $T=0.3$ s and return periods of 50 and 100 years. Kitagawa and Matsushima (1980) illustrated their method by an example of an application to Sendai City, in terms of the ductility factor μ for a degradated bilinear SDF system with different initial stiffness under given ground motion acceleration a_{max} and velocity v_{max}. Zhou *et al.* (1984) completed the response spectrum microzonation of Beijing, with maps in Figure 12.9 for peak acceleration a_{max} and predominant periods corresponding to three kinds of typical structures, namely, four- and five-story brick buildings with period $T=0.28$–0.32 s; 12- and 13-story framed buildings with $T=0.65$–0.75 s; and 30–45 m tall brick chimneys with $T=1.0$–1.2 s. It is seen that the high-frequency spectral ordinate and peak acceleration maps are quite similar, as they should be.

12.6.5 Multiple microzonation

The USGS suggested a multiple microzonation approach in the first International Microzonation Conference in 1972, modified later in 1975 and again in the second International Microzonation Conference in 1978. The

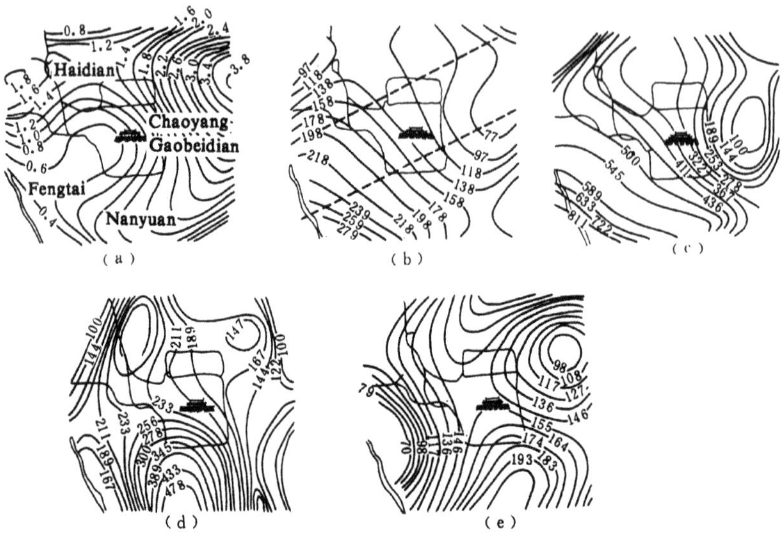

Figure 12.9 Seismic microzonation maps of Beijing: contours of (a) predominant period of ground; (b) peak acceleration (cm/s^2); (c) acceleration spectrum for $T = 0.28$–0.32 s; (d) acceleration spectrum for $T = 0.65$–0.75 s; (e) acceleration spectrum for $T = 1.0$–1.2 s.

first microzonation map (Figure 12.10) in China was completed for Lanzhou City in 1981 (Sun and Chen, C1982).

This type of seismic microzonation requires three sets of microzonation maps as follows for a city to satisfy different requirements of the engineering field.

(a) Site condition maps

These maps show the distribution of geological data within the area, which includes:

1. depth of recent deposits together with blow counts of standard penetration and/or shear wave velocity;
2. depth of liquefiable sandy soil;
3. location of old rivers and lakes;
4. predominant period of site;
5. contour lines.

(b) Ground motion microzonation maps

Similar to response spectrum microzonation, this approach provides microzonation maps of several ordinates of the site response spectrum corresponding to several periods to satisfy the design of different structures, or to provide maps of site acceleration or velocity together with a microzonation map of spectrum shapes.

(c) Ground failure microzonation maps

Several maps of different possible ground failures are required to satisfy the requirements of foundation design and site selection. The following maps are usually required:

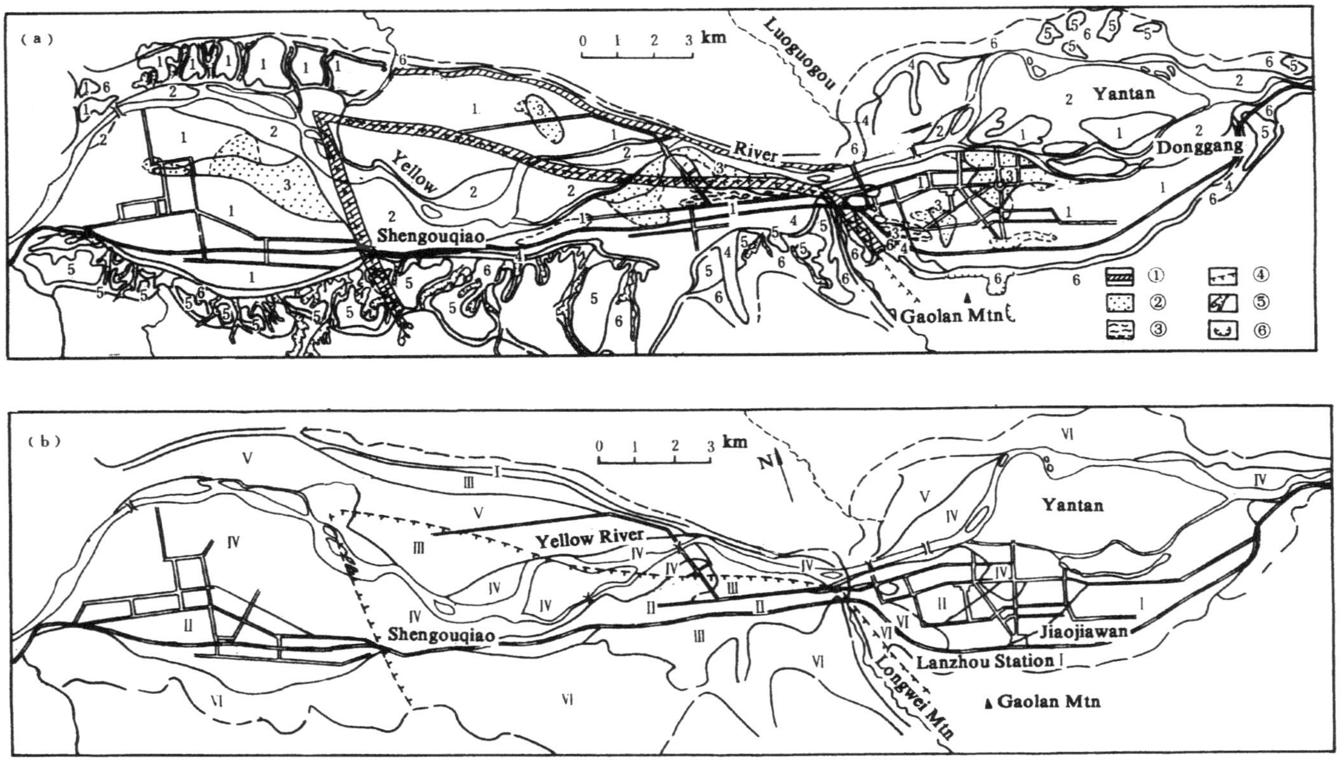

Figure 12.10 Microzonation map of Lanzhou City.

1. active fault types, their location, width and expected relative displacement;
2. liquefaction potentials;
3. slope slides.

13 Principles of seismic design of structures

Optimization goals in engineering design are approached through empirical methods, and the same is true of seismic design. In large scale structural design the results of different trials are compared and the best selected. The optimum quantity should be least cost or maximum benefit. In practice, however, there is no specific optimum quantity or function, and 'optimum' is mostly a subjective and fuzzy idea based on past experience. Engineers seem satisfied with the safety level of their present design, which is usually taken as a standard in their evaluation of safety of new designs. Perhaps because of this implicit nature, safety level is generally not stated in quantitative terms and the users of design codes do not deal with it explicitly.

Another point neglected by the users of design codes is the approximation of some quantities they use to control the design, which may be simply a tradition, but actually these quantities may vary because of the factors not considered in the code. For example, most codes specify base shear and a standard response spectrum for the design level of earthquake motion and allowable strength and deformation for the structure; but actually, according to current understanding, the spectrum shape may change greatly for different earthquakes and a specification of a maximum acceleration in base shear and a standard spectrum cannot cover the important changes in real cases. Also, the effect of duration has not yet been considered in any design code, but it is an important factor in nonlinear behavior and collapse of structures. Similarly, one set of allowable strengths and deformations cannot describe different levels of the capacity of a structure to satisfy different functions of the structure. For example, the functioning state of a structure is controlled by the elastic limit but the ultimate state by the ductility of the structure. To design a structure safely and economically, the designer needs to realize the factors which affect both the load and the resistance.

13.1 Developments and trends in seismic design theory

According to the considerations of both the earthquake motion and dynamic properties of structure, theory of seismic design of structure may be divided into three stages: static which covers the first half of this century; response spectrum, which started in the 1930s; and dynamic, from roughly 1970 onwards. A review of these stages help understanding of the trend of development.

13.1.1 Static theory

Although activity may be traced back to the nineteenth century, it is commonly considered that earthquake engineering is a 20th century development (Housner, 1984). The static theory of seismic design started in Europe and developed in Japan. European engineers studied strong earthquakes early in the 19th century and introduced terms like intensity, seismology, epicenter etc. After field investigation of some strong earthquakes in Italy, some European engineers suggested designing structures for a static horizontal force equal to a few percent of the weight of the structure (Housner and Jennings, 1984). This idea was brought to Japan and the Japanese engineers put it into practice. Omori, at the end of the 19th century, in his report 'Investigation of Overturning of Brick Columns and Column-Type Objects', pointed out that the earthquake force on a structure might be written as

$$P=\frac{W}{g}a_{\max}=kW \tag{13.1}$$

where W is the weight of the structure and $k=a_{\max}/g$ the seismic coefficient, which was taken as one-twelfth or one-eighth. Japanese engineers Sano and Naito applied this method to the design of some buildings in Tokyo before the 1923 earthquake and showed success. Martel designed a building the same way in the late 1920s in Los Angeles. Since then, both in Japan and the United States, official codes adopted this approach, and became popular worldwide.

This fraction of the weight of a structure is the current seismic coefficient $k=a/g$, where a is the earthquake acceleration. The static method and the seismic coefficient k both originated from field experience, and the acceleration is an effective value. For example, Omori from Japan obtained the conclusion from analysis of field damage data that the threshold value of k for structural damage is about 0.3, and if a factor of safety for structures under ordinary load is 3, a design seismic coefficient $k=0.1$ is appropriate.

Although the dynamic nature of earthquake response was realized at the beginning, the difficulty in dealing with the problem of complex irregular vibration meant that the static method was adopted for simplicity, and because of the rigidity of most structures at that time. This method is based on the assumption of a rigid structure, so that the design coefficient k is the same for all kinds of structure.

At this stage, European and Japanese engineers made important contributions to earthquake engineering development; but in the early 1930s, engineers in the United States started to work on strong earthquake motion observation, an important step in the development of seismic design theory.

13.1.2 Response spectrum theory

Although Ishimoto (J1931a–c, J1932, J1935) had started strong earthquake motion observation in Japan at the beginning of the 1930s, the United States government set up a steady and well organized program a year after Suyehiro gave his famous lectures on engineering seismology in the USA, with special emphasis on strong motion observation, and the first acceleration record was obtained in the USA in 1933 at Long Beach (Matthiesen, 1980). The most famous record was obtained at El Centro in 1940 (section 5.2.1).

Up to the 1940s, there were a considerable number of strong motion records of significant relevance in the engineering field obtained in the USA, which led to the application of the response spectrum theory of structural analysis and design in the 1950s.

The response spectrum theory was the product of a group of engineers at the California Institute of Technology (CIT). Although it had been suggested for the analysis of structures by Biot and Benioff in the early 1940s, it was some years before Housner and his colleagues computed response spectra using an analog torsional pendulum instrument from real earthquake accelerogram records. This idea was adopted in the 1952 seismic design specification of lateral forces (section 14.2.2), and this may be considered as the beginning of the response spectrum stage of seismic design theory.

The response spectrum theory combines the spectral properties of ground motion and the dynamic properties of the structure into a simple form:

$$V_0 = k\beta(T)W \tag{13.2}$$

where $\beta(T) = S_a(T)/a_{max}$ is the standard or average acceleration amplification spectrum of the ground motion and V_0 the base shear of the structure.

At the beginning of this stage in the 1950s, seismic design was performed on the basis of linear theory of structural response; but in the 1960s, the nonlinear theory of structural analysis followed the realization of the importance of nonlinear behavior of structures, especially the energy absorption capacity of a structure after initial damage but before its collapse, and limit design was popular in static design of structures, which had a strong effect on response spectrum theory. A group of engineers at the University of Illinois, headed by Newmark, modified the ground motion peak values, such as acceleration, velocity and displacement into the nonlinear response spectrum by introducing a new factor μ, the ductility factor (Section 9.7.2). This study continued up to the 1980s when Newmark passed away.

Another important advance in the 1960s is the design consideration of site conditions in terms of response spectral shape. This led to the formulation of the following ideas.

1. Foundation or ground failure of weak sites should be considered separately from the superstructure design.
2. Ground motion should be described by a spectrum, and different spectral components have different effects from site conditions; e.g. soft sites increase the amplitude of the longer period components.

A typical example is the specifications of these two conclusions in the 1964 Seismic Design Code (Draft) in China (Section 14.2.3) and also in the later codes.

13.1.3 Dynamic theory

In the 1960s, there were other important developments in earthquake engineering in addition to response spectrum theory, such as the following.

1. Wide application of computer techniques and experimental methods to find the nonlinear behaviour of structures in strong earthquakes helped the understanding of failure mechanisms of structures.
2. Many critical and special engineering works such as nuclear power plants, offshore platforms and pipelines were newly constructed; there was therefore no field damage experience for these structures, and their safety depended mainly on analysis.
3. Earthquake motion records were rapidly accumulated.
4. Several strong earthquakes had occurred in the 1950s and 60s which provided some strong motion records, together with new experience on tall buildings, industrial buildings, dams, foundations and pipelines, some of which were heavily damaged.

The above gave a large impetus to the development of a dynamic theory of structural design in the 1970s.

The seismic design or analysis of a structure considers the following four factors: input ground motion, a dynamic model of a structure and its members, a practical dynamic response method of analysis and a design principle. These four factors are greatly simplified in static theory to:

1. a simple peak acceleration for complicated motion without consideration of spectrum and duration;
2. a simple elastic model;
3. linear analysis for a nonlinear structure;
4. a simple allowable strength for safety checking.

Response spectrum theory brings the following modifications.

1. Input motion is still a peak acceleration, but with a standard spectrum which does not vary with earthquake environment.
2. The structure is considered simply as linear with a global ductility factor, and two-dimensional.
3. The members are checked by elastic or limit-state design.

Finally, dynamic theory introduces the following.

1. The input ground motion is a one- or three-dimensional time history $a(t)$ with probabilistic definition.
2. Models of structure and its members may show nonlinear hysteretic behaviour.
3. Dynamic analysis considers all the responses in the time domain with nonlinear deformation and energy dissipation.
4. The design requires consideration of multiple levels of no damage, functioning and no collapse, and the safety of the structure is finally in terms of probability.

The consideration of these four factors is complicated but is closer to real cases; the result is more reliable.

In structural modeling, the dynamic method models the nonlinear behavior of each member on the basis of test results, in contrast to other methods which model only the whole structure.

13.2 Principle of selecting the design level

The principle of seismic design, similar to other structural design, is to design a structure for maximum safety with minimum cost. Ideally, this principle may be expressed in the following way:

$$E = \text{Benefit} - \text{Initial cost} - \text{Probable loss} = \text{max.} \tag{13.3}$$

where benefit includes the direct and indirect gains from the structure; initial cost includes not only the cost of the design structure but also the gain which may be obtained if the money is spent in other ways; loss includes the expected loss from the building, casualties and other effects. Because of the difficulty in evaluating the indirect loss and in combining casualty and monetary losses, equation 13.3 is not easy to apply.

If nonstructural loss is not considered, the principle may be written in another way:

$$C = \text{Initial cost} + \text{Repair cost} = \text{min.} \tag{13.4}$$

Figure 13.1 plots this relation. It is seen that the repair cost is decreased if the structure is made stronger with more initial constructional cost and the expected cost C has a minimum or an optimum. In the figure,

$$\text{BD} = \text{AB} = \text{constructional cost}$$

$$\text{BC} = \text{DE} = \text{repair cost}$$

$$\text{BE} = \text{BD} + \text{DE} = \text{C} = \text{total cost}$$

The state corrcsponding to (or near) point E on the curve of total cost will give the optimum choice. Because the occurrence of an earthquake is a random event, the damage and thus the repair cost is also uncertain. Suppose P_y is the probability density function of ground motion y, and B_y the cost to repair the structure to the original strength corresponding to y; the expected repair cost is then

$$\int_0^{y_{\max}} B_y P_y \mathrm{d}y \tag{13.5}$$

where $y_{\max}$ is the maximum ground motion possible at the site.

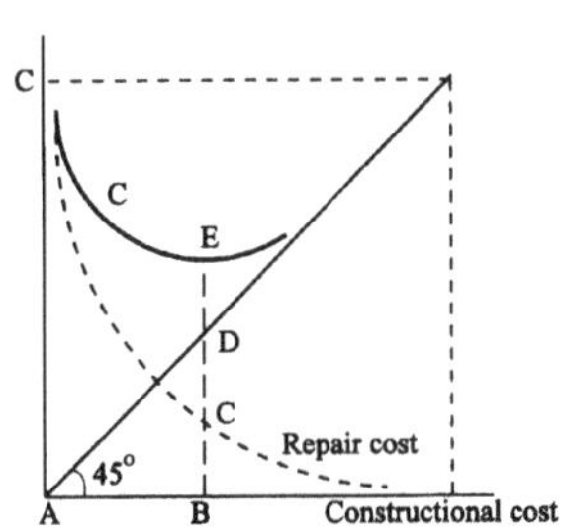

Figure 13.1 Total constructional cost.

The principle mentioned above sounds nice but is not always preferable, for the following two reasons. Firstly, money cannot cover all the important factors of concern. For example, it is not easy to express the benefit of, or society's need for, a structure in monetary terms, especially the need 10 or 20 years later. Secondly, seismic design is not always a controlling factor.

Present thinking is to assume that the current level is acceptable, and optimization of monetary benefit is not performed.

13.3 General guidelines for seismic design

On the basis of past experience, the following general guidelines are usually agreed to be of vital importance; violation usually means greater earthquake risk.

13.3 Site selection

Least seismic hazard and stable ground should be the first considerations in the choice of a construction site. Firstly, sites with potential ground failure, such as those on or below a slope which may be subject to slope slide, should be avoided. Secondly, differential settlement of foundations may easily occur on loose and soft sites, such as ground of saturated fine sand and silt, very soft clay and recent fill. If a structure is built on such ground, it is difficult or very expensive to provide the superstructure with sufficient safety.

13.3.2 Regular and uniform structure layout in plan and elevation

Both in plan and in elevation, the structural layout of dimensions, mass, stiffness and ductility should ideally be regular, uniform and symmetric, to avoid sudden changes. Nonsymmetric shapes in plan, such as L, winged shapes, such as T, H, U and Y, and irregular features, such as the penthouse, stepped roof and flexible first floor in elevation, will increase the torsional response, local vibration and/or stress and deformation concentration, and thus introduce damage to the structure.

13.3.3 Sufficient ductility or energy absorption capacity

Damage to a structure comes from the energy transmitted to it in the form of vibration. In order to increase the safety of structures against earthquakes, one possible measure is to prevent or reduce the energy transmitted into the structure by designing the structure to have a natural period away from the predominant period range of input motion, by means of base isolation or selection of structural layout. Another measure is to increase the damping or the ductility of the structure to absorb the energy already input to the structure, by means of artificial damping devices or construction materials, or by structural detailing to prevent local failure of joints and members to assure large plastic deformation with no strength degradation. Simply increasing strength with no attention to deformation is not good seismic design practice.

13.3.4 Design of equal safety

Ideally, designs should have all members and joints of nearly equal resistance to collapse so that no part of a structure is much stronger or weaker, in terms of final global failure. A design involving weak beams and strong columns is preferred, because when a column reaches its yield point, the P–Δ effect will produce a fast strength degradation which brings the whole structure to failure. Equal resistance does not mean all members of a structure will fail at the same time, but means that all members have nearly the same resistance to the worst possible attack.

13.3.5 Multiple lines of defence

In terms of seismic resistance, a statically indeterminate structure is better than a statically determinate structure because failure of one defence line does not mean the failure of the whole structure; local failure dissipates some part of the energy input to the structure and saves the structure from total collapse.

13.3.6 Prevention of brittle and instability failure

Brittle and instability failure is the opposite of ductility. Ductility means to deform with no significant decrease of strength, while brittle or instability failure means to deform with a large instantaneous drop, or complete loss, of strength. Common brittle failures are cracks in masonry, shear failure of reinforced concrete beams and short columns and failure of unconfined concrete in compression; common instability failures are buckling of main reinforcements in reinforced concrete columns with insufficient confining stirrups, local buckling of thin-walled steel members and buckling of columns under large lateral drift of upper floors.

13.4 Special requirements of various types of engineering works

13.4.1 Buildings

Because of the design principle 'no damage to occur in frequent earthquakes, repairability under probable maximum earthquake and no collapse in maximum possible earthquake', the inherent requirement of building is then nonlinear considerations in design.

An important achievement in nonlinear structural analysis in the 1960s was to design buildings for ductility or energy absorption capacity, in the form of a structural coefficient in the code, which specified a structure of large energy absorption capacity designed for a smaller earthquake load, which thus has a lower yield limit.

13.4.2 Highrise structures

The highrise structure has a height much more than its horizontal dimensions, a fairly simple symmetrical plan construction and a long fundamental natural period, say a few seconds. For this type of structure, the long-period components of the input motion require special attention. In the following, the example of chimneys is given and most of the features are applicable to other highrise structures.

(a) Field damage experience

There is a great amount of field experience of chimneys (Qian, C1983a, b), and the following may be concluded.

1. Unreinforced brick smokestacks are easily damaged.
2. There are various possible patterns of damage. For unreinforced smokestacks, these include: falling to one side in a long segment or to all sides in pieces; some with the top portion fallen but some with it twisted and remaining on top; some with several horizontal cracks, some with inclined cracks and some cracked irregularly (Figures 13.2(a) and (b)).
3. Damage to smokestacks is common not only in mesoseismal areas but also away from the epicenter, with the top portion fallen or badly cracked in mesoseismal areas, but mostly cracked, especially with several closely spaced horizontal cracks, in areas far away.
4. Damage occurs mostly in the upper portion of the smokestacks, and mostly with horizontal cracks.
5. The elevation where damage mostly occurs seems correlated with ground stiffness.
6. Witnesses have observed smokestacks to twist and vibrate violently in all directions like dancing snakes, sometimes with repeated opening and closing (or even hammering) of the top portion on the bottom one.

(b) Effect of higher-modal responses (Hu and Hu, C1983; Hu, Hu and Chen, C1985)

Under ordinary conditions, the predominant period of ground motion is close to the second or third natural period of the smokestacks in the horizontal direction but always to the first in the vertical direction. The higher modal response of the smokestacks is then of more importance in

Figure 13.2 Damage to unreinforced brick smokestacks: (a) horizontal offset; (b) overturning of top portion.

the horizontal direction than in the vertical. If consideration is given to the fact that the horizontal ground motion is usually greater than the vertical, the higher modal response in the horizontal direction is relatively more important than the fundamental modal response in the vertical direction. Numerical results support this analysis (Liu and He, C1983; Liao and Wang, C1983; Zhou *et al.*, C1983).

13.4.3 Bridges

(a) Field experience (IEM, C1979; ATC, 1979; Liu, C1985–86)

Many bridges have been damaged by earthquakes; highway bridges have shallower foundations and are thus more heavily damaged. The following experiences have been repeatedly confirmed in strong earthquakes.

Slide of abutments and piers. Abutments are located near river banks, and some piers on the slopes of the river channel. Where there are liquefiable soil layers underground, part of the foundation near the bank may slide due to gravity, together with the abutment and pier, toward the river channel. This kind of damage is typical on alluvial plains.

Figure 13.3 shows a sketch of the damaged bridge in Tangshan City over the Dou River. Slope slide and disturbance of the river bed were found near the bridge; evidence shows that the banks slid toward the river channel and the distance between abutments shortened by 3.5 m. Analysis identified that the damage to the bridge was initiated by the liquefaction of the saturated fine sand layer underneath the abutments.

Figure 13.4 shows the damage to a railway bridge over the Dou River. This bridge of three simple spans of 16 m each and piers of plain concrete, with piles 24 m long, was located in area of intensity X. After the earthquake, two abutments moved 3.7 m toward the river channel but the distance between piers was not changed. Pier no. 2 broke at a constructional joint and the bridge was close to collapse.

Girder drop. Drop of superstructures, girders or beams, is another common and serious type of bridge damage. The cause of girder drop is the

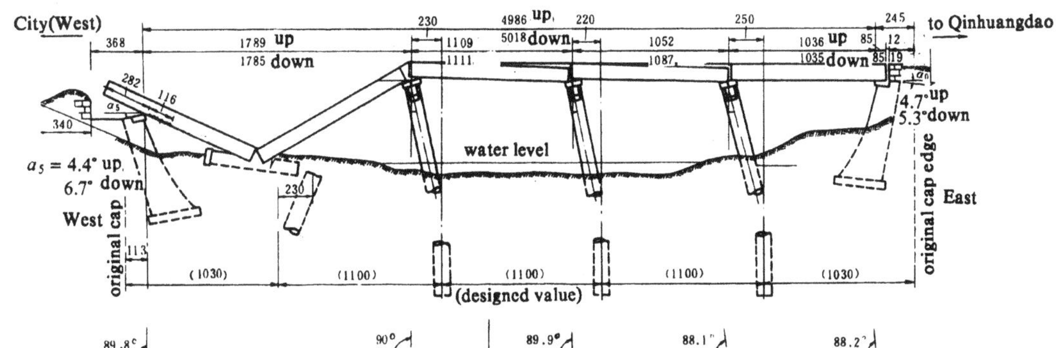

Figure 13.3 Failure of Victory Bridge in Tangshan City.

Figure 13.4 Failure at pier no. 2 of a Railway Bridge over the Dou River.

large relative displacement of the superstructure from its support, mostly due to permanent displacement of the abutment toward the river channel which pushes the superstructure, breaks the pier and the girder drops (Figure 13.4), or anchor bolts shear off and the other girder is pushed off the pier (Figure 13.5), as in the case of railway bridge no. 55 over the Ji Canal in the Tangshan earthquake. Although less common, girder or beam drop may also be induced by dynamic relative displacement in the longitudinal direction. A highway bridge over the Luan River of 35 simple spans of 22.2 m each, made of reinforced concrete T-beams, with piles bearing on hard sandy gravel layers, was in the intensity VIII area and perhaps not seriously damaged by the 1976 main Tangshan earthquake, as it was opened for traffic; but it was badly damaged the same day by a strong aftershock, although still in an area of intensity VIII; spans nos 2–24 all dropped off their piers but no slope or foundation slide was found (Li, 1982).

Damage to piers and columns. Damage as shown in Figure 13.6 is common on hard ground, which is similar to the failure of ends of building columns, with concrete inside the main reinforcements crushed because of weak confinement, and followed by the buckling of the main reinforcements.

(b) Design features of bridges

Bridges are important transportation links, and play a vital role after a strong earthquake in ensuring that disaster relief measures can be carried out. It is then a special design requirement that temporary transportation over the bridge is guaranteed, either immediately after a strong earthquake, or as soon as possible if damage cannot be prevented. Emergency repairs, such as temporary piers, for relief transportation is possible if the superstructure does not fall.

Effective measures to prevent the collapse of the superstructure are:

1. to consider the possibility of liquefaction of the foundations and, if necessary, to consider an adequate type of foundation, such as deep or pile foundations;
2. to increase the continuity of the superstructure, widen the pier cap by at least one meter, and provide additional flexible connections between girders of neighboring spans.

(c) Current code specifications (ATC, 1979)

United States (ATC, 1979, pp. 115–40) The 1977 California code specifies clearly that the effects of the following factors should be considered: distance from the causative fault, maximum credible earthquake (MCE) on the active fault, site soil conditions, dynamic response of the bridge and measures to reduce design forces and increase ductility. It requires also the flexibility to incorporate possible future developments in earthquake engineering.

The code specifies the following four ground motion parameters:

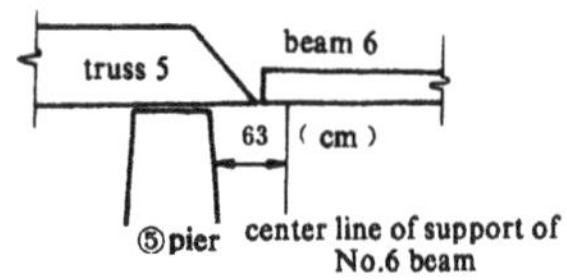

Figure 13.5 Damage over Pier no. 5 of a railway bridge over the Ji Canal.

1. bedrock peak acceleration A_a, determined from the seismicity of the active fault and attenuation of ground motion; the code gives the map of

Figure 13.6 Damage to a bridge pier in the 1971 San Fernando earthquake.

contours of the MCE bedrock acceleration A_a of California, where $A_a \leqslant 0.7g$;

2. amplification response spectrum R of bedrock acceleration, based on actual records, given in Figure 13.7;
3. site soil spectrum S or amplification of different thickness (10–80 ft (3–24 m), 81–150 ft (24.3–45 m), and > 150 ft (45 m)) of alluvial deposit and different acceleration levels (0.1g and 0.7g) for different period T, also determined from actual records;
4. hazard/ductility factor Z, derived from damage data, given in Figure 13.8.

The product $A_a SR(T, A_a)$ of the first three factors gives the elastic response spectrum of the site from the MCE on the closest active fault, which remains constant when $T \geqslant 5$ s; the product of all four factors is the design earthquake spectrum.

There are several new features of this code, but the consideration of the nonlinear effect of acceleration level from 0.1g to 0.7g is the first in any code.

This code emphasizes strongly the importance of detailing for seismic design; it requires checking each detail and evaluating its behavior after the yield limit. It emphasizes also the importance of ductility, and warns that

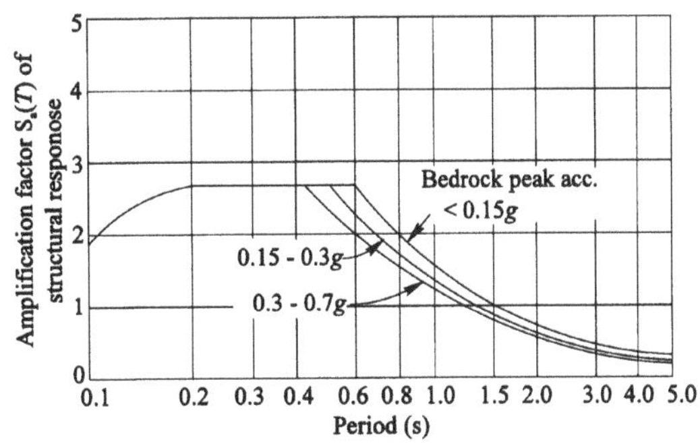

Figure 13.7 Bedrock spectra S_a (T, A_g).

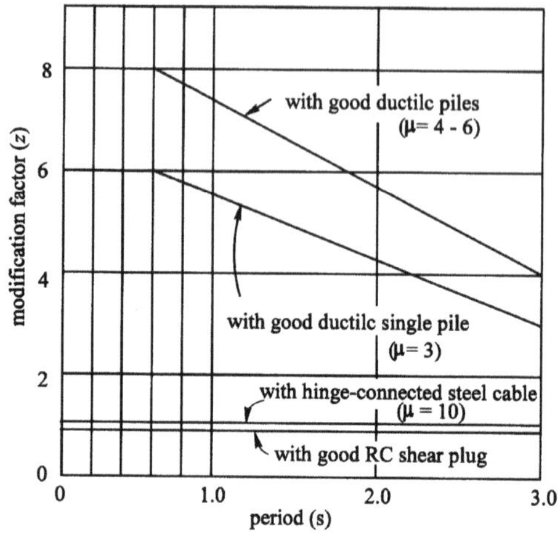

Figure 13.8 Hazard/ductility factor Z.

simply to increase the design force may result in a brittle and weak design; it requires that a bridge should not collapse like a mechanism, but serious damage may be allowed. The following three items are to be checked carefully:

1. control of excessive displacement at expansion joints;
2. detail of columns, with increased stirrups at location of maximum moment;
3. detail of the shear dowels.

In 1977, the Federal Highway Administration Bureau supported ATC to prepare a new code for the United States and a draft outline was completed in 1979 on the basis of the California highway code and that of New Zealand, but more specific; for example, it specifies the minimum length of support of beam overlap. This draft was later adopted.

Japan (ATC, 1979, pp. 43–65). The Japan highway code before 1979 followed closely the Japan building code; the 1979 code made some important modifications. Firstly, a detailed specification on sand liquefaction: scope and formula of liquefaction checking and a reduction of bearing power of zero to two-thirds for liquefiable soil. Secondly, the time history method and response spectrum method of analysis to be used for bridges of special type, on very soft ground or when a special study of ductility is required. When the time history method is used, the ground motion may be evaluated in one of two ways:

1. estimation of the earthquake coefficient from the lifetime of bridge and the return period of the earthquake;
2. estimation of the ground motion from the potential source and magnitude of the earthquake.

13.4.4 Storage tank on ground

(a) Field experience

The most important form of damage caused by storage tank failure is the indirect disaster induced by the liquid content. For example, during the 1964 Niigata earthquake in Japan, major fires were caused by the petroleum from damaged storage tanks. During the Alaska earthquake in the same year, fires continued for three days. During the 1978 Argentina earthquake and the 1980 central California earthquake, hundreds of storage tanks were damaged. Invaluable lessons may be learned from these cases (Guo, C1982). Local buckling of the thin cylindrical wall of a storage tank is the most common type of failure. There are two types: elephant foot and diagonal chain of diamonds. The former occurs mostly at the bottom of the tank and in a ring or a section of a ring (Figure 13.9(a)); the latter usually near the bottom and goes up diagonally in two branches (Figure 13.9(b)). The tank shown in Figure 13.9(a) was damaged in the 1971 San Fernando earthquake, and had a height of 15 m and diameter of 12.3 m. There were 18 tanks in that area, and those damaged were of a height/diameter ratio not less than 1 and with full liquid content.

During the central California earthquake of 24 January 1980 (magnitude 5.5), the Wente Brothers Winery at Livermore had 177 wine tanks, with diameters $D = 1.8–6.6$ m and height/diameter ratio $H/D = 0.8–3.0$, anchored to concrete pads of height 0.6–1.2 m with a few anchor bolts; some had cooling covers made of steel plate on the outside of the tank near the bottom. The earthquake epicenter was about 13 km away. During the earthquake, 10 were not damaged; 24 slightly damaged and 70 moderately damaged. Statistical analysis shows the following.

1. Empty tanks are not or only slightly damaged;
2. Damage is related to ratio H/D, full elephant foot failure for smaller H/D ratio (<1.5), various patterns of failure for moderate H/D ratio (1.5–2.0) and no, or only slight, damage on walls for large H/D ratio (>2.0).
3. A cooling cover helps to prevent wall from buckling.

Elephant foot. This type of failure is quite common and a storage tank is a not very complicated structure, but the mechanism of this type of local buckling is not yet clear. Many believe it comes from the horizontal vibration of the tank together with its liquid content, i.e. excessive vibration causes uplift of the tank and the compression edge is subject to large vertical compression which is the main cause of local buckling. Because the tank is vibrating in a constantly changing direction, and the local buckling of the wall will shift the compression to its neighboring part, the local buckling will shift and form a ring or part of a ring. Field experience indicates that elephant foot buckling occurs only:

1. when the tank is full of liquid;
2. on tanks placed freely on the ground.

Figure 13.9 Damage to storage tank: (a) elephant foot (b) chain of diamonds.

Chain of diamonds This second pattern of local buckling is seen quite often in the field, and was produced in a shaking table test carried out at EERC in 1980 when a tank from the Wente Brothers Winery was tested to the maximum capacity of the shaking table. The inclined chains of diamonds were clearly shown dynamically, appearing and disappearing on the upper half of the tank alternatively during vibration, but disappearing when the test was stopped. The tank was simultaneously under vibration in one horizontal and one vertical direction. It is possible that chains of diamonds occur when a tank is vibrating primarily in one horizontal direction, but elephant foot buckling occurs when the tank is vibrating in a changing horizontal direction.

(b) Seismic response analysis of storage tanks

Dynamic analysis of tanks with rigid walls. The most popular method, introduced by Housner, deals with an ideal liquid and assumes that part of the liquid content of mass m_1 moves with the tank and another part of mass m_2 moves relative to the tank. The latter part is then simplified to an SDF system of mass m_2 attached to the tank by a spring of spring constant k, representing the effect of the first mode of the liquid oscillation. This model has been widely used because of its simplicity and practicability.

Storage tank placed freely on the ground. Here special consideration has to be given to uplift and local buckling of the wall. Many studies have been carried out in the last two decades, including experiments. Shih (1981) obtained a solution for the compression stress in cylindrical walls under the assumption that the contact area of the bottom plate and foundation during uplift may be simplified as circular, as shown in Figure 13.10:

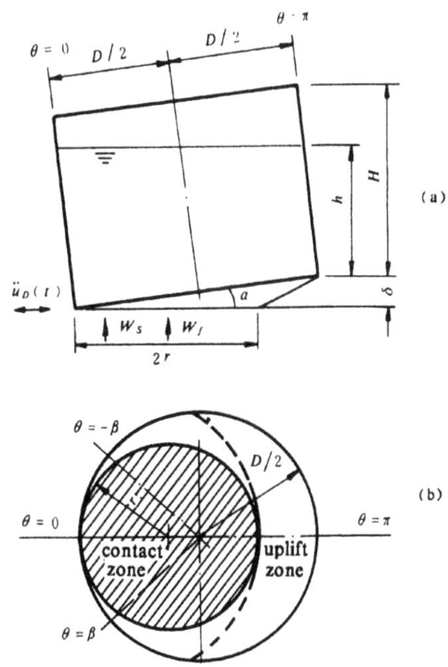

Figure 13.10 Uplift of tank.

$$\sigma_c=\frac{N_c}{t_s}=\frac{W\beta}{Dt_s(1-\cos\beta)}-\left[\frac{2M}{WD}-\frac{4\beta[2-\beta/\pi)}{\pi(\pi-\beta)}\left(\frac{(\pi/2)-1}{\pi}+\frac{\cos\beta+\beta-(\pi/2)}{\pi-2\beta}\right)\right] \quad (13.6)$$

where N_c and t_s are respectively the compression and thickness of wall; $W=\rho gh\pi D^2/4$ the total weight of liquid; the effective contact angle β or radius r depends on the uplift angle $\alpha=\tan^{-1}(\delta/D)$ and the stiffness of tank, especially that of the bottom plate of the tank; the overturning moment at the bottom is

$$M=\frac{8D^2N_c(1-\cos\beta)}{\beta}+16D^2N_t\left(\frac{\pi/2-1}{\pi}+\frac{\cos\beta+\beta-\pi/2}{\pi-2\beta}\right) \quad (13.7)$$

Table 13.1 shows a comparison of the compressive stress in walls from a test, from API Standard 650, and from equation 13.6. The test was made on a tank of height $H=4.57$ m, diameter $D=2.36$ m and wall thickness $t_s=2.29$ mm. $\beta=28°$ according to test. It shows that the API code results are too small, about one-fifth to two-fifths of the test, and that equation 13.6 gives acceptable results, about 0.95–1.75 times of the test.

13.4.5 Underground pipelines

(a) Field experience

Many engineers have published good reviews and summaries (Wang and O'Rourke, 1977; Ariman and Muleski, 1981; Ariman, Lee and Chen, 1987; Kubo, 1975) and the following points are important.

1. Damage is much more serious in weak and/or complicated ground than in hard and uniform ground.
2. Damage occurs mostly at pipe joints or bends.
3. Damage is due to ground deformation, not inertial force.
4. Flexibility and ductility control the earthquake resistance capacity of pipelines, especially the joints.
5. Effect of axial deformation is greater than that of bending. The results of strong motion observations (Ariman and Muleski, 1981) support these conclusions.

Table 13.1 Comparison of maximum compressive stress in walls

Liquid depth h (m)	α (deg.)	Compressive stress σ_c (kg/cm^2)		
		Test	API 650	Eq. 13.6
3.96	4.2	63.3	25.1	112.6
3.96	5.6	140.6	33.5	167.5
3.96	6.5	179.3	38.9	202.8
3.96	8.5	295.3	61.9	281.5
3.96	6.0	160.2	35.4	183.2
3.96	6.0	77.2	23.4	111.4
2.13	6.0	35.3	10.4	40.2

(b) Theoretical analysis

Static theory. A long underground pipe may be simplified as a beam on an elastic foundation, whose equation of motion is

$$\rho A_0 \frac{\partial^2 w}{\partial t^2} + EI \frac{\partial^4 w}{\partial x^4} + k_b w = k_b w_g(x, t) \quad \text{(transverse)} \tag{13.8a}$$

$$\rho A_0 \frac{\partial^2 u}{\partial t^2} - EA_0 \frac{\partial^2 u}{\partial x^2} + ku = ku_g(x, t) \quad \text{(axial)} \tag{13.8b}$$

where ρ, E, A_0 and I are respectively the material density, elastic modulus, sectional area and moment of inertia of cross-section of the pipe; k_b and k are the spring constants of the pipe foundation in the transverse and axial directions; w and u are the transverse and axial displacements of the pipeline; and w_g and u_g are the transverse and axial ground displacements. The solutions of equation 13.8 are

$$w(x, t) = \frac{1}{D_b} w_g(x, t) \tag{13.9a}$$

$$u(x, t) = \frac{1}{D_u} u_g(x, t) \tag{13.9b}$$

where $1/D$ are amplification factors. If the stiffness of the pipe is large, the natural frequency $\omega_0 = (k/\rho A_0)^{1/2}$ of the pipe on elastic springs will be much greater than the predominant frequency of ground motion for pipes in soft soils, and because the damping of soils surrounding pipes is large, the dynamic response of the pipe can be neglected. A static solution is then obtained as follows:

$$D_b = 1 + \left(\frac{EI}{k_b}\right)\left(\frac{2\pi}{L}\right)^4 \quad \text{(transverse)} \tag{13.10a}$$

$$D_u = 1 + \left(\frac{EA_0}{k}\right)\left(\frac{2\pi}{L}\right)^2 \quad \text{(axial)} \tag{13.10b}$$

$$L = vT, \; T = \frac{2\pi}{\omega}$$

where v is the wave velocity along the pipeline and T the period of ground motion. The bending and axial strains of the pipeline due to ground motion are then respectively

$$\varepsilon_b = \left(\frac{2\pi}{L}\right)^2 \frac{1}{D_b} u_0 r_0 \tag{13.11a}$$

$$\varepsilon = \left(\frac{2\pi}{L}\right) \frac{1}{D_u} u_0 \tag{13.11b}$$

where

$$u_0 = \frac{a_{max}}{\omega^2} = \frac{a_{max} T^2}{4\pi^2} \tag{13.12}$$

and a_{max} are respectively the amplitude of displacement and accelerations of ground motion; and r_0 is the external radius of the pipe. For long pipes of very small radius, from equations 13.10, $D_b = D_u = 1$, and finally

$$\varepsilon_b = \left(\frac{2\pi}{L}\right)^2 u_0 r_0 \tag{13.13a}$$

$$\varepsilon = \frac{2\pi}{L} u_0 \tag{13.13b}$$

which means that the pipe moves together with the free ground, i.e. the static theory. The ratio of axial to bending strain is

$$\frac{\varepsilon}{\varepsilon_b} = \frac{L}{2\pi r_0} \tag{13.14}$$

It is seen that, for a large ratio L/r_0, the axial strain will be much greater than bending, which is consistent with damage experience.

Under the assumption that the pipeline moves together with the free ground, the foregoing relations can be derived as follows:

axial strain, $$\varepsilon = \frac{\dot{u}}{v_p} = \frac{2\pi u_0}{vT} = \frac{2\pi u_0}{L} \tag{13.15a}$$

bending strain, $$\frac{\varepsilon_b}{r_0} = \frac{\ddot{u}}{v_s^2} = \left(\frac{2\pi}{T}\right)^2 \frac{u_0}{v_s^2} = \left(\frac{2\pi}{L}\right)^2 u_0 \tag{13.15b}$$

axial strain between two points, $$\varepsilon_L = \varepsilon L \tag{13.15c}$$

angular strain between two points, $$\varepsilon_\theta = \varepsilon L \tag{13.15d}$$

where $\dot{u} = (u_0 T)/(2\pi)$ and $\ddot{u}$ are respectively particle velocity and acceleration of ground motion, and L is the distance between the two points. These results are exactly the same as equation 13.13.

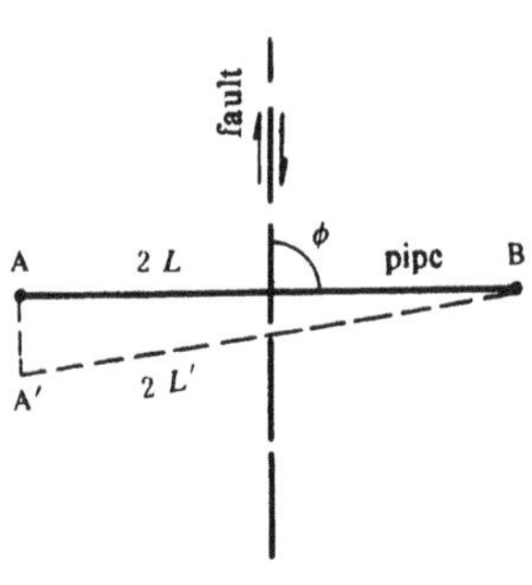

Figure 13.11 Displacement over fault.

Underground pipeline over active fault. When studying oil lines over Alaska, Newmark and Hall (1975) suggested a simple but practical method of design to take very large displacement. As shown in Figure 13.11, suppose that there is a pipeline of length L on each side of an active fault, supported at A and B; the offset of two sides of the fault is D, and the angle between the pipeline and the fault is ϕ. The elongation of segment AB due to fault offset is

$$\Delta(2L) = D\cos\phi \text{ or } \varepsilon = \frac{\Delta(2L)}{2L} = \frac{D}{2L}\cos\phi$$

When $\phi = 90°$, the pipe will lengthen to

$$(2L')^2 = (2L)^2 + D^2$$

or

$$\varepsilon + \frac{\varepsilon^2}{2} = \frac{1}{2}\left(\frac{D}{2L}\right)^2 \tag{13.16}$$

In the general case,

$$\varepsilon + \frac{\varepsilon^2}{2} = \frac{1}{2}\left(\frac{D}{2L}\right)^2 + \frac{D}{2L}\cos\phi \tag{13.17}$$

A first approximation of the solution can be obtained by neglecting the higher order terms of this equation, i.e. $\varepsilon = (D/2L)\cos\phi$, which is then substituted in the original equation for the higher order term, and the final approximate solution is

$$\varepsilon = \frac{D}{2L}\cos\phi + \frac{1}{2}\left(\frac{D}{2L}\right)^2 \sin^2\phi \tag{13.18}$$

which can be used to find the average pipeline strain or the minimum unsupported length $2L$.

(c) Design consideration

Many water and sewage pipeline codes require the following principles to be adopted.

1. Keep away from active faults, or at least avoid the direction parallel to that of maximum deformation.
2. Avoid steep slopes.
3. Use several pipelines of small diameter instead of one pipeline of large diameter to reduce the stiffness and to include redundancy to safeguard against complete loss of functioning.
4. Select materials of good ductility, such as steel, ductile iron, copper or plastics which will take large deformation.
5. Use ductile joints and very long sleeves for sliding joints.
6. Use special flexible joints and design the main pipeline for large ground motion at locations over active faults.

13.4.6 Special requirements of critical or special structures

For nuclear power plants (NPP), offshore platforms, large bridges, high pressure liquid storage tanks, and high dams, special considerations are usually required because of the serious consequences of failure, its special nature and the lack of field damage experience for these structures. In the next few sections, the requirements of some specific structures will be discussed; here only some common features are briefly stated.

Because the safety requirements for these structures are in general high, corresponding design measures are necessary. The first measure of safety is to adopt a two-level design: the lower level is to design the structure to operate normally during and after the largest probable earthquake motion in its lifetime; and the higher level is to design the structure to be safe to society for the largest probable earthquake motion in its lifetime. The lower level is referred to as safe operational or design basis earthquake (DBE, OBE or S1) or earthquake motion, while the higher level is called safe-shut-down, safe-ultimate, or maximum credible earthquake (SSE, MCE or S2). Although a two-level design is considered by many as a reasonable approach, the specific definitions of the two levels are not yet unified. Roughly speaking, the lower level is similar or slightly higher than that for ordinary buildings, say with annual probability of exceedance 10^{-2}–10^{-3}, and the higher level is a design value of ground motion, twice that of the lower level, or with annual probability of 10^{-4}.

Two-level design was first considered in the late 1960s in the United States for nuclear power plants and other kinds of structure soon after. The Federal Bureau of Reclamation requires the design of high dams to be for three levels of earthquake motion: operational basis (safely operational during earthquake), design basis (safely operational after earthquake) and maximum credible (no collapse).

The second measure for high safety design is to consider various possible characteristics of the earthquakes. For example, in the Los Angeles area, important structures such as highrise buildings, offshore platforms, nuclear power plants, large bridges and high dams have been designed according to the following procedure (Housner, 1979).

1. Identify the possible active faults which may affect the site.
2. Estimate the maximum possible magnitude M and closest distance R of the earthquakes on each fault.
3. Select four typical earthquakes from these faults from large distant earthquakes ($M = 8^+$, R = 55 km) to small close earthquakes ($M = 5$, $R = 0$).
4. Find the time histories $a(t)$ for ground motion at sites from these four earthquakes.
5. Calculate the structural responses for each of these four ground motions, and design the structure to be safe for all of them.

The third measure is to adopt a high-accuracy method of analysis, such as the response spectrum and/or step-by-step integration method. The modern trend is to require a direct dynamic method of analysis for the second or third level of design.

13.4.7 Concrete gravity dam

Developments in the last two decades have improved the design procedure for dams, such as: consideration of the interaction of dam, water and foundations; nonlinear analysis; computer programming; and consideration of reservoir earthquakes. This section discusses only the seismic design of concrete gravity dams.

(a) Evaluation of the routine design procedure

Routine design method. The routine method considers a combination of static seismic force, static and dynamic water pressure, and weight. The seismic force on the dam is taken as the product of the dam's mass and acceleration. The dynamic water pressure is obtained from dam motion under the assumption of incompressible water and a rigid dam. In the analysis stage, the dam is assumed to be a simple beam in order to compute the stresses in the dam, which are then compared with the allowable values; no consideration is given to vertical ground motion. In safety checking to prevent sliding and overturning, all forces are considered static.

In the design stage, usually no tension is allowed in the dam; when allowed in a few special cases, the possibility of cracking is not seriously considered. The routine method emphasizes checking to prevent sliding and overturning; for gravity dams, the routine method requires that these safety factors should not be less than 1.

Earthquake damage to the Xinfengjiang and Koyna dams. The Xinfengjiang concrete buttress dam, during the 1962 Heyuan earthquake in China, and the Koyna gravity dam, during the 1967 Koyna earthquake in India, were damaged in a very similar manner. Both earthquakes are considered to be reservoir earthquakes, occurring in a region considered otherwise as of low seismicity. The two dams were in the epicentral areas. The earthquake magnitudes were moderate, being 6.1 and 7.0 respectively, and the intensity was VIII and VIII–IX respectively. Both dams are located in regions where the intensity was considered low; the Chinese one was not seismically designed, while the Koyna was designed for 0.05*g*. The cross-sections are very similar; the damage was therefore also similar. During the earthquake, the maximum acceleration recorded on the Koyna dam near the base was 0.5*g*, and a long horizontal crack occurred at the sharply turning section near the top. Although there was no collapse or loss of water, the crack was so serious that the dam had to be strengthened by concrete buttresses after the earthquake. The Chinese dam was also strengthened, by filling the lower part of the open space between buttresses to make it close to a gravity dam.

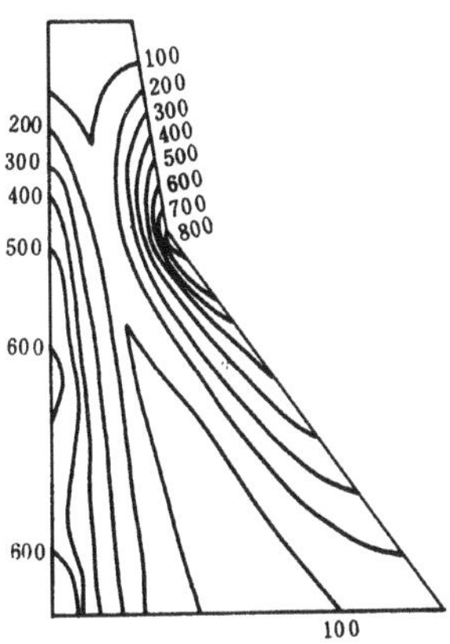

Figure 13.12 Tensile stress (lbf/in^2) in Koyna Dam (1 lbf/in^2 = 6.89 kPa). (After Chopra, 1978.)

Dynamic analysis, including dam–water interaction under both horizontal and vertical ground motion, reveals that for the Koyna dam, under recorded motion, the tension stresses near the section of sudden change of cross-section are rather large, about 40 kgf/cm^2 at upstream face and 60 kgf/cm^2 at downstream face, as shown in Figure 13.12, much more than the tensile strength of concrete, 25 kgf/cm^2.

Limitations of the routine design method. There are at least three defects in the routine method.

1. The design value for the earthquake motion is much lower than the actual.
2. The design principle is not suitable for dams. The routine static method was derived from the experience of building damage and the empirical coefficients may not be applicable to dams.

3. The dynamic water pressure given by the routine method is too low. If the compressibility of the water and the elasticity of the dam are considered, the dynamic water pressure will increase considerably.

(b) Improved method of analysis

The current idea of a suitable method of analysis should consider the effect of the following three factors:

1. horizontal and vertical ground motion compatible with the specific conditions;
2. the dynamic properties of the dam–foundation interaction system, including natural period, mode and especially damping;
3. dam–water interaction.

An accurate method is available on computers software, but is complicated. A simplified method is also available (Chopra, 1978), which uses the ordinary response spectrum, scaled by maximum acceleration, and only the first mode of the dam is considered.

(c) Simplified dynamic method of design

The principle of design for a gravity dam is to allow it to remain in the elastic stage for probable moderate earthquakes; but allow some local cracking, providing the storage of water is not affected and repair is easy, for very strong possible earthquakes. A simplified dynamic method of design is as follows.

Earthquake design levels. Two levels of hazard for the specific earthquake environment are used, one probable maximum and one possible maximum in the lifetime of the dam. Time histories of horizontal and vertical ground motion should be given, either by generating or by modifying some recorded histories. The smoothed horizontal response spectrum may be used only in the initial design stage.

Elastic design. A two-step method is used. In the first step, a simplified method (Chopra, 1978) may be used for the initial design and the allowable values of tensile and compressive stresses are not over. The second step is a dynamic check by an accurate method. The routine analysis of sliding and overturning of dams has no meaning in dynamic analysis and should be omitted.

The tensile and compressive strength of concrete depends on loading rate; for earthquake motion, the compressive strength may be increased by 50%, and tensile strength may be taken as 10% of the compressional strength if a tension test is not performed.

Design for strong earthquake: Minor cracks not endangering the storage function of dam are permitted because a design which does not allow cracking in a very strong earthquake is not only difficult but also uneconomical. A nonlinear analysis of the dam with consideration of

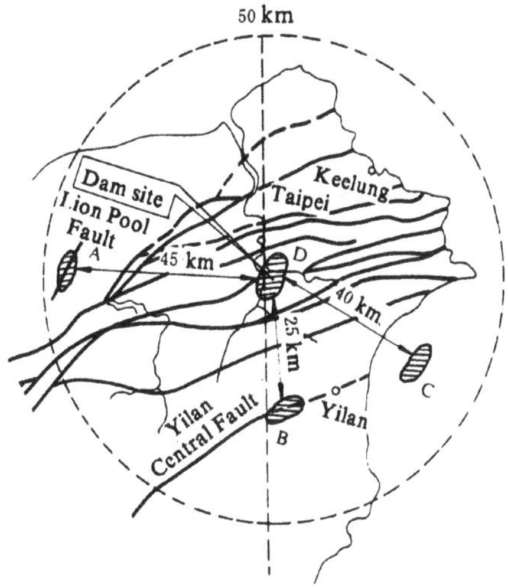

Figure 13.13 Potential earthquake sources of Feitsui arch dam.

cracking is possible but the result is sensitive to the properties of the concrete considered; very careful and dependable determination of these properties is necessary.

(d) Seismic design of the Jadeite (Feitsui) arch dam in Taiwan, China

This dam was designed in the late 1970s; because it is close to Taipei City (Figure 13.13), it was designed with a high factor of safety and with a good design method, which will be discussed below as an illustration of the modern design of dams (Chinese Institute of Civil and Hydraulic Engineering, 1981, Lecture 31, Seismic Design of Concrete Dams, by Y. Chen).

The principle and procedure of evaluation of the design ground motion at the site of the Jadeite dam is shown in Figure 13.14. For dam design, application of both the traditional static method and the dynamic method is required. For the dynamic analysis, the following three levels of ground motion are specified.

1. MCE is defined as the maximum possible earthquake motion at the site, under which minor damage is allowed but with no sudden or uncontrollable loss of reservoir water and with some main facilities maintaining their functioning.
2. DBE is defined as the earthquake motion that may occur once in the economic lifetime of the dam which is taken as the return period of 100 years, under which the dam is repairable after damage and the main facilities can maintain their functioning.
3. OBE is taken as the return period of 25 years, under which there is no loss of function of the dam and other main facilities.

The following design considerations are of interest.

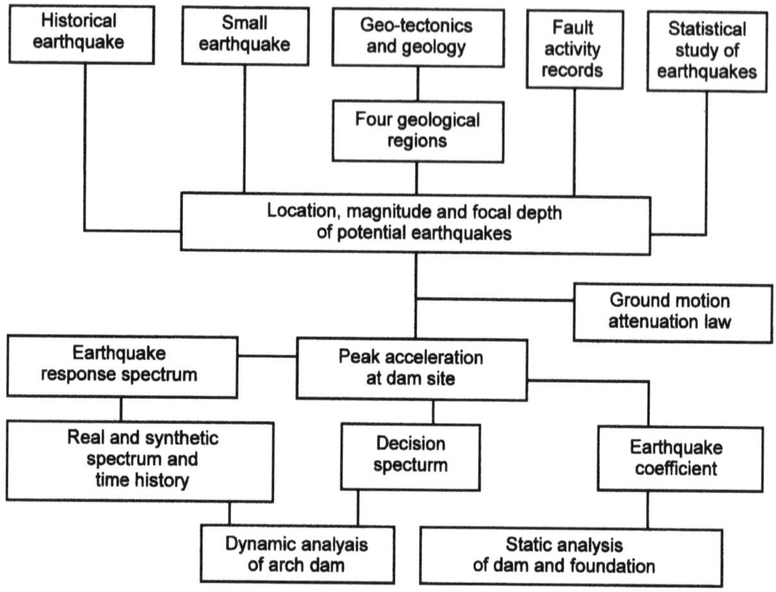

Figure 13.14 Procedure of evaluation of ground motion at the site of Feitsui Arch Dam.

Selection of the static earthquake coefficient. The peak ground acceleration (PGA) obtained for the dynamic analysis is reduced by 50% for static design for three reasons.

1. Real damping is much greater than that used.
2. The dimensions of the dam are very large and the global acceleration is much smaller than the peak value at one point.
3. Phase differences in the motion between points at the base may exist because the dam bottom is at different elevations.

Design allowable values. For concrete in compression, $0.25f_c'$ is taken for normal loads; $0.33f_c'$ (OBE), $0.5f_c'$ (DBE) and f_c' (MCE) for dynamic analysis of earthquake loads; $1.3\times0.25f_c'$ (DBE) for pseudo-static analysis of earthquake loads. Here f_c' is the concrete compressive strength reacted in 365 days. Factors of safety against shear sliding of foundations are taken as 4 (normal loads, OBE and DBE) and 1.33 (MCE) for pseudo-static analysis. Factors of safety against sliding of foundations are taken as 1.5 for normal loads, 1.3 for OBE and DBE, 1.1 for MCE for pseudo-static analysis.

13.4.8 Nuclear power plant

(a) Fundamental principles of seismic design and special features.

The fundamental principles of the seismic design of a nuclear power plant (NPP) concern extreme conservation measures to protect the nearby community from the danger of harmful leakage of radioactive material from the reactor. The disaster of such an event is not only large in terms of economy and human lives but also its long-term effect; the guarantee in the design stage to achieve this requirement by all means is the special consideration in nuclear design. The seismic design has to prevent the failure of radioactive-related structures and piping systems and thus the occurrence of leakage of radioactive material, and to ensure the normal functioning of the parts housing radioactive materials, such as the supply of cooling water, and the implementation of emergency measures against radioactive diffusion, such as safe shutdown devices.

Compared with ordinary buildings, an NPP is different both in terms of seismic experience and design requirements. Firstly, there is no seismic experience of NPPs; buildings can be designed primarily from experience, but NPPs cannot. Secondly, in the seismic design of NPPs the safety of both the structure and the facilities inside the structure must be considered. Thirdly, seismic failure of redundant or spare facilities may be a common-mode failure, i.e. they may fail at the same time.

(b) Earthquake load and design

Seismic design practice in the USA, UK and European countries adopts OBE and SSE discussed in section 13.4.6 as the design load, which are evaluated from the earthquake environment of the specific region in terms of peak accelerations. A standard spectrum is usually specified and scaled according to the accelerations corresponding to OBE and SSE. The one currently used in RG 1.60 of the Federal Regulation 10CFR 50, Appendix A, is the Newmark–Blume–Kapur spectrum of 1973, shown in Figure 13.15. This spectrum is for hard sites only; the vertical spectrum is taken as half of the horizontal. The Newmark–Blume–Kapur spectrum has been used since then universally. Figure 13.15 compares four spectra used for NPPs. It is seen that they became progressively more conservative. The RG 1.60 spectrum is about double the TID 7024.

Although it is required to evaluate the site's specific spectrum, a standard spectrum is commonly used.

In the above method, the structures and systems are classified into three categories according to their importance to nuclear safety:

I. related to nuclear safety;
II. those related to nuclear safety but not included in category I;
III. not related to nuclear safety.

The Japanese method is somewhat different from that of the United States. A recent important change is the adoption of the earthquake-related response spectrum in the Seismic Design Guidelines JEAG 4601 in 1987, in which the Ohsaki spectrum of bedrock motion is specified to change with magnitude and distance, instead of the too-conservative standard spectrum. Another important difference of the Ohsaki spectrum from the standard spectrum is that the former is scaled according to velocity instead of acceleration (Japan Association of Electricity, 1987).

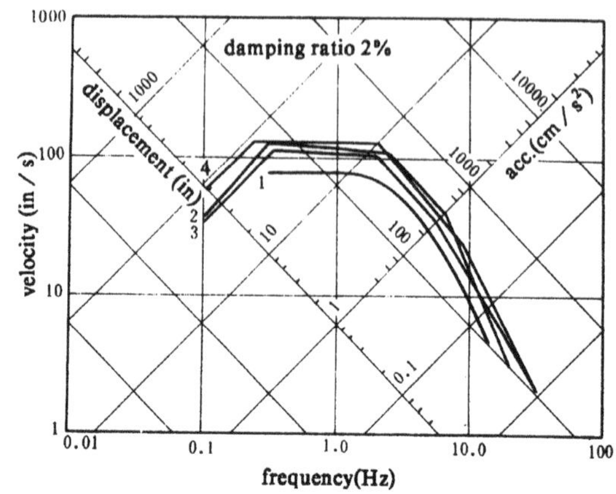

Figure 13.15 Development of nuclear power plant design spectra in the USA: 1 = Housner TID 7024 (1959); 2 = Newmark and Hall (1969); 3 = Newmark and Hall (modified) (1972); 4 = Newmark *et al.* (1973).

In China, a draft code for seismic design of NPP specifies both a standard spectrum and an earthquake-related spectrum.

13.4.9 Offshore platforms

(a) General features

There are two kinds of fixed offshore platform, and the most commonly used one is a three-dimensional truss-type structure with pile foundations, as shown in Figure 13.16; the members are usually steel tubes. The other kind is concrete gravity type, built from a few large cylindrical tubular columns supported on large hollow concrete tank foundations, placed freely on the sea bed.

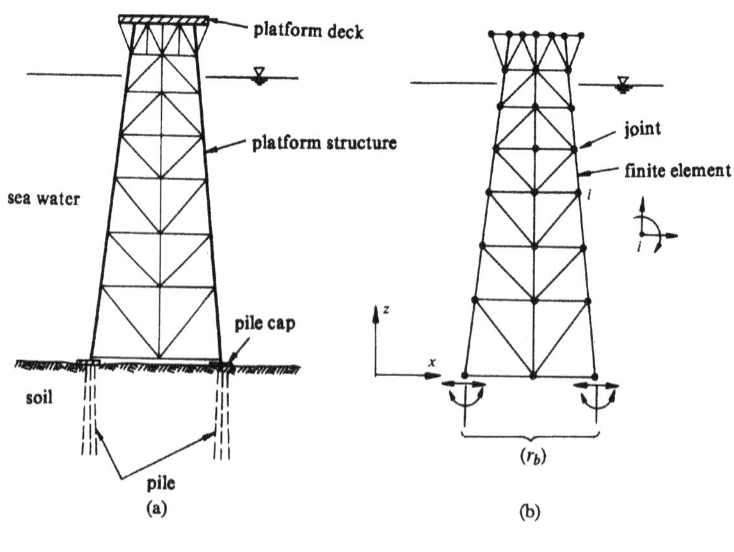

Figure 13.16 (a) Steel truss-type platform; (b) its analytical model.

(b) Design principle

The seismic design of offshore platforms is modern in evaluating design ground motion and in consideration of seismic capacity of the platform. A two-level design principle is used: the first is for no significant damage in a design-specified earthquake, and the second for no serious damage, heavy casualities or serious pollution to the environment in a rare strong earthquake. The API RP2A code, revised each year, includes the following steps:

1. seismicity study
2. site response
3. selection of design criterion
4. dynamic analysis (iteration: steps 4–8)
5. stress analysis
6. evaluation of failure pattern
7. check of ductility requirement
8. detail seismic design.

In evaluation of design ground motion, it is required to include the effect of causative fault type, maximum magnitude, rate of occurrence, ground motion attenuation and site soils; but use of general conditions is also permitted when data are not available. The so-called general condition is taken from the building code to evaluate an effective acceleration according to a seismic zoning map, and select a design response spectrum for the specific site conditions. In analysis, simultaneous action of two horizontal and one vertical component motions of a relative magnitude $1:\frac{2}{3}:\frac{1}{2}$ is required (API, 1982).

(c) Design of steel truss platform with pile foundation

Code API RP2A requires redundant stress transmitting paths to secure a multiple defence line, so that the damage is slowly developed to avoid a sudden collapse when one location is in its plastic stage. Global and detail design should follow this principle.

Dynamic analysis. The response spectrum method and direct integration method are both applicable. Where there are no data, the design spectrum shown in Figure 13.17 may be applied which is normalized to a peak acceleration of 0.1*g*. It should be studied particularly for ground of soft clay, loose sand or silt. Three-dimensional motion should be considered and at least two important modes in each direction together with significant torsional mode.

Ductility check. The API code considers a rare strong earthquake during which the structure can absorb, without collapse, at least four times the energy corresponding to the maximum strain of the soil–structure system during the design-specified earthquake.

Detail design. Similar to ordinary structures, it is also important to design the details to ensure the global ductility of the structure; to design the joints

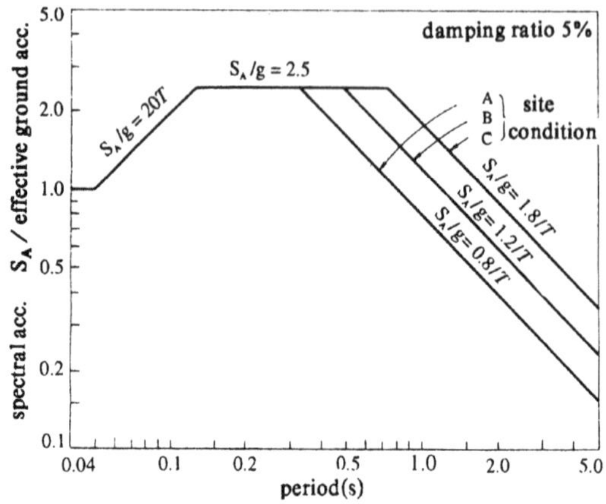

Figure 13.17 Design response spectra for offshore platform ($S_a = (2\pi/T)S_r = (2\pi/T)^2 S_d$).

so that the initial failure starts in members, not at joints, and in cross bracings, not in vertical columns or piles; and to have redundant systems to transmit the forces through other paths when one part fails.

13.5 Seismic constructional measures

There are three categories of seismic measures. In the widest sense these may include social and economic measures, such as the policy of earthquake considerations and disaster reduction planning, and earthquake prediction, such as middle- and short-term prediction. More specifically, these refer only to engineering measures, which include drafting and implementation of design codes, earthquake zonation and seismic constructional measures; only the last item will be discussed in this section.

13.5.1 Lateral bracing systems

Because the major action of earthquake motion is in the horizontal direction, constructional measures for the horizontal direction are the most important. The function of the lateral bracing system is to take a horizontal load or action and is the most important constructional measure. Commonly used lateral systems include the rigid shear wall, the flexible rigid frame and a combination of these such as frames with infill walls, as shown in Figure 13.18. The vertical layouts of lateral systems include continuous, separate and intermittent (Figure 13.21). Different systems have different functions and characteristics. In the following, the properties and seismic capacity of some important systems will be discussed, together with field damage experience. Bracings in the horizontal plane will not be discussed here.

(a) Shear wall

A shear wall is suitable for low and medium-sized buildings; its advantage is that it is rigid, easy to construct and effective in limiting large horizontal

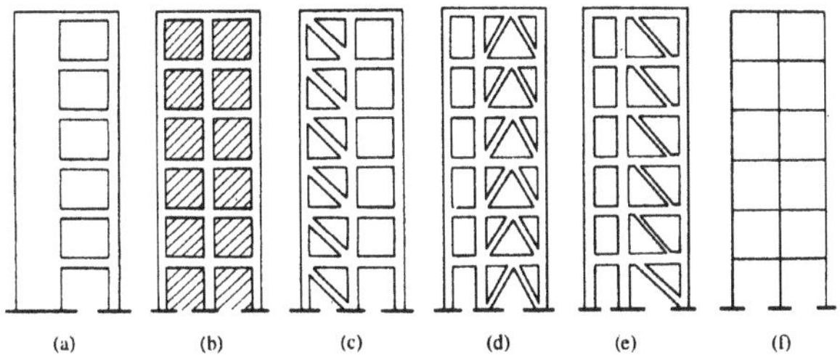

Figure 13.18 Lateral bracing system.

displacement and damage to nonstructural members. A shear wall is also a bearing wall, which takes vertical loads as well as wind load. It is usually made of reinforced concrete and reinforced masonry with good strength and ductility.

Design considerations for shear walls. In resisting horizontal loads, a low shear wall deforms mainly in shear, but a tall one deforms also in bending. If there is no opening or only small openings, the shear deformation is usually small.

For shear walls, especially masonry, some codes include allowable stress design, but this may not be a good procedure. A shear wall has large stiffness and a greater part of the lateral load will be acting on it; the specified acceleration is usually an effective one which is much smaller than the actual. For example, under motion such as that which occurred in the 1940 El Centro earthquake, a shear wall may undergo an actual acceleration of $1.0g$, i.e. $0.3g$ with an amplification of 3, which is too large to expect a shear wall to stay in its elastic range. The design of shear walls should be based on inelastic theory with due consideration to the actual acceleration and the required ductility.

Earthquake capacity of shear walls. This is controlled by two parameters, the strength and the ductility; it is better to have the shear strength somewhat greater than the compressional to prevent shear failure, which is brittle, sudden and of serious consequence.

The distribution of reinforcements should be concentrated at edges with minimum reinforcement over the whole wall to increase both the bending strength and the ductility, but attention should be paid to preventing bond and buckling failure of the reinforcements.

The following is known from the results of tests on reinforced masonry shear walls.

1. There is obvious energy-dissipating capacity below theoretical yield displacement.
2. The strength and stiffness degradations are small and the wall behaves like a reinforced concrete shear wall when $\mu \leqslant 2$.
3. The wall still behaves well when $\mu \geqslant 4$ if shear is not large and there is confinement in the compression zone.
4. Degradation will occur quickly when $\mu > 2$ if there is no confinement.

(b) Moment-resistant rigid frames

Pure moment-resistant rigid frames are generally used in steel structures where members are mainly in bending. When under lateral loads, a well-designed moment-resistant rigid frame having a sufficient number of plastic hinges will fail as a global mechanism. If not properly designed, or under special conditions such as poor workmanship, very low temperature, three-dimensional stress states, low-cycle fatigue and large P–Δ effects, the ductility may greatly reduce and a frame may fail before forming a global mechanism, due to local failure.

Figure 13.19 shows a set of test results for a beam–column under cyclic loading. It is seen that the low-cycle fatigue strength decreases when axial load increases; and local instability (in other cases, perhaps local cracking) is an important factor in inducing low-cycle fatigue.

Compared with a shear wall, a rigid frame has much smaller stiffness, and thus greater lateral displacement. Displacement comes from shear and bending of columns, rotation at beam ends, deformation of beam–column connections, deformation inside the beam–column joint panels, axial deformation of columns and deformation at foundations. For strong-column and weak-beam design, displacement comes mainly from the plastic hinges at ends of beams and bottoms of columns.

(c) Diagonal bracing systems

Because a rigid frame actually has very low overall rigidity and thus very large deformation if not reinforced, it is combined usually with diagonal bracing or a gravity wall, whose function is to resist lateral forces and to provide enough lateral rigidity to the structure, and the lateral moment on the structure is taken by the axial forces in columns. There are two kinds of inclined bracings: diagonal bracing (Figure 13.18(c)) and K-bracing (Figure 13.18(d)). To avoid the difficulty of detailing at the joints and to avoid brittle or instability failure inside the joint panels, the eccentric form of diagonal bracing (Figure 13.18(e)) is suggested, which provides better ductility to the structure and is consistent with the design idea of strong columns and weak beams; but it is important to prevent the beams failing in shear at their ends.

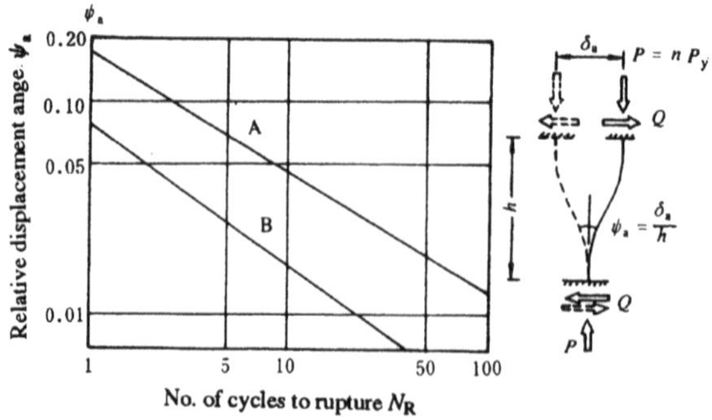

Figure 13.19 Low-cycle fatigue of an I-beam.

(d) Frames with infill walls

Frames with infill brick walls are commonly used in China. If properly designed, they have considerable seismic capacity. The seismic capacity of an infill wall is much better than a wall with no frame. The confinement provided by the frame to the wall increases considerably the ductility of the wall itself.

Stiffness. At the elastic stage, shear beam theory can be used to estimate the shear stiffness of the wall; when deformation increases, the frame in bending and the infill wall in shear do not act as a homogeneous material and local separation may take place (Figure 13.20), and the wall acts like a diagonal strut. The effective width w depends on the relative stiffness of frame and wall, the stress–strain relation of the wall and the magnitude of deformation. When 50% of the ultimate strength of the infill wall is reached, $w=0.25L$, where L is the length of the diagonal.

Strength. The strength of the frame with infill wall is the minimum of the following strengths.

1. Axial strength of columns: under lateral forces, the columns will be stressed by bending moments, especially the corner column, in tension or in compression.
2. Shear strength of infill wall: the ultimate shear strength V_u of each panel is the sum of the horizontal frictional strength V_f of the wall and the shear strength of the columns,

$$V_u = \frac{M_t + M_c}{(h_e/2)} + V_f \tag{13.19}$$

   where M_t and M_c are respectively the ultimate bending strength of the columns in tension and in compression, and h_e is the height between the plastic hinges of each column, or roughly h/2.
3. Compressive strength of the effective diagonal bracing, or

$$R_c = \frac{2}{3} h_\theta t f_m' \sec\theta \tag{13.20}$$

   where f_m' is the compressive strength of the wall, θ the angle of the diagonal with beam, t the thickness of the wall and h_θ the contact length of wall and columns (Figure 13.20)

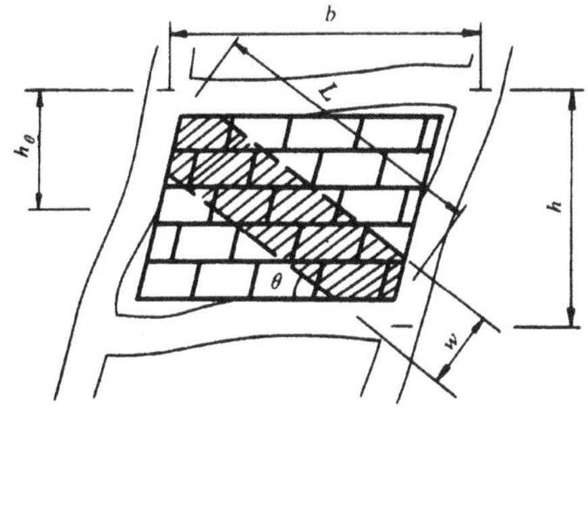

Figure 13.20 Equivalent diagonal bracing on infill wall.

$$h_\theta = \frac{\pi}{2}\left(\frac{4E_f I_f h_m}{E_m t \sin 2\theta}\right)^{1/4} \tag{13.21}$$

where E_f and I_f are respectively the elastic modulus and moment of inertia of the columns, and E_m and h_m are those of the wall.

Ductility. Because of the confinement of the frame, the infill wall can maintain a certain amount of bearing power after cracking, which gives the frame with infill wall a certain amount of ductility. For an ordinary frame with brick infill wall, constructed with no defects, its ductility may be taken as 2 (Rosenblueth, 1980).

(e) Constructional columns

It was widely accepted in China after the 1976 Tangshan earthquake that the use of constructional columns is a simple and practical way of increasing the seismic capacity of brick buildings. Constructional columns are not designed to take loads but are placed in the outside walls at the junction with transverse walls or at the midpoint of very long walls; the reinforcements in the constructional columns are the minimum for constructional requirements. Test results show that the constructional columns, together with spandrel beams, provide support or confinement for brick walls and thus increase the ductility of the brick building.

(f) Layout of lateral bracing systems

When designing the layout of the lateral bracing system, it is necessary to consider the functional and architectural requirements and also the effectiveness of the arrangement. The fundamental principle of a reasonable layout is the provision of short, effective and continuous paths of force transmission.

Figure 13.21 shows several possible layouts, of which (a), (b) and (c) are continuous and the stress flow is smooth, but (d), (e) and (f) are intermittent, and stress and deformation will concentrate at points of abrupt change. Field experience has shown heavier damage for the latter case.

The layout in Figure 13.21 (a) is usually used in frame–shear-wall systems, but is not good for very tall buildings because the deformation of the wall in shear and the frame in bending are not well matched and will introduce stress concentration. For very tall buildings, the layout in Figure 13.21 (b) is better, but attention should be given to the ductility required for the tie-beams between the shear walls.

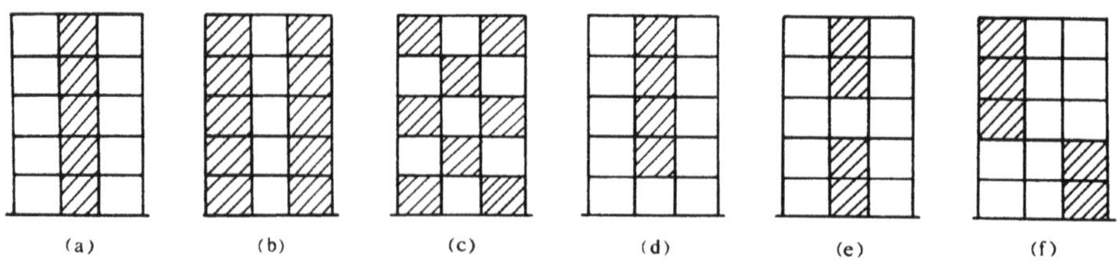

Figure 13.21 Stirrups in column.

It is equally important for the layout in plan of the building and the fundamental principle is the same. The primary concern is to reduce the torsional vibration of the building by keeping the center of stiffness and the center of mass as close as possible and to couple the lateral system as much as possible to the outer frames and to form a closed loop in each floor to increase its torsional resistance and rigidity.

13.5.2 Constructional detailing

Experience shows without doubt that a good detail design is vital to the survival of a structure during a strong earthquake. Neglect of detail design is certainly not wise. Details are different for different structures but their common requirements are to ensure the ductility capacity by avoiding brittle failure.

(a) Reinforced concrete (RC) structures

One common experience with RC structures is that most damage starts at inadequate detailing. One disadvantage of RC is its brittleness in tension and shear. The structural detailing has then to overcome this disadvantage.

Stirrups. Confinement of concrete in the compression zone will increase both the strength and ductility. There are at least four factors in ensuring good confinement: the spacing and continuity of the stirrups, the coupling of the stirrups to the main reinforcement, and the strength of the stirrups. Closely spaced spiral stirrups are thus the best. The stirrup layout in Figures 13.22(b) and (c) is adequate but that in Figure 13.22 (a) is no good.

Anchorage and overlap of the main reinforcement. These should be in areas of compression to ensure the bond between the concrete and the reinforcing bar, and effort should be made to bend the bar in the compression zone instead of the tension zone. The length of anchorage should be enough to develop the tensile strength of the bar.

Beam–column joints. Many RC-framed structures have weak joints. Ordinary structural analysis can give only stresses in beams and columns, not inside the panels of the joints; joint panels are designed according to some rough assumptions with no particular attention to the possibility of failure inside the panel.

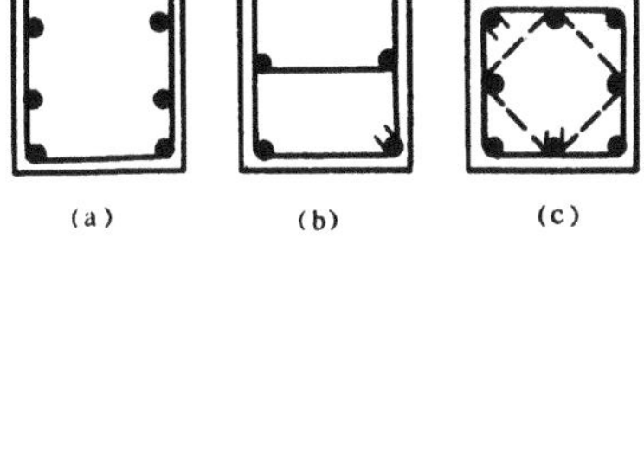

Figure 13.22 Layout of lateral bracing system.

(b) Brick structures

Field damage experience has always shown that masonry buildings are vulnerable to earthquakes. In developed countries, they are not used without reinforcement. There are many brick buildings without reinforcement in China and it is believed that unreinforced brick buildings with adequate design and construction can be built to resist earthquakes with intensity of VIII or less with no collapse, but that great care should be taken in areas of intensity IX.

The important design consideration for brick buildings is the integrity of the structure. Because brick buildings are constructed from small elements, the weak parts are the connections: the connection between bricks, between walls and between walls and floors. Although integrity is also an important factor in other buildings, brick buildings require special attention to the following points.

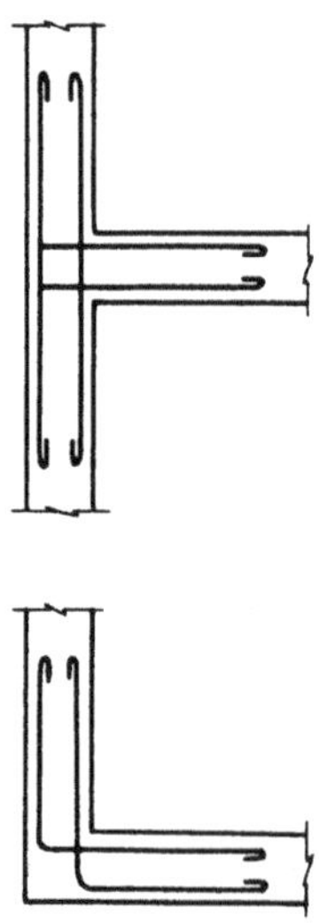

Figure 13.23 Horizontal reinforcing in brick walls.

1. Connections between transverse and longitudinal walls (Figure 13.23): in addition to the requirement of overlapping of bricks at the joints, horizontal reinforcement may be added at the joints.
2. Spandrel beams should form a closed loop at floor level and should not have long lateral unsupported lengths.
3. Weak parts such as the parapet and gable walls and the small chimneys on top of buildings should be strengthened.
4. The performance of brick walls depends on the standard of constructional skill.

13.6 Vibrational control

Apart from within this section, so-called seismic design, as discussed in this book, is actually a safety check of the ordinary design to minimize earthquake excitation, with additional consideration of lateral bracing systems. It is only a process of proportioning members to be strong enough to take the forces induced by an earthquake. Structural control is quite different from the routine meaning of design; it tries to prevent or reduce the energy transmitted into the structure, to absorb the energy in the structure by specially designed energy dissipation devices, and/or to modify the structural property to minimize the structural vibration with or without energy input. Although this was only an unrealized idea some decades ago, it is, at least partially, practical or feasible in the 1990s with the progress in the rubber industry and computer technology (Liu, Lagorio and Chong, 1991; Soong, Masri and Housner, 1991; Kelly, 1981, 1986).

13.6.1 General features of structural control

Structural control may be passive or active. Routine design makes use of the parts in the structure to absorb the energy input to the structure. Passive control designs special devices to achieve the same goal; active control designs also some special devices but introduces energy source or force into the structure to:

1. act against the response; and/or
2. change the dynamic property of the structure in a way to reduce the response.

Because earthquake motion changes randomly, active control uses a computer to control such forces instantaneously and continuously so that the structural responses are within the required limit. Base isolation is a popular passive control.

13.6.2 Base isolation

(a) Early ideas of base isolation

The earliest record may be traced back to 1909 when a medical doctor designed a base isolation system with a layer of talc on the ground to separate the building from ground. The first real building constructed using the idea of base isolation was the Imperial Hotel designed by Wright and built in Tokyo in 1921, which made use of an 18–21 m layer of soft mud under the building. This building, tied with the top hard layer, floated on the soft layer and performed extremely well in the devastating 1923 great Kanto earthquake while similar buildings were generally badly damaged.

(b) Modern base isolation designs

The rubber bearing is the most popular type of base isolation and is easy to manufacture. It can be made very stiff in the vertical direction to take the vertical load and very flexible in the horizontal direction to isolate the horizontal vibration. This kind of base isolation is very good in reducing the high acceleration, or the high frequency motion. The main idea of this system is to make the natural period of the structure very long, say 1s to avoid the resonance of the structure with the strong high frequency components of ground motion.

In the modern design of rubber bearings, auxiliary measures are usually added to meet different requirements, such as large damping and sliding support. Strong motion observation and test results all show that base isolation, when properly designed, is reliable and effective, reducing high-frequency motion to a half or one-fifth (Hisano, Nagashino and Kawamura, 1990).

The first base-isolated building in the United States was the Foothill Communities Justice Center in San Bernardino County, completed in 1985. It is 126 m × 33.5 m in plan and four stories high, with basement and sub-basement for the isolation system which consists of 98 isolators of laminated rubber bearings reinforced with steel plates. The site of the building is 20 km from the San Andreas fault and the building was designed for a 5% damping constant velocity spectrum of 1.27 m/s over the 0.8 s to 4.0 s period. Conventional design could not provide a reasonable solution and a base-isolation plan was decided upon. After completion, the record obtained during an earthquake of $M = 4.9$, 30 km away, showed the satisfactory result of base isolation design.

Base isolation has also been used in other countries in the past two decades. The French nuclear construction company Framatome (EDF) used

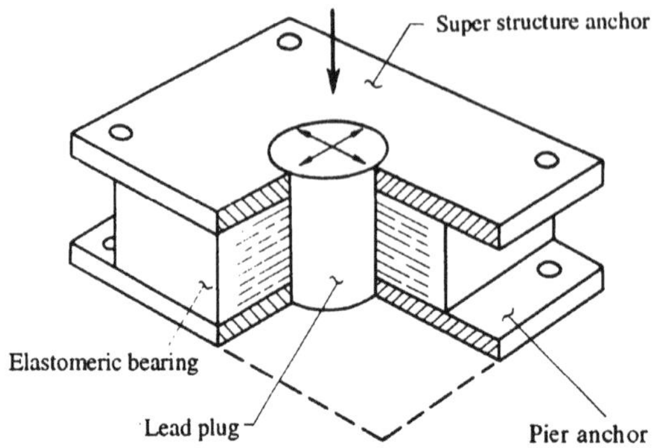

Figure 13.24 A base isolator used in New Zealand.

a standard nuclear power plant design for 0.2 g acceleration with a base isolation system for a site with higher acceleration. This isolator uses laminated neoprene bridge bearings with lead–bronze–stainless steel slip plates on top of the bearings. The neoprene bearings are conventional and cannot accept large displacements because they have only a few layers of elastomer; sliding takes place on the slip plates when a large displacement is needed.

Figure 13.24 shows an isolator of New Zealand design, which has been applied to a four-story reinforced-concrete-framed building in Wellington. The isolators are made of laminated natural rubber bearings with a cylindrical plug of lead in the central hole of each bearing to increase damping and control the displacement of the system.

There are other designs of rubber-isolated systems used, such as multiple rubber sheets bonding with thin steel reinforcing plates, together with additional damping and/or restraint against wind, such as the lead plugs, mild steel energy absorbing devices and frictional elements.

Base isolation is very attractive and will find many practical applications. Its main advantages are:

1. very small relative deformation in the superstructure which avoids the damage to structural and nonstructural elements;
2. economical and easy design of the superstructure;
3. possibility of building structures for the very worse conditions such as very high seismicity;
4. safety of important structures or systems during unexpected strong earthquakes.

But when the building is very tall, attention should be given to the effectiveness of the isolation and the possible uplift of foundations from rocking motion of the building.

13.6.3 Other passive control systems

In addition to the commonly used base isolation system, there are other passive seismic control systems, which have also been used in actual constructions, such as:

1. sand base isolation
2. suspended structure
3. energy absorbers
4. passive tuned mass damper.

13.6.4 Active control

An active control system consists of:

1. sensors, which measure earthquake motion and/or the structural responses in real time;
2. a controller, which analyzes the response data from the sensors using a specially designed control program, computes the compensating force and generates commands for them to be added to the structure;
3. an actuator, which produces the required force from an external energy source.

If only some structural responses are measured and used, it is a closed-loop control; if only the external excitations are measured, it is an open-loop control; if both are measured, it is an open–closed-loop or a combined control. This idea has been widely used in electronics and other fields of engineering. If passive structural control is in a stage of design and construction, active structural control is still in a stage of research and initial application, but current results show great potential of practical applications.

The equation of motion for a controlled structure of n degrees of freedom may be written as (Soong, Masri and Housner, 1991)

$$\boldsymbol{M}\ddot{\boldsymbol{u}}(t)+\boldsymbol{C}\dot{\boldsymbol{u}}(t)+\boldsymbol{K}\boldsymbol{u}(t)=\boldsymbol{D}\boldsymbol{f}(t)+\boldsymbol{E}\boldsymbol{p}(t) \tag{13.22}$$

where the bold variables are matrices or vectors; **f** and **p** are respectively the control force and external excitation, $\boldsymbol{D}$ and $\boldsymbol{E}$ define respectively the locations of $\boldsymbol{f}$ and $\boldsymbol{p}$. The control force may be written as

$$\boldsymbol{f}(t)=\boldsymbol{K}_1\boldsymbol{u}(t)+\boldsymbol{C}_1\dot{\boldsymbol{u}}(t)+\boldsymbol{E}_1\boldsymbol{p}(t) \tag{13.23}$$

where $\boldsymbol{K}_1$, $\boldsymbol{C}_1$ and $\boldsymbol{E}_1$ are control gains. If $\boldsymbol{D}=\boldsymbol{0}$, the structure is not controlled. Combining these two equations,

$$\boldsymbol{M}\ddot{\boldsymbol{u}}(t)+(\boldsymbol{C}-\boldsymbol{D}\boldsymbol{C}_1)\dot{\boldsymbol{u}}(t)+(\boldsymbol{K}-\boldsymbol{D}\boldsymbol{K}_1)\,\boldsymbol{u}(t)=(\boldsymbol{E}+\boldsymbol{D}\boldsymbol{E}_1)\,\boldsymbol{p}(t) \tag{13.24}$$

It is clearly seen that, by selecting proper gains or control forces, the damping, stiffness and external excitations can all be changed, within practical limitations by the designer to arrive at a suitable design.

Passive control modifies the structure itself using ordinary design techniques, such as adding a base isolator or an energy absorber which changes the structure permanently. Active control modifies the structural responses in a predetermined way by adding energy sources to the structure which apply forces, varying dynamically in the time domain with different magnitude and direction according to the earthquake motion and structural responses, in terms of a performance function

$$J = J(\boldsymbol{u}, \boldsymbol{f}) = \min. \tag{13.25}$$

J may be simply the maximum displacement, maximum acceleration or maximum shear of the response, or a function of statistical values of the responses. If the control forces $\boldsymbol{f}(t)$ are included in the target function, it is then a closed-type, and its solution requires iteration. Choice of $\boldsymbol{f}(t)$ and J differs according to the method of control and purpose.

A common example of active control is the tuned mass damper (TMD). The TMD is a comparatively small vibration system itself, which consists of a mobile mass block, a spring and a damper. The mass block is controlled by an actuator and is connected to the structure. The position of the mass block is controlled or tuned by external energy according to the design to minimize the selected response quantity. If there is no energy input, the system is then a passive TMD. The active TMD was first applied in mechanical engineering systems and then to wind and earthquake design of structures in Australia, Canada and Japan, mostly to tall structures.

The first structure with an active TMD was built by the Kajima Institute of Construction Technology. The damper consists of two mass blocks, of weight 4t and 1t respectively, about 1% of the weight of the structure. The controller accounts for about 4–5% of the cost of the structure.

The efficiency of the TMD depends on the ratio of the mass of the mobile block and the structure; the larger the ratio, the more efficient the control, but more control power is needed. Reduction of cost and the power needed to operate the controller during a strong earthquake remains a problem.

14 Seismic design codes

14.1 General features

A seismic design code consists usually of three parts:

1. general purpose, scope of application, general philosophy and the goal to be achieved;
2. computational analysis of structural behavior, such as the method of evaluation of the global and local earthquake loads and responses, specification of allowable stress and deformation, and some necessary requirements on site and foundation;
3. earthquake-resistant constructional measures.

14.1.1 Computational analysis

Some aspects of seismic design practice, after theoretical generalization, may be expressed in terms of mechanical principles and used to evaluate quantitatively the force and deformation or other indices related to the safety of structures induced by earthquake ground excitation. Such design practice is specified in the code in terms of response analysis. The main content includes the following items.

(a) Computation of seismic loads

The computation of earthquake loads acting on the structure is given in terms of response sepctra in most of the current codes.

Specification of base shear–equivalent static approach. This specifies first the procedure and necessary numerical data for the computation of base shear and bending moment at each floor either directly or through the computation of the intertial forces on the structure. Take the former as an example; the commonly used format is as follows:

$$\text{Horizontal base shear } V_0 = \alpha W \tag{14.1a}$$

$$\alpha = ZISC\beta(T, \zeta)\, k_0 \tag{14.1b}$$

where α is the horizontal earthquake load coefficient or earthquake coefficient; $k_0 = a/g$ is the basic earthquake coefficient and a the horizontal acceleration of ground motion. The basic coefficient is modified by a seismic zoning coefficient Z and the product Zk_0 is the zone earthquake coefficient. I is the importance factor of the structure; C is the structural coefficient; S is the site effect factor; $\beta(T, \zeta)$ is the design response spectrum, a function of the natural period T and damping ratio ς of the structure; W is the weight of structure, including the total dead load and partial live load. The damping ratio ς of an ordinary structure is usually taken as 0.05.

The earthquake inertial force F_m acting at the elevation H_m is usually taken as

$$F_{\mathrm{m}}=\frac{(V_0-F_{\mathrm{d}})\,W_m H_m}{\sum(W_j H_j)} \qquad (14.2a)$$

$$F_{\mathrm{D}}=C_{\mathrm{D}}V_0 \qquad (14.2b)$$

where W_m is the weight concentrated at elevation H_m, $\sum W_j=W$; F_{D} is the force concentrated at the top and the concentration coefficient C_{D} takes values usually in the range 0–0.15, but sometimes is a function of the fundamental period T of the structure. The larger the T value, the larger the coefficient.

Modal combination approach. The base shear of the response of the jth mode is

$$V_{0j}=ZIC\beta(T_j,\varsigma_j,S)\,k_0W_j \qquad (14.3)$$

where the subscript j indicates the quantity of the jth mode; $\beta(T,\varsigma,S)$ is the amplification response spectrum of acceleration, considering the effect of site S; W_j is the weight of the jth mode.

When the inertial force on the structure is directly specified, the inertial force at location m of the jth mode is usually

$$F_{mj}=ZIC\beta(T_j,\varsigma_j,S)\,k_0\phi_{mj}\gamma_j W_j \qquad (14.4)$$

where γ_j is the participation factor of the jth mode; ϕ_{mj} is the value of the jth mode at point m; and W_m is the weight at point m. From the base shear or the inertia forces computed according to the procedure given above, the internal forces and deformation Q_j anywhere in the structure can be found for the jth mode response first, where $j=1, 2\ldots$, and the total response quantity Q is finally obtained from responses of all modes ($j=1, 2, \ldots$) by the method of square-root-of-sum-of-squares (SRSS),

$$Q=\left(\sum Q_j^2\right)^{1/2} \qquad (14.5)$$

(b) Structural coefficient C or its reciprocal $1/C$

When designing according to the method of allowable stress or limit strength, it is required that the response Q be not over the corresponding strength R,

$$Q\leqslant R \qquad (14.6)$$

If the structural coefficient C is not considered in the computation of response Q, this requirement will be

$$CQ\leqslant R \quad \text{or} \quad Q\leqslant\frac{R}{C} \qquad (14.7)$$

The coefficient C is used to consider the effect of the energy absorption capacity of the structure; the greater the energy absorption capacity, the smaller the response and thus the C value. In the United States, a notation R is used for $1/C$. The ratio of the smallest to the largest values of C is roughly in the range of 2–5.

(c) Effect of site on ground motion

For simplicity, codes usually classify the sites into a few classes or types according to the stiffness of the ground, and specify different ground motion factors for them. In the 1950s, different acceleration values were specified with higher values for softer sites. Up to the 1970s and 80s, different response spectra, instead of accelerations, were specified with spectra of greater responses at longer periods for softer sites. This trend started first in Mexico for very soft sites and in China in 1964 for cases in general; the United States and Japan followed about ten years later, and the former Soviet Union in 1982.

(d) Local seismic force

The above-mentioned seismic force or load is referred to as the global structural response. A single member or feature in the structure, such as a penthouse, appendage, parapet or partition, and mechanical or electrical parts supported on the structure may be subject to much larger instantaneous local seismic forces due to the local amplification or high mode effect. These local inertial forces have little effect on the global structure because their peak values occur at different times and act in different directions; but they do have important effects on the individual parts and their supports or the connections with the structure. Therefore, codes usually specify forces several times larger for their design, to ensure their safety.

(e) Torsional vibration

Most codes specify two torsional effects: one from the static eccentricity e_s or the horizontal offset between the stiffness center and the mass center of each story and its dynamic amplification $e_d = 1.5e_s$, and the other from accidental eccentricity due to random errors in stiffness, mass and geometric dimensions, which is usually specified as a certain percentage, such as 5% or 10%, of the corresponding horizontal dimension B of the structure. The final eccentricity e in some codes is whichever of the following gives the larger response:

$$e = 1.5e_s + 0.05B \tag{14.8}$$

$$e = e_s - 0.05B \tag{14.9}$$

(f) Limitation of sidesway and story displacement

To ensure sufficient stiffness for a structure, to avoid collision with neighboring structures or to guarantee the safety of nonstructural parts, design codes generally specify limiting values for the global and story displacement δ or rotation θ in the following form,

$$\delta < \theta h$$

where h is the story height. Many codes specify $\theta \leq 0.005$.

(g) Vertical seismic load

For cantilevered parts on the structure, the vertical seismic load is generally treated as the local earthquake force; for those structures, such as gravity dams, whose stability depends on their own weight, or high chimneys, which are sensitive to vertical vibration, a vertical seismic coefficient of half the horizontal coefficient is generally specified in the code.

14.1.2 Earthquake-resistant constructional measures

In addition to the design practice specified in terms of computational analysis described above, the third part of the design code specifies earthquake-resistant constructional measures or limitations, which come mostly from field experience of past strong earthquakes. The following principles are the most important.

(a) General design considerations

Many codes list some criteria to be followed in design, such as: in site selection, to avoid sites of ground failure potential and to select flat and stable sites; in structural configuration, to avoid sudden changes or irregularities in plan and elevation and to have simple and regular shapes; in construction, to strengthen measures of quality control.

(b) Structural layout

Many codes specify limits on heights of buildings, stories and parapet walls, and lower limits on widths of walls between openings; some codes specify requirements on ring beams and constructional columns.

(c) Structural details

Many codes have specific requirements on structural details, such as the main reinforcements and stirrups, the supported length of precast members, strength of mortar on masonry etc.

14.2 Development of Codes in Japan, USA and China

Review and analysis of the history of seismic codes may be of help to accelerate the application of research results to engineering practice and to strengthen the tie between code and experience.

14.2.1 Development in Japan

(a) Chronology of codes and related issues in Japan

1915 Mr Sano suggested a static approach using a seismic coefficient k for earthquake-resistant design of structures, which became the foundation of seismic design in Japan for several decades.

1922 Mr T. Naito retrofitted some buildings according to the seismic coefficient method, with $k=\frac{1}{15}$.

1923 In the Kanto earthquake, many buildings were heavily damaged, but retrofitted buildings with $k=\frac{1}{15}$ withstood it successfully.

1924 The city building law required seismic design, and a seismic coefficient $k=0$ was specified. This was the first such specification in a code.

1931 The code specified an upper limit of 31 m for buildings.

1941–43 The Earthquake Construction Subdivision published a book, *Important Requirements on Earthquake-Resistant Construction*, in which limit design was specified and the seismic coefficient for upper stories was increased.

1944 Limit design was required with $k=0.15$.

1947–48 In the Japanese Building Association Standard, JES building 3001, the coefficient was increased to $k=0.2$, and the allowable stress increased accordingly at the same time.

1950 The Building Standard Law was issued to replace the codes used since 1924, to be applied nationwide, in which the seismic code adopted was JES 3001, and $k=0.2$. For buildings over 16 m, coefficient k was increased at a rate of 0.01 for each 4 m increase; $k=0.3$ for buildings on soft ground and for penthouses.

1951 In the recommendation on 'reduction of seismic coefficient' by the first sub-committee of the Basic Law Research Committee, the whole of Japan was divided into three seismic hazard zones with zone-reduction coefficient $C_1=0.8$, 0.9 and 1.0 respectively; a site-reduction coefficient C_2 as given in Table 14.1 was specified for four classes of site and four kinds of structures. The rest remained the same as in the 1950 Building Standard Law.

1952 Notice No. 1074 of the Ministry of Construction adopted the reduction coefficients C_1 and C_2, and specified $k=0.2$ and $C_1C_2 \geqslant 0.5$. This specification was used until 1981.

1955 Standard (draft) of Design of Civil Engineering Constructions of Japan Railways divided the country for seismic design into three zones A, B and C, with horizontal seismic coefficients specified according to site class, and the vertical seismic coefficient was taken as half the horizontal coefficient.

1956–59 Similar specifications made for highway bridges, dams, ports and harbors.

1964 Review Committee of Highrise Buildings founded in the Ministry of Construction. *Technical Guidelines on Highrise Buildings* published by the Japan Architectural Association.

Table 14.1 Ground-reduction coefficient C

Ground	C			
	Wood	Steel	Reinforced concrete	Masonry
1. Bedrock; tertiary	0.6	0.6	0.8	1.0
2. Diluvium	0.8	0.8	0.9	1.0
3. Alluvium	1.0	1.0	1.0	1.0
4. Fill; thick alluvium	1.0	1.0	1.0	1.0

1967 *Technical Guidelines on Highrise Buildings*, for buildings taller than 45 m, adopted explicitly the response spectrum design method. The main specifications were as follows:
- base shear $V_0 = ZIC\alpha_1 W$;
- dynamic amplification factor $\alpha_1 = 0.18/T$ to $0.36/T$ (or $0.15/T$ to $0.30/T$) $\geqslant 0.05$, depending on ground and structure;
- base shear is distributed throughout the height of the building but with a certain amount concentrated at the top.

1972 A five-year plan was laid down for the development of a new earthquake-resistant code, which was carried out in 1976, primarily by the Building Research Institute and Civil Engineering Research Institute.

1977 *New Earthquake-resistant Building Design Code (draft)* and *Seismic Safety Appraisal of Existing Reinforced Concrete Buildings* published.

1980 *Enforcement Order of Modified Building Standard Law* published by the Research Committee on Earthquake-resistant Building Standard announced.

1981 *New Earthquake-resistant Building Design Code* became officially effective in June.

(b) The 1981 design code

Background to drafting the code. It is seen from the chronology above that 22 years passed between the announcement of the 1950 Standard Law, and the laying down of the five-year plan in 1972. During this long period of time, great advances were made, including the following.

- The response spectrum approach was widely accepted in design codes, and more advanced dynamic methods of analysis used for the seismic design of critical or special structures.
- Records of strong earthquake ground motion were rapidly accumulated in both the USA and Japan.
- Many strong earthquakes occurred in Alaska, Niigata, Tokachi-oki and San Fernando, which provided much invaluable experience for modern earthquake-resistant constructions.
- In 1967, the dynamic method of analysis was adopted for reinforced concrete highrise buildings of height over 45 m.

All of these highlighted the inadequacy of the static approach.

Essentials of the code. In the 1977 draft, the code consisted of two main parts, building and civil engineering work, but only the building part was adopted in the Building Standard Law and only this part will be discussed in the following.

Expected seismic capacity. There are two design levels of earthquake:

1. small earthquake with ground motion expected several times within the lifetime of the structure;

2. large earthquake with ground motion expected less than once within the lifetime of the structure.

Seismic lateral shear

- First level: story shear of the nth story,

$$Q_n^e \geqslant ZR_t A_n k_1 W_n, \quad k_1 = 0.2 \tag{14.10}$$

- Second level,

$$Q_{un} \geqslant CC_{es} ZR_t A_n k_2 W_n, \quad k_2 = 1.0 \tag{14.11}$$

Coefficients

- Seismic zoning coefficient $Z = 1$, 0.9, 0.8, 0.7 respectively for zones A, B, C and D.
- Response spectrum $R_t = \beta(T)/\beta_{max}$,

$$R_t = 1.0 \qquad T \leqslant T_0 \tag{14.12a}$$

$$= 1.0 - 0.2\left(\frac{T}{T_0} - 1\right)^2 \quad T_0 \leqslant T \leqslant 2T_0 \tag{14.12b}$$

$$= 1.6\frac{T_0}{T} \qquad T \geqslant 2T_0 \tag{14.12c}$$

$T_0 = 0.4$ s (site I); 0.6 s (site II); 0.8 s (site III).

- Story shear distribution coefficient (Figure 14.1),

$$A_n = 1 + \frac{(\gamma_n^{-1/2} - \gamma_n)\,2T}{1 + 3T} \tag{14.13}$$

$$\gamma_n = \frac{W_n}{W_N}$$

  where W_n is the weight of the building above the nth story and W_N the total weight of the building.
- Structural coefficient,

$$C = 0.25 \text{ (large ductility) or } 1.0 \text{ (no ductility)}$$

- Eccentricity-stiffness coefficient,

$$C_{es} = 1.0 \text{ (regular) or } 2.25 \text{ (very irregular)}$$

- Seismic coefficient of underground part of building

$$k \geqslant 0.1\left(1 - \frac{H}{40}\right)Z \tag{14.14}$$

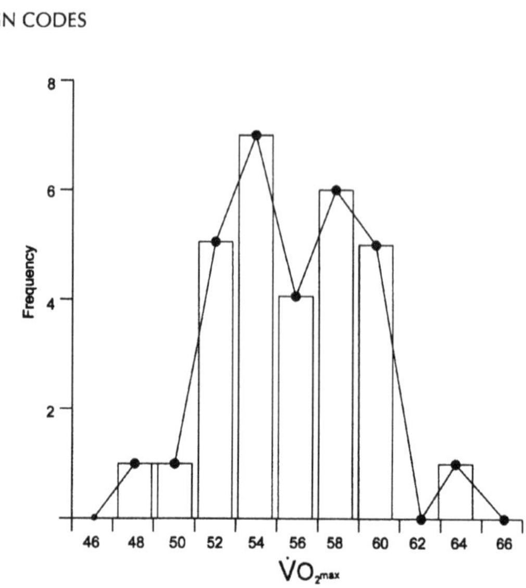

Figure 14.1 Distribution factor of story shear A_n.

where H = depth (m) from ground level to bottom of basement; H is taken as 20 for $H>20$.

- Local seismic coefficient $k=1.0$; if no casualties, $k=0.5$.

Story lateral drift limitation. This should be 1/200 for the first design level.

Design procedure

- First level design: necessary for all buildings, and to carry out elastic analysis and detailing. For group 1 buildings, it is not necessary to go to the second level; group 1 includes masonry buildings with less than three stories above ground; wooden houses of one or two stories and with total floor area less than 500 m^2; some reinforced concrete buildings with $H \leqslant 20$ m.
- Second level design: necessary for group 2 and 3 buildings. Group 2 includes those of height not more than 30 m, except group 1, and group 3 buildings of height 31–60 m.
- Special design: necessary for group 4 buildings, which are those of height greater than 60 m. This comprises dynamic analysis with input of given time history $a(t)$ and two-level design.

Special features of the code.

1. The two-level design method was adopted.
2. Damping and deformation capacities were considered specifically by decomposing the structural coefficient C into two factors, the damping factor C_ς and the ductility factor C_μ:

$$C = C_\zeta C_\mu \tag{14.15}$$

$$C_\zeta = \frac{1.5}{1+10\zeta} \tag{14.16}$$

$$C_\mu = \frac{Q_y}{Q_e} = \frac{1}{\mu} \text{ (for stiff structures)} \tag{14.17}$$

$$C_\mu = (2\mu - 1)^{-1/2} \text{ (for flexible structures)} \tag{14.18}$$

3. There should be continuous vertical distribution of inertial forces.

In addition to these three features, the 1981 Japan code changed also the effect of site conditions by modifying the design spectrum; and it added local seismic specification.

Influence of the code. After the approval of the code, several specifications and regulations were modified accordingly. The published seismic codes are for foundations, building facilities, storage tanks and chimneys. The 1989 modified building seismic design code GBJ 11–89 for China adopted the two-level design principle and there is such a trend also in the USA (Whitman, 1989).

14.2.2 Development of codes in the USA

In the United States, each state may have its own law and seismic code; and because earthquakes are frequent and strong on the west coast, seismic codes were therefore first considered there, especially for cities in California.

(a) Chronology of codes and related issues in the USA

1927 The appendix to the first edition of the UBC (Uniform Building Code) was applied to seismic design with $k=0.075$–0.10.

1933 The Field Act, as an emergency measure, became effective on 1 April, and specified $k=0.02$–0.05 for public school buildings and 0.1 for masonry buildings. The Riley Act became effective on 26 May, and specified $k=0.02$ for ordinary structures. Los Angeles city building regulations specified $k=0.08$ for ordinary buildings and $k=0.10$ for school buildings. Long Beach and other southern Californian cities adopted these regulations.

1935 UBC specified $k=0.08$ (hard ground) or 0.16 (soft ground).

1937 The Field Act adopted $k=0.06$–0.10 for buildings of height less than three stories or with no moment-resisting rigid frames, and $k=0.02$–0.06 for others.

1941 The Field Act adopted $k=0.06$–0.10, depending on site conditions.

1943 The Los Angeles building code specified the story shear coefficient $\alpha=0.06/(N-n+4.5)$ and $N-n \leqslant 13$, where n is the story number (counting the bottom) and N the total number of stories.

1946 The UBC specified k values according to seismic zones.

1952 The Lateral Force Joint Committee recommended a code as follows.

- Base shear $V_1 = \alpha W$, where $\alpha = k/T$, $k=0.015$ for $0.02 \leqslant \alpha \leqslant 0.06$ (buildings); $k=0.025$ for $0.03 \leqslant \alpha \leqslant 0.010$ (others).

- Seismic inertial force at nth floor

$$F_n=\frac{V_0 W_n H_n}{\sum W_j H_j}$$

where H_j is the height of the jth story; $W=j$ is the weight of the jth floor.

1953 The Field Act adopted $V_n=\alpha_n W$, and $\alpha_n=0.06/(N-n+4.5)$. The Riley Act adopted $k=0.03$ for buildings of height less than 13.3 m and $k=0.02$ for higher buildings.

1956 The San Francisco code adopted the 1952 Lateral Force Committee recommendations, but with some changes in numerical values.

1957 The Structural Engineers Association of California (SEAOC) appointed an earthquake committee to draft a seismic design code for California.

1959 Los Angeles code removed the limitation of $N-n\leqslant 13$ and modified the story shear coefficient to

$$\alpha_n=\frac{0.0046N}{N-n+0.9(n-8)}$$

when $N>13$, take $N=13$.

SEAOC published its first edition of the *Recommended Regulations of Lateral Forces*, which were revised every few years and have had important academic influence on other legal codes internationally, although these regulations are not themselves legal requirements. The important features of the first edition were as follows.

- Base shear $V=C\alpha W$, where $\alpha=0.05/T^{1/3}$, $T\geqslant 0.1$ s and $C=0.67-1.33$.
- Inertial force at nth floor

$$F_n=\frac{V_0 W_n H_n}{\sum W_j H_j} \quad \text{(ordinary buildings)}$$

$$\left.\begin{aligned} F_n&=\frac{0.9V_0 W_n H_n}{\sum W_j H_j} \\ F_D&=0.1V_0 \end{aligned}\right\} \quad \text{(highrise buildings)}$$

- Base overturning moment $M=J\sum W_j H_j$, where reduction factor $J=0.5/T^{1/3}$.
- nth story overturning moment $M_n=M(H_N-H_n)/H_N$.
- Ductile moment-resisting space rigid frames should take at least 25% of the total horizontal seismic load when the height of building is greater than 53 m or 13 stories.
- There should be an accidental eccentricity between centers of gravity and stiffness of 5% of the building width.

1961 The UBC adopted the 1959 SEAOC recommendations together with a seismic zoning factor $Z=0.25$, 0.5 and 1.0.

1963 The SEAOC recommendations decided to remove the limitation of building height for buildings with a ductile moment-resisting space frame taking at least 25% of the total seismic load.

1966 SEAOC made the following modifications.

- $\alpha_{max} = 0.1$ (when $T_{min} = 0.1$ s).
- $F_D = 0.004V(H_N/D)^2$ where D is the dimension of the building along the direction of vibration.
- The lower limit of $J_{min} = 0.33$ was removed.
- Overturning moment

$$M_n = J_n \left[F_D(H_N - H_n) + \sum_{n+1}^{N} F_j(H_j - H_n) \right]$$

$$J_j = J + (1 + J)\left(\frac{H_n}{H_N}\right)^3$$

- Ductile moment-resisting space frames, including both steel and those reinforced concrete frames satisfying certain requirements, may be used for buildings higher than 53 m.
- Buildings with coefficient $C = 0.67$–0.80 should have a ductile moment-resisting space frame which takes no less than 25%–100% of the total seismic load.

1967 SEAOC suggested the following modifications.

- $J = 0.6/T^{1/3}$ (buildings).
- $J_{min} = 0.45$ (other structures).
- When dynamic analysis is performed, the specification regarding setback may be disregarded.
- UBC adopted these recommendations.

1969 SEAOC suggested removal of the specification of reduction factor on overturning moments.

1970 SEAOC appointed a speciality committee to review the fundamental principles of seismic design.

1971 SEAOC recommended the following modifications.

- All seismic resistant space rigid frames should be ductile moment-resisting.
- Strength of light concrete should be limited.
- The spacing of stirrups should be reduced.
- Precast members should be allowed.

UBC adopted these recommendations.

The SEAOC speciality committee published a report on the fundamental principles of seismic design with the following important conclusions.

1. The currently used equivalent static method based on dynamic analysis of regular structures is a practical format.

2. Dynamic analysis is recommended for important and irregular structures to discover and strengthen the weak links.
3. Design seismic loads may be raised.
4. The structural coefficient C may be used but numerical values changed.
5. It is highly recommended that soil–structure interaction be considered in the future code.
6. For important structures, it is necessary to evaluate the current recommendations from the criterion of public safety.

1972 The City of Los Angeles issued a tentative seismic regulation with specification of dynamic method of analysis, similar to the 1971 SEAOC recommendation, but stating that static analysis should be used for regular buildings lower than 53 m; for important buildings such as hospitals and communication centers, the seismic load should be increased by 50%.

The SEAOC Speciality Committee published a report on fundamental design criteria.

1973 The ATC suggested the ATC-3 program on 'Comprehensive Seismic Design Regulations'.

1974 The SEAOC recommendations were modified on the basis of the 1971 recommendations of the Speciality Committee, and made important modifications in the following three respects.

1. $V=ZIC\alpha SW$, where Z is the seismic zoning coefficient; I is the important factor, $I=1$ for ordinary structure, 1.5 for important structures; S is the site soil coefficient in the range 1.0–1.5, depending on the predominant period of the site.
2. The seismic coefficient α was increased to $\alpha=1/(15T^{1/2})\leqslant 0.12$; when compared with the 1959 specification; it had increased by 60% at $T=0.7$ s and 13% at $T=2.5$ s.
3. Dynamic analysis should be used for important structures. For highly irregular buildings or buildings with greatly different strength and stiffness in adjacent stories, the effects of structural dynamic properties on the lateral force distribution should be considered.

1976 The UBC code adopted the 1974 SEAOC recommendations.

1977 The ATC published the model code ATC 3–06.

1979 The FEMA founded and supported the ATC-3 program to be carried out under the BSSC (Building Seismic Safety Committee), whose duty was to modify the code on the considerations of safety.

1980 The following modifications were made to the Los Angeles code.

- Dynamic analysis should be used for all structures higher than 53 m.
- The input motion $a(t)$ should be evaluated on the basis of soil, geology and earthquake reports for the specific site.

1986 *Seismic Design Guidelines for Essential Facilities* published by the US Army.

1988 NEHRP (National Earthquake Hazard Reduction Program) started by FEMA.

1990 SEAOC recommendations adopted the new zoning map of California based on EPA/g (0.4, 0.3 and 0.2 respectively for zones 4, 3 and 2) and, when dynamic analysis is used, three acceleration site amplification spectra $\beta(T)$ for three site types, as suggested in ATC-3. An appendix on seismic-isolated structures was included for the first time in a design code.

(b) Essentials of the ATC-3 model code

Purpose of the code. The ATC-3 program was initiated by the ATC, financially supported by NSF (National Science Foundation) and NBS (National Bureau of Standards) and started at the end of 1973, with the purpose of providing building design professionals with a comprehensive and up-to-date knowledge base to write the tentative design provisions applicable to all probable earthquake areas of the United States and to all buildings, including existing buildings.

Special features of the code. This model code made the following nine important changes.

Design value of ground motion. This is specified through two parameters: the effective peak acceleration (EPA) A_a and the acceleration expression of the effective peak velocity (EPV) A_v. It is widely believed that the most practicable representation of a response spectrum is its simplified formation, reflecting not only the magnitude but also the frequency content of the ground motion. The code drafting committee believed that the effects of distant earthquakes on flexible buildings, were significant, and would be easier to take into consideration by adopting a response spectrum with variable shapes. According to the results obtained by McGuire (1977) EPA represents the short-period ground motion, roughly proportional to the response spectral ordinates in the range $T=0.1$–0.5s, and EPV represents the moderate-period ground motion, roughly proportional to the response spectral ordinates near $T=1.0$ s. The following definitions were used:

$$\mathrm{EPA}=\frac{S_a}{2.5}, \quad \mathrm{EPV}=\frac{S_v}{2.5} \tag{14.19}$$

where S_a are the average acceleration response spectral ordinates in the range $T=0.1$–0.5 s and S_v the average velocity response spectral ordinates in the range $T=0.8$–2.5 for damping $\zeta=0.05$.

For simplicity in computation, EPA and EPV were normalized into nondimensional variables as follows,

$$A_a=\frac{\mathrm{EPA}}{g}, \quad A_v=\frac{\mathrm{EPV}}{30} \tag{14.20}$$

where the unit of EPV is inches per second and its physical meaning will be explained in the following section.

Contour maps of A_a and A_v. The contour map of A_a was that given by Algermissen and Perkins (1976) for bedrock sites of the United States. This map has the following features.

1. It is for hard sites.
2. The highest contour is 0.4, not 0.6 as given by Algermissen and Perkins.
3. It has a probability of exceedance of 10% in 50 years.

The contour map of A_v was obtained as follows.

1. The innermost contour of A_v was chosen to coincide with the innermost contour of A_a and have the same value by using a normalizing constant of 30 in/s (76 cm/s) because the maximum value of EPV was found to be 12 in/s (30.5 cm/s) which implies that the spectra at epicentral areas do not change.
2. For the western part of the USA, the velocity attenuation law was obtained from actual records of ground motion; for eastern USA, the approximate velocity attenuation was obtained by comparison of intensity attenuation laws of east and west.

Velocity contours were then obtained in a similar way to those for acceleration. The velocity contours attenuate more slowly than the acceleration contours, which means a spectrum with higher A_v for distant earthquakes, such as that shown in Figure 14.2.

Effect of site soil. This code was the first in the USA to adopt different design spectra for different site soil conditions, as shown in Figure 14.2, instead of a soil modifying factor depending on period, on the basis of results obtained by Seed (1976). The modificatons of design spectra for site conditions was the trend worldwide, but the special point of the ATC code was to consider at the same time a reduction factor of 0.8 for the soft site S_3 as shown in Figure 14.3.

Specification of seismic design-stages for various seismic performance categories. In this code, a new term seismic performance is defined as a measure of the degree of protection provided against the earthquake ground motion on the structure. This protection should be higher for important structures under potentially severe ground motion. The strength of ground

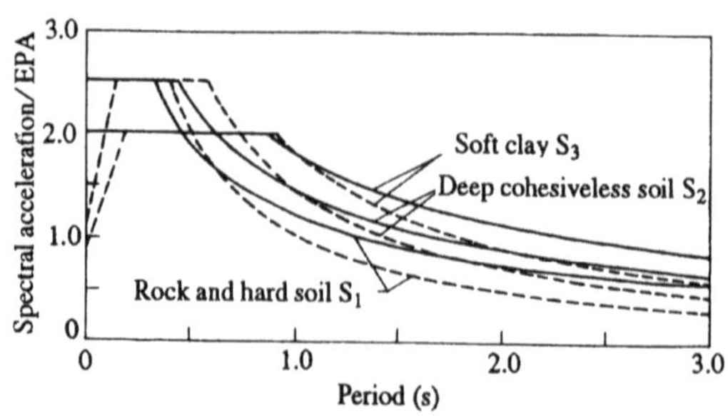

Figure 14.2 Design spectra (solid lines) for different site types. Dashed lines show free-field spectra.

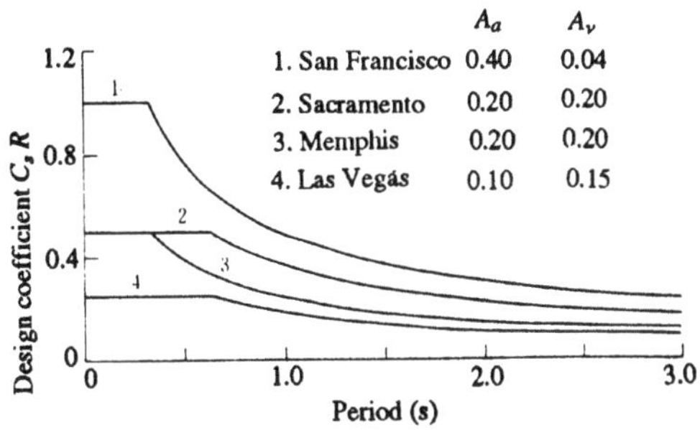

Figure 14.3 Design spectra for various cities.

motion is classified into four seismicity indices according to the value of A_V as shown in Table 14.2 and the importance of structure into three seismic hazard exposure groups, group III being essential facilities, whose functioning is needed after a strong earthquake, group II, important structures and group I ordinary structures. The seismic performance category is also defined in Table 14.2. The difference from the current codes is that for very important structures, if the seismicity index is low, the seismic protection is not necessarily high.

Many important stages in seismic design, such as the selection of framing system, analysis procedure, design and detailing requirements and foundation design, are then specified according to the seismic performance category defined in Table 14.2. Category A is the lowest protection specified and D the highest.

Soil–structure interaction. Quantitative specification of soil–structure interaction is another new feature. Because of the compliance of the foundations, the natural period of the combined system lengthens, and the equivalent damping increases, mostly from radiation damping. This code provides formulas to compute these values together with formulas for lateral deflection of the combined system.

P–Δ effects. Formulas are given for the modified story shear, overturning moment and internal forces for *P*–Δ effects.

Inspection, evaluation and retrofitting of existing buildings before or after earthquakes. It was quite a new feature to include seismic problems of existing buildings in a seismic design code. The ATC-3 code specifies both qualita-

Table 14.2 Seismicity index and seismic performance category

A_a, A_V	Zone no.	Seismicity index	Seismic hazard exposure group		
			III	II	I
0.20–0.40	5–7	4	D	C	C
0.15–0.20	4	3	C	C	B
0.05–0.15	2–3	2	B	B	B
0.05	1	1	A	A	A

Table 14.3 Ground motion (Z_a in g, Z_v in cm/s) zoning of Canada

	Z_a/Z_v for Zone no.						
	0	1	2	3	4	5	6
Zone boundary	0	0.04	0.08	0.11	0.16	0.23	0.32
Zone middle	0	0.05	0.10	0.15	0.20	0.30	0.40

tive and quantitative methods to evaluate the seismic strength of existing buildings.

Influence of ATC-3 code. Since the publication of the ATC-3 code in 1978, its influence has been obvious over the world. The Tri-Services Manual of the US Army *Seismic Design Guidelines for Essential Facilities* was completed in 1984 and published in 1986; Columbia code Norma AIS 100–83, the Canadian NBCC 1985 code and the 1979 UBC code have adopted some important features from the ATC-3. We will take the 1985 Canadian code as an example.

This code has the following features.

1. Following ATC-3, two parameters Z_a and Z_v, instead of EPA and EPV, were used for seismic hazard and zoning maps published for Canada. As can be seen from Table 14.3, the variation of parameters is in a ratio of 2 before zone 2, but in a ratio of 0.707 after zone 2.
2. The exceedance probability changes from 40% in the 1980 code to 10% in 50 years, in keeping with the world trend.
3. The design spectrum is defined by Z_a and Z_v (Figure 14.4).

14.2.3 Development of codes in China

(a) Chronology of codes and related issues in China

1955 A translation of the 1951 Soviet code 'Building Code in Seismic Regions PSP-101–51', was published.

1956 The first edition of the China Earthquake Intensity Zoning Map was published, which has not been officially used.

1957 The State Construction Commission entrusted the Institute of Civil Engineering and Architecture of the Academic Sinica to be the chief editor of a seismic design code for China.

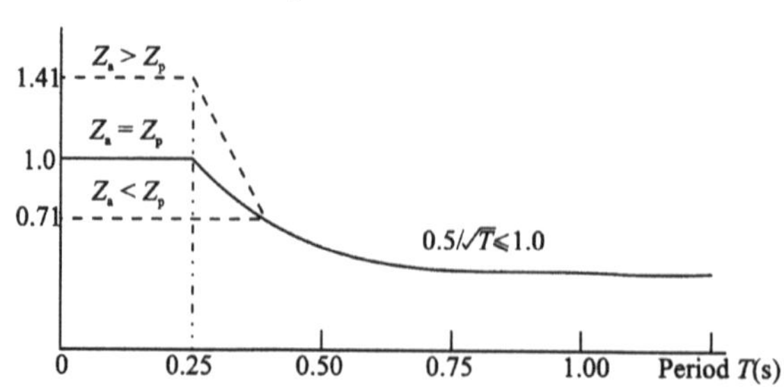

Figure 14.4 Design spectra in 1985 NBCC.

1959 The first seismic design code (draft) for China was completed, applicable to buildings, bridges, dams, water supply systems, etc. and was used in practice. It is comparable to the 1957 Soviet code CH-8-57. It remained a draft because the State Construction Commission was rescinded soon afterwards.

1962 The Institute of Civil Engineering and Architecture was reorganized under its present name, the Institute of Engineering Mechanics (IEM). The State Construction Commission returned and entrusted the IEM once more to edit the seismic design code jointly with other institutions.

1964 The second seismic design code (draft) 'Building Design Code for Seismic Region (draft)' was completed, which included buildings, water supply systems and bridges, but not hydraulic works. The code had been changed extensively; it had the following features.

1. Earthquake intensity was determined by referring to intensity of nearby cities.
2. Site effects: the idea of modifying the intensity, used in the 1959 draft, was discarded; instead the design spectrum was to be modified. Site spectra are given in Figure 14.5. Similar specifications in the USA and Japan appeared about ten years later.
3. Site classification: a hybrid method of combined qualitative and quantitative definitions was given (Table 14.4).
4. Seismic coefficient: The coefficient k_c in the 1959 draft was split into two coefficients C and k but $k_c = Ck$ on average; $k = a/g$ inherited the name seismic coefficient and represents the real peak acceleration, as given in Table 14.5; the structural coefficient C represents the capacity of absorbing energy or ductility, varying in the range $\frac{1}{3} - 1$.

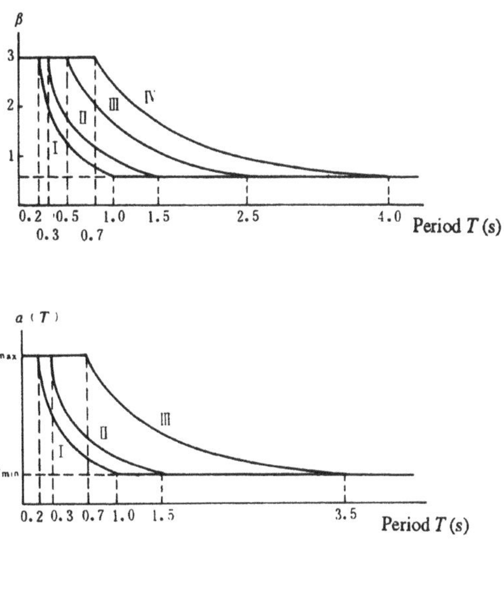

Figure 14.5 Design spectra for 5% damping in 1964 China code. ($0.6 \leq \beta = 3T_0/T < 3.0$)

Table 14.4 Site classification in 1964 code draft

Site type	Qualitative description	Bearing power	Density (T/m^2)		Predominant period (s)	V_P(km/s)
			Sand	Clay		
I	Stable rock	> 40	–	–	< 0.2	> 2
II	Ordinary soil, such as dense sand and stiff clay	15–40	> 1.5	> 1.4	0.15–0.50	1–2
III	Soft soil, such as loose sand and soft clay	8–15	1.4–1.6	1.2–1.5	0.35–0.70	0.3–1
IV	Very soft soil, such as deep and very soft clay and fill	< 8	< 1.4	< 1.2	> 0.6	< 0.3

5. Analytical methodology: the following two methods were specified.

- Equivalent static method, applicable to cantilevered structures:

$$\text{base shear } V_0 = Ck\beta qW \tag{14.21a}$$

$$\text{base moment } M_0 = Ck\beta mWH \tag{14.21}$$

  where H is the height of the center of gravity of the structure, q and m are coefficients of equivalent shear and moment whose values are specified for simple structures.
- Response spectrum method, applicable to all structures: the SRSS (square root of sum of squares) method was specified for computing the total response from modal responses; the first three modes are generally considered.

1974 The first legal code in China, Seismic Design Code of Industrial and Civil Buildings TJ 11–74 (Tentative), was published (chief editor was Research Academy of Building Science), which included only buildings. It followed closely the 1964 draft, but with the following major differences.

1. Noncausative faults: intensity of a site on a noncausative fault was not modified, instead of raising by one intensity level as in the 1964 draft.
2. Site and site spectra: site type III of the 1964 draft was deleted and the corner period T_0 of the spectrum for the new site type III was changed to 0.7 s.

Table 14.5 Seismic coefficient k in 1964 draft

Design intensity	k	
	Horizontal	Vertical
VII	0.075	0.038
VIII	0.15	0.075
IX	0.30	0.15
X	0.60	0.30

3. Seismic coefficient: a new term $\alpha(T)=k\beta(T)$ was introduced in place of k and $\beta(T)$.
4. Liquefaction: it was specified for the first time in a seismic code that a sand layer may liquefy if the blow count of standard penetration at the layer is less than the critical value N_c given by the formula

$$N_c=N_0[1+0.125(d_s-3)-0.05(d_w-2)] \qquad (14.22)$$

where d_s is the thickness of the soil above the sand layer, d_w the water level below the surface and N_0 is 6,10 or 16 respectively for design intensity VII, VIII or IX. Taiwan zoned her province into three zones, namely strong, moderate and weak, with the relative seismic coefficients 1.25, 1.0 and 0.75 respectively.

1977 The first legal inspection standard in China, and perhaps the world, Seismic Inspection Standard of Industrial and Civil Buildings TJ 22–77 (Tentative), was published. It included buildings of multistory brick wall bearing, inner frames, single-story reinforced concrete frames or brick columns, old-style timber frames, rural style, chimneys and water storage towers.

1978 The 1974 edition of the design code was revised on the basis of the experience from the 1975 Haicheng and 1976 Tangshan earthquakes and a new edition TJ 11–78 published. Seismic Design Code of Hydraulic Structures SDJ 10–78 (Tentative) was published and the chief editor was the Research Academy of Water Conservation and Electric Power. It has the following main features.

1. Intensity zonation is included.
2. The following two seismic analysis methods are specified.
   - Static method, applicable to dams of height less than 150 m:

     Inertial force $F_n=Ck\alpha_n W$ (earth dam and rockfill dam)

     overall coefficient $C=\frac{1}{4}$; earthquake coefficient $k=0.1$, 0.2 and 0.4 for intensities VII, VIII and IX respectively; earthquake acceleration coefficient $\alpha_n=1$ at base and 2.5 at top.

     Base shear $V_0=Ck\alpha W$ (concrete dam)

     $\alpha=1.0$ at base and 3.0, 5.0 and 6.0 at top respectively for height $H\leqslant 30$ m, $H=30$–70 m and 70–150 m

     $$\text{inertial force } F_n=\frac{V_0 W_n \Delta_n}{\sum W_j\Delta_j}$$

     Δ is specified in graphs and tables.
   - Spectrum method, applicable to dams higher than 150 m: the response spectrum is shown in Figure 14.6.

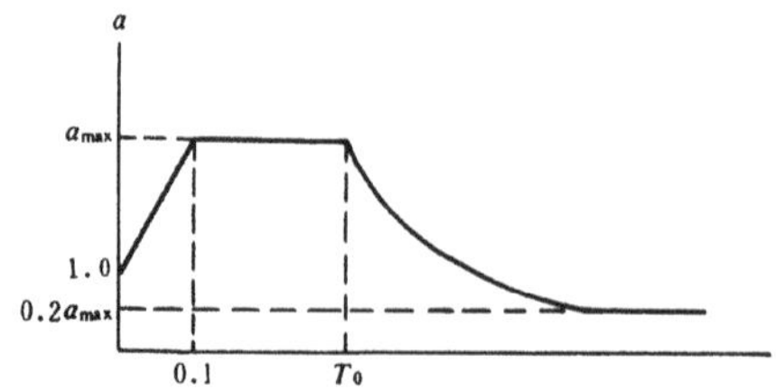

Figure 14.6 Design spectrum SDJ 10–78: T_0(s) = 0.2 for site I; 0.3 for site II; 0.7 for site III.

3. The soil liquefaction specification in TJ 11–78 was modified to include the effect of relative density D_r for sand and plastic index PI for silty sand.
   A tentative seismic design code for the coal industry was published.

1980 Modifications of seismic-related items of the 1974 edition of *Building Technical Guidelines* were suggested by the Division of Seismic Design Codes of the Earthquake Engineering Research Center in Taiwan; the main points are as follows.

1. Base shear: $V_0 = Zk\alpha IW$.
2. Seismic zoning factor: $Z = 1.0$, 0.8 and 0.6 respectively for strong, moderate and weak seismicity zones and the zonation was also modified.
3. Design spectrum: $\alpha = 1/(8T^{12}) \leqslant 0.15$.
4. Importance factor: $I = 1.5$, 1.25 and 1.0 respectively for important, moderately important and ordinary buildings.
5. Structural factor: $K = 0.67$–2.5 for buildings of various ductilities.
6. Vertical distribution of inertial forces

$$F_n = \frac{(V_0 - F_D)\, W_n H_n}{\sum W_j H_j}$$

$$F_D = 0.07TV_0 \leqslant 0.25$$

$$= 0 (T \leqslant 0.7\,\text{s})$$

1982 Revision to code BJ 11–79 was begun, with the Research Academy of Building Science as the chief editor.

1987 The draft of the revised edition of BJ 11–78 was completed. The State Seismological Bureau (SSB) assigned a group of specialists to edit a new earthquake zoning map of China.

1988 The Earthquake Zoning Group of the SSB decided to have two zoning maps, one for intensity of ordinary site and another for peak acceleration on rock sites, both with probabilistic definition.

1990 A new code, 'Seismic Design Code of Buildings GBJ 11–89', was approved to be the only legal code from 1991. The Earthquake Zoning Group of the SSB completed the intensity zoning map of China which was approved in 1991.

(b) Background to the revision of GBJ 11–89

The earthquakes at Haicheng in 1975 and Tangshan in 1976 warned the Chinese of the need for better structural design in earthquake areas. At the same time, two important seismic design codes were published with many important new features: ATC-3 in the United States and the 1981 building code in Japan. These two codes are milestones in earthquake engineering in the 1970s, a period of the following developments.

1. Several strong earthquakes provided important experience on modern structures, together with many strong motion observation data.
2. Laboratory tests accumulated many data on dynamic performance of elements and structures.
3. Theoretical work achieved success in seismic hazard and ground motion assessments, nonlinear analysis and reliability design methods.
4. Many nuclear power plants and other special and critical structures were constructed.

On the other hand, theory and practice of structural design for ordinary loads had undergone a great change from deterministic theory of safety factor to reliability theory. In China, a draft of Uniform Standard of Building Design was completed, which set up criteria, of which reliability design is one, to be followed by design codes for all kinds of buildings.

(c) Main features of GBJ 11–89

After the efforts of at least 50 specialists (a consultant group of nearly 20 specialists and a drafting group of more than 30) working for four years, a draft was completed in 1985 and distributed for comments; the final copy was completed in 1989 and approved in 1990. This edition had many important new features as follows.

Site classification. Thickness of deposit, bearing capacity and shear wave velocity were considered in site classification. The soil types were classified into four types: stiff, moderately stiff, moderately soft and soft, according to either shear wave velocity or qualitative description and bearing capacity. Soil type and thickness of the deposit were then considered together in the site classification, as shown in Table 14.6.

Soil liquefaction. Four major changes were made: a two-step judgement, consideration of silt liquefaction and effect of duration, and countermeasures against liquefaction.

A simple initial judgement should be made first; if one of the following conditions is satisfied, liquefaction may not be considered:

$$d_u > d_0 + d_b - 2 \tag{14.23a}$$

$$d_w > d_0 + d_b - 3 \tag{14.23b}$$

$$d_u + d_w > 1.5d_0 + 2d_b - 4.5 \tag{14.23c}$$

Table 14.6 Site classification in GBJ 11–89[a]

Soil type	Classification for d_{0V} =				
	0	0^{+}–3	3^{+}–9	9^{+}–80	> 80
Stiff	I	–	–	–	–
Moderately stiff	–	I	I	II	II
Moderately soft	–	I	II	II	III
Weak	–	I	II	III	IV

[a] d_{0V} = thickness of deposit in meters.

Table 14.7 Characteristic depth of lignefield soil

Intensity	Depth (m)	
	Silt	Sand
VII	6	7
VIII	7	8
IX	8	9

where d_w is the water table below the surface; d_u is the depth of unliquefied soil above; d_b is the depth of foundation bottom below surface; and d_0 is the characteristic depth of sand layer as given in Table 14.7. All depths are in meters. If the initial check does not indicate that there is no liquefaction, another check by standard penetration is then carried out. It will liquefy if the number of blows is less than

$$N_{\mathrm{cr}} = N_0[0.9 + 0.1(d_s - d_w)]\left(\frac{3}{\rho_c}\right)^{1/2} \tag{14.24}$$

where ρ_c is the precentage of the clay particle content, and the nominal blow counts $N_0 = 6$, 10 and 16 for intensities VII, VIII and IX for a nearby earthquake, 8 and 12 for intensities VII and VIII for a distant earthquake take account of the influence of duration or number of cycles of the dynamic action.

Another modification consisted of countermeasures against liquefaction added for different levels of liquefaction potential. A liquefaction index is defined first as

$$I_{\mathrm{1E}} = \sum\left(1 - \frac{N_i}{N_{\mathrm{cr}i}}\right)d_i w_i \quad (i = 1, 2, \ldots, n) \tag{14.25}$$

where n is the total number of tests of standard penetration made within 15 m depth of one boring; N_i and $N_{\mathrm{cr}i}$ are respectively the measured and the critical number of blows of the ith test with $N_i \leqslant N_{\mathrm{cr}i}$; d_i is the thickness of soil represented by the ith test; w_i is the weighting function for the ith test, with $w_i = 10$ for mean depth of the ith test less than 5 m and zero for mean depth 15 m, with unit $1/m$. Liquefaction level is then defined as light for $0 < I_{\mathrm{1E}} \leqslant 5$, moderate for $5 < I_{\mathrm{1E}} \leqslant 15$, and heavy for $I_{\mathrm{1E}} > 15$. Various countermeasures were finally specified for different liquefaction levels and different structural importance levels. Countermeasures varied from complete replacement of liquefiable soil to condensation, deep foundations and piling.

Design spectrum (Figure 14.7). There were two purposes in modifying the design spectrum. One was to consider the effect of distance or magnitude of the earthquake on the spectrum and the other to reduce the big jump between the spectra of the original types II and III. For close earthquakes, curves A(site I), B(site II), C(site III) and D(site IV) should be used; for distant earthquakes, curves B(site I), C(site II), D(site III) and E(site IV), should be used. No minimum is given for spectra but the range of applications of the specified spectra is limited to a maximum period of 3.0 s, as shown in Figure 14.7.

Deformation checking – two-level design. At the beginning of the code it is stated that buildings designed according to the code should need no repair to maintain functioning when subject to frequently occurring earthquakes, should maintain functioning with no or minor repair when subject to an earthquake of design intensity, and should not collapse when subject to rarely expected earthquakes. The code takes the ground motion of a small earthquake to be approximately one-third that of the design intensity and that of the large earthquake as twice that of the design intensity, or four to six times the ground motion of a small earthquake as specified in the code. For small earthquakes, elastic analysis is used and the cross-section of structural elements proportioned. Only for the following cases is second-level design required:

1. large and tall single-story factories of reinforced concrete columns on type III and IV sites in a region of intensity IX;
2. framed structures and brick buildings with a framed first story and a story yield strength coefficient $\xi_y = V_y/V_e$ less than 0.5 in a region of intensity VII to IX;
3. reinforced concrete structures of importance level A.

For regular buildings not over 12 stories and not belonging to importance level A, a simple method is given to compute the inelastic deformation.

Probabilistic format of elastic strength checking. For first-level design, a fractional loading factor format was adopted on the basis of failure probability, which is required by the Uniform Standard of Building Design in China.

Among the special features of the 1989 code, distance- and site-dependent design spectra, liquefaction and some constructional measures such as constructional columns for brick walls are particular to China; but two-level design is similar to the Japanese 1981 code and the principle of adjustment of design spectra as a function of distance may have been influenced by the ATC-3 code. The adoption of a probabilistic format of partial loading factors in a seismic code of ordinary building structure was for consistency with the design code for ordinary loads in China, which was already in a probabilistic format. However, all these features reflect modern world trends.

Earthquake distance	Characteristic period (s) for site			
	I	II	III	IV
Near	0.20	0.30	0.40	0.65
Far	0.25	0.40	0.55	0.85

Intensity	α_{max} for strength check
VI	0.04
VII	0.08
VIII	0.16
IX	0.32

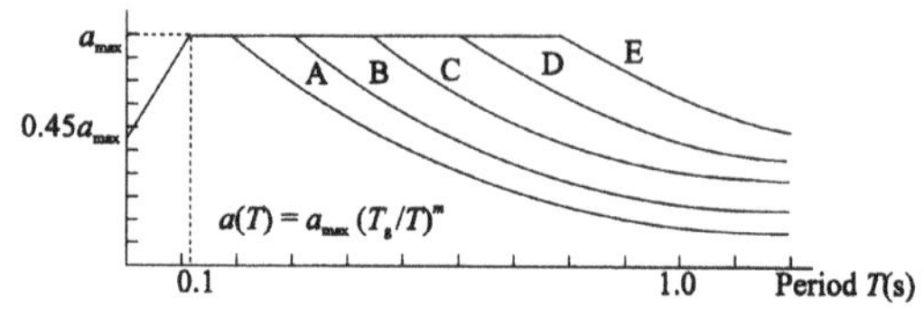

Figure 14.7 Design spectra of China code GBJ 11–89.

14.3 Trends in seismic codes

In present codes, response theory has been widely applied and real dynamic theory and nonlinear analysis are applied only for special cases; peak acceleration and standard average spectra are widely used in place of earthquake-related design spectra; and seismic design is mainly a form of checking rather than actual design.

Comparison of the development of earthquake engineering theory and code specifications shows their close relationship, but also that the latter tends to lag some way behind. It is therefore possible to predict the near future trends in code specifications on the basis of the present, and on the trends in theory. The following predictions may be useful.

Direct application of ground motion. It is already clear that design codes increasingly use ground motion parameters directly instead of specifying ground motion through intensity, defined in terms of a qualitative intensity scale. There are 14 codes which still use the term intensity, some of them only for zonation, not for design, for example the 1979 UBC code of the United States, which may be changed very soon to something similar to the zoning map of ATC-3. The code of the former Soviet Union is perhaps representative of the group defining ground motion through intensity and using it for design. An important disadvantage of this old approach is the need to tolerate a large abrupt jump in peak acceleration at the boundary of two intensity zones, especially for high intensities from VIII to IX, and a constant design spectrum for earthquakes of different magnitudes, both of which are neither scientific nor economic.

Because engineering design relies increasingly on quantitative analysis and because it is almost universally accepted that there is no adequate way to define ground motion from intensity, it is almost certain that intensity will soon lose its place in design codes, though it will still be useful in damage evaluation and prediction for simplicity reasons.

Consideration of site- and earthquake-dependent spectra. Most codes consider only site-dependent spectra; only a few consider both site-dependent and earthquake-dependent spectra, for example ATC-3, Romanian, Canadian and the building code for China. The draft of the 1981 Japan building code had this idea but it was not adopted in the final edition. The present method of specifying two parameters EPA and EPV to scale the spectrum is an acceptable simplification in this direction; if long-period motion is to be considered, it is easy to extend this approach by introducing a third parameter EPD for displacement. This idea may be traced back to the work done by Newmark's group in the 1970s. This is one case which shows the degree of delay between application and theory.

Protection as a function of hazard and importance. ATC-3 was the first code to require clearly the protection of a structure according to seismic hazard expected and the importance of the structure, which may be accepted by other codes.

Long-period portion of the spectrum. The accuracy of response spectra in the long-period portion is much lower than in the short-period portion, because errors from recording instruments and data processing in the former are not known. However, because many highrise structures such as buildings, towers and long-span bridges were built in the last two decades, the need for accurate specification of spectra in the period range 2–10 s is increasing. Newly available ground motion data from digital accelerographs may be used to check the accuracy and to give more dependable spectra for long-period sites.

Multilevel design. A two-level design principle has been present in design codes for many years, but the 1981 Japan building code first specified it explicitly, followed by the 1989 China code and others. In the United States, a research program is underway to introduce multilevel design into seismic design codes (Whitman, 1989). However, it may take many years to gain wide acceptance.

Probabilistic method of seismic design. The 1989 code for China was the first to adopt the probability format in order to be consistent with other design codes currently used in China. In principle, the probabilistic method is reasonable in the sense that there are so many uncertain factors involved in making design decisions and a probabilistic treatment provides an adequate and uniform base for all of them.

Programming of design process including complex analysis. Computer-aided design is already used for part of the design procedure; it is reasonable to expect programming of the whole design process to become the practice in design offices. Therefore it is likely to be applied increasingly to complex analysis such as nonlinear, torsional and dynamic analysis, as well as to detailing and drawings.

Other inclusions. It can be expected that seismic evaluation and retrofitting of existing buildings and structural control will be included in codes.

Part Five

New Analytical Tools in Earthquake Engineering

15 Stochastic methods

Since earthquake occurrence, ground motion and thus the structural responses to it, are highly random, stochastic methods are a natural tool for their consideration. Although recent theory classifies uncertainties into random, fuzzy and incomplete knowledge, the stochastic approach is still the main approach and deserves a whole chapter of discussion.

To comply with structural design principles used in some countries, a reliability-based design is also required for earthquakes. To complete such a design procedure, it is necessary to consider uncertainties from sources using:

1. seismic hazard assessment;
2. random earthquake ground motion modeling;
3. stochastic structural response analysis;
4. reliability-based design principles.

All of these four topics will be discussed in this chapter.

15.1 Stochastic models of earthquake occurrence

Because of the uncertainties associated with the underlying physical process of earthquake occurrence, its characterization must be probabilistic in nature. There have been many stochastic models proposed to describe earthquake occurrence. As can be readily expected, there is not one model which can be accepted to be applicable all over the world. In this section, only those models used in seismic hazard assessment are presented, which can be divided into Poisson models, renewal models, Markov models and semi-Markov models.

15.1.1 Poisson models

A Poisson process is a continuous-time, integer-valued counting process. For t_0, $t \geqslant 0$, let $N(t_0, t_0+t)$ be the number of events occurring in (t_0, t_0+t) and set $N(t)=N(0, t)$ and $N(0)=0$. It is called a Poisson process if it satisfies the following conditions.

1. The process has stationary increments, that is

$$P[N(t_0+t)-N(t_0)=k]=P[N(t)-N(0)=k] \qquad (k=0, 1, 2\ldots) \tag{15.1}$$

   in other words, the number of events occurring in the interval (t_0, t_0+t) depends only on the length of the interval t.
2. The process has independent increments, which means the probability of an event occurring in (t_0, t_0+t) is independent of the past history and can be expressed as

$$P[N(t_0+t)-N(t_0)=k \mid N(t_0)=m]=P[N(t_0+t)-N(t_0)=k] \tag{15.2}$$

3. The distribution of the number of events in (t_0, t_0+t) is

$$P[N(t_0+t)-N(t_0)=k]=\frac{\exp-(\nu t)(\nu t)^k}{k!} \qquad (k=0.1,\ldots) \qquad (15.3)$$

in which $\nu(\nu>0)$ represents the earthquake occurrence rate.

Besides, the Poisson process also assumes that the probability of two or more events occurring simultaneously is negligible. On the basis of the above assumptions, it is not difficult to deduce that the expected value and variance of the number of events in the time interval (t_0, t_0+t) are the same and equal to νt, and the density function of time interval between events is

$$f_T(t)=\nu\exp-(\nu t) \qquad (\nu,\ t\geqslant 0) \qquad (15.4)$$

According to the definition of hazard rate,

$$h(t)=\frac{f(t)}{1-F(t)} \qquad (15.5)$$

where $F(t)$ is the cumulative distribution function of the time interval between events. Here for the Poisson process, the following equation can be easily obtained:

$$h(t)=\nu$$

That is to say the Poisson process has constant hazard rate.

The Poisson process is perhaps the most commonly used in practice. Cornell (1968) and Der Kiureghian and Ang (1977) have used it in seismic hazard assessment, and Shi *et al.* (C1991) have used the segmental Poisson model in the preparation of the new seismic zonation map of China. There are several reasons for its popularity (Cornell and Winterstein, 1988), which include:

1. some successful comparisons of its predictions with observations (Gardner and Knopoff, 1974; McGuire and Barnhard, 1979);
2. rather broad acceptance that, lacking evidence to the contrary, the model is not unreasonable (especially for the less-large events that may govern the hazard);
3. the fact that the sum of non-Poissonian processes may be approximately Poissonian.

Apart from the above, the most important reason is its simplicity. It needs only one parameter ν, which can be obtained from the Gutenberg–Richter magnitude–frequency relationship, to provide a complete description of earthquake occurrence.

15.1.2 Renewal models

In the last section it was shown that the Poisson process has constant hazard rate and is independent of the past history of occurrence of events.

This 'memorylessness' characteristic of the Poisson process can not reflect the property of occurrence of large earthquakes and is contradictory to the well-known elastic rebound theory in seismology, which assumes that the longer the time since the last earthquake, the more probable the occurrence of the next earthquake. According to the elastic rebound theory of earthquake occurrence, when an earthquake occurs, it releases accumulated strain energy, which may in turn decrease the possibility of occurrence of another earthquake until the strain energy has again accumulated in the same region. The use of Poisson models may be appropriate for regions with relatively uniform seismic activity. However, for regions where infrequent but large earthquakes occur after long quiescence, it is not suitable. Therefore a variety of different models have been proposed to try to overcome this deficiency.

A counting process $\{N(t),\ t \geqslant 0\}$ is called a renewal process if the interval times between events are independent and identically distributed with arbitrary distribution. For a renewal process, the assumption of distribution of interval times between events is important. According to the definition of a renewal process, a Poisson process is just a renewal process, but with the difference that the distribution of interval times between events for a Poisson process is exponential, whereas that of a renewal process is arbitrary.

Generally, the distributions of interval time between events are assumed to be Weibull distributed (Savy *et al.*, 1980; Cornell *et al.*, 1993) or gamma distributed (Wang, C1987). Here, as an example, assume that the distribution of interval time between events is gamma distributed:

$$f_T(t) = \frac{\beta}{\Gamma(\alpha)} (\beta t)^{\alpha - t} \exp(-\beta t) \tag{15.6}$$

in which α and β are the parameters of the gamma distribution. The hazard rates for t when $\beta = 1$ corresponding to $\alpha > 1$, $\alpha = 1$ and $\alpha < 1$ are shown in Figure 15.1. It is not difficult to see that the hazard rate is the increasing function of t when $\alpha > 1$; this means that the longer the time since the last earthquake, the more probable is the next earthquake, which is what the elastic rebound theory implies. When $\alpha = 1$, the gamma distribution reduces

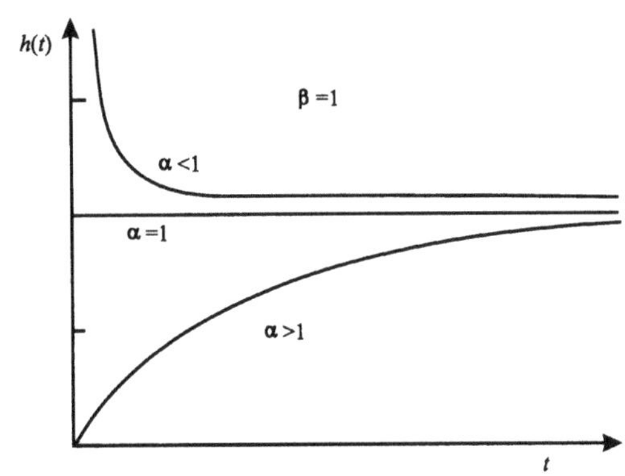

Figure 15.1 Hazard rate function for gamma distribution.

to a Poisson distribution, so the hazard rate is constant. When $\alpha < 1$, the hazard rate is the decreasing function of t, which means that the longer the time since the last earthquake, the less probable is the next earthquake. This property can also be used to model 'clustering' earthquakes after some period of quiescence.

It is not difficult to see from the above discussion that in contrast to the Poisson model, renewal models are dependent on the time of occurrence of the last earthquake.

15.1.3 Markov models

A stochastic process $\{Y(t): t \geqslant 0\}$ with state space E is a continuous-time Markov process if for t_0, $t \geqslant 0$ and $j \in E$

$$P[Y(t_0+t)=j \mid Y(u): u \leqslant t] = P[Y(t_0+t)=j \mid Y(t)] \qquad (15.7)$$

That is, the probability of the process being in state j at time t_0+t depends only on the state at time t_0 and is independent of the history up to t_0. In general, the transition probability function

$$P[Y(t_0+t)=j \mid Y(t_0)=i] = P_t(i, j) \qquad (15.8)$$

depends on t. When the conditional probability in equation 15.8 is independent of t ($t \geqslant 0$) for all $i, j \in E$, the process $\{Y(t): t \geqslant 0\}$ is a time-homogeneous Markov process. It is not difficult to prove that the transition probability satisfies:

$$\begin{gathered} P_t(i, j) \geqslant 0 \\ \sum_{j \in E} P_t(i, j) = 1 \\ \sum_{k \in E} P_{t_0}(i, k) = P_{t_0+t}(i, j) \end{gathered} \qquad (15.9)$$

for any $i, j \in E$ and $t \geqslant 0$. The sequence of states visited in the Markov process form a discrete-time Markov chain.

Many Markov-based models have been proposed to describe earthquake occurrence. Vere-Jones (1966) used one to model aftershock sequences; Knopoff (1971) used one to model the occurrence of main events; Chinese researchers (Wang, C1973) have used them to model the spatial migration of large earthquakes.

15.1.4 Semi-Markov models

A stochastic process $\{(Y_n, T_n): n \geqslant 0\}$ is called a semi-Markov process with state space E for all i, $j \in E$ if

$$\begin{aligned} &P[Y_{n+1}=j,\ T_{n+1}-T_n \leqslant t \mid Y_0, \ldots, Y_n;\ T_0, \ldots, T_n] \\ &\quad = P[Y_{n+1}=j,\ T_{n+1}-T_n \leqslant t \mid Y_n=i] \end{aligned} \qquad (15.10)$$

in which T_n takes values in $[0, \infty)$ and $0 \leqslant T_0 \leqslant T_1 \leqslant T_2 \ldots$. The above equation implies that the joint probability of the $(n+1)$th transition of the process to state j and of the time between the nth and $(n+1)$th transitions (also called holding time) being less than or equal to t depends only on the state at the nth transition. In other words, the semi-Markov process makes transitions according to a discrete-time Markov chain $\{Y_n: n \geqslant 0\}$ and has a random holding time which depends on the present state as well as the next state. Compared with a Markov process, it is easily seen that a semi-Markov process is more general.

The semi-Markov process was first used to model earthquake occurrence by Patwardhan *et al.* (1980) and has been applied to areas of the circum-Pacific belt (Patwardhan *et al.*, 1980) and the Wasatch fault zone, Utah (Cluff *et al.*, 1980). In its application, the related model parameters such as transition probability and holding time are estimated from historical and geological data as well as subjective assessments. Though the semi-Markov model is physically more appropriate than others, it needs a relatively large number of data for the reliable estimation of model parameters; such data are generally not available. Therefore many researchers propose some specific models which are somewhat simpler but physically still appropriate such as slip-predictable and time-predictable models (Kiremidjian and Anagnos, 1984; Anagnos and Kiremidjian, 1984), and the updated slip-predictable, time-predictable and combined slip- and time-predictable models (Cornell *et al.*, 1993), in which the correlation between the earthquake magnitude and the holding time is considered.

15.1.5 Summary

In this section, various models of earthquake occurrence used in seismic hazard assessment are discussed. As stated above, there is not one model which is applicable to the whole world. In practice, one should select an appropriate model according to the characteristics of the problem concerned. Besides, only some of the possible models are included here; many other models have been proposed which can be found in various journals and conference proceedings.

15.2 Stochastic models of earthquake response

Both stochastic modeling of ground motion and stochastic response analysis aim at finding some stochastic estimates, which include mean and variance, correlation function, power spectrum or maximum values with given probabilities of exceedance. Fundamental theories of stochastic processes and stochastic representation of earthquake motion are given in Chapter 3.

15.2.1 Stochastic representation of a linear transformation of a stochastic process

Let $x(t)$ represent a stationary stochastic process and $y(t)$ its linear transformation, i.e. $y(t) = Lx(t)$. The operator L may be linear differentiation, integration or a differential equation such as equation 9.3, the equation of motion of an SDF system, where input $p(t)$ corresponds to $x(t)$ and output $u(t)$ corresponds to $y(t)$. As defined by equations 3.18, $X(\omega)$ and $Y(\omega)$ are respectively the Fourier transformations of $x(t)$ and $y(t)$. Let $H(\mathrm{i}\omega)$ be the transfer function of operator L; it will be $H(\mathrm{i}\omega, t)$ if the nonstationary effect of the initial condition of the system is considered. The output process will be

$$y(t)=\frac{1}{2\pi}\int_{-\infty}^{\infty} H(\mathrm{i}\omega, t)\, X(\mathrm{i}\omega) \exp(\mathrm{i}\omega t)\, \mathrm{d}\omega \tag{15.11}$$

The correlation function of $y(t)$ is then

$$R_y(t, t')=\frac{1}{2\pi}\int_{-\infty}^{\infty} S_x(\mathrm{i}\omega)\, H(\mathrm{i}\omega, t)\, H^*(\mathrm{i}\omega, t')\cdot \exp(\mathrm{i}\omega(t-t'))\mathrm{d}\omega \tag{15.12}$$

For stationary conditions, $H(\mathrm{i}\omega, t)$ reduces to $H(\mathrm{i}\omega)$, and the correlation function and power spectrum of $y(t)$ will be

$$R_y(\tau)=\frac{1}{2\pi}\int_{-\infty}^{\infty} H(\mathrm{i}\omega)\, H^*(\mathrm{i}\omega)\, S_x(\omega) \exp(\mathrm{i}\omega\tau)\, \mathrm{d}\omega \tag{15.13a}$$

$$S_y(\omega)=H(\mathrm{i}\omega)\, H^*(\mathrm{i}\omega)\, S_x(\omega) \tag{15.13b}$$

For a differentiation operator, $L=\mathrm{d}/\mathrm{d}t$, $H(\mathrm{i}\omega)=\mathrm{i}\omega$ and

$$S_y(\omega)=\omega^2 S_x(\omega) \tag{15.14}$$

For an integration operator, $L=\int(\cdot)\mathrm{d}t$, $H(\mathrm{i}\omega)=(\mathrm{i}\omega)^{-1}$ and

$$S_y(\omega)=\omega^{-2} S_x(\omega) \tag{15.15}$$

It is important to realize that the conditions of differentiability and integrability of the random process $x(t)$ are respectively the existence of the following integrals:

$$\int_{-\infty}^{\infty} \omega^2 S_x(\omega)\, \mathrm{d}\omega \quad \text{or} \quad \int_{-\infty}^{\infty} \omega^{-2} S_x(\omega)\mathrm{d}\omega \tag{15.16}$$

15.2.2 Stochastic representation of earthquake ground motion

A popular stochastic model of earthquake ground acceleration is white or filtered noise with the following power spectrum:

$$S_{\mathrm{a}}(\omega)=\frac{c\omega^2+d}{\omega^4-a\omega^2+b} \tag{15.17}$$

For example, Barstein (Hu and Zhou, C1962) used the following form:

$$S_{\mathrm{a}}(\omega)=2R_{\mathrm{a}}(0)\frac{(\alpha-\mu\beta)\omega^2+(\alpha+\mu\beta)(\alpha^2+\beta^2)}{\omega^4-2(\beta^2-\alpha^2)\omega^2+(\alpha^2+\beta^2)^2} \tag{15.18}$$

and its related correlation function is

$$R_a(\tau) = R_a(0)\exp(-|\tau|)(\cos\beta\tau + \mu\sin\beta|\tau|) \qquad (15.19)$$

Kanai and Tajimi used the idea of site filter of an SDF system defined by equation 9.3 and suggested the following model for ground acceleration:

$$S(\omega) = S_0 \frac{1 + \zeta_g^2(\omega/\omega_g)^2}{[1-(\omega/\omega_g)^2]^2 + 4\zeta_g^2(\omega/\omega_g)^2} \qquad (15.20)$$

$$R(\tau) = \sigma_g^2 \exp(-\alpha|\tau|)\cos\beta\tau \qquad (15.21)$$

$$\sigma_g^2 = \frac{\pi S_0 \omega_g (1 + 4\zeta_g^2)}{2\zeta_g} \qquad (15.22)$$

where ω_g and ζ_g are respectively the natural frequency and damping ratio of the site, subjected to a white noise input from the bedrock (Clough and Penzien, 1975).

It was pointed out by the first author (Hu and Zhou, C1962) that equations 15.17, 15.18 and 15.20 will give unbounded displacement because the condition given in equation 15.16 after double integration is not satisfied; the following modification was suggested for the ground acceleration model:

$$S_a(\omega) = \left(\frac{\omega^2}{\omega^4 - a\omega^2 + b}\right)\left(\frac{\omega^2}{\omega^2 + \gamma^2}\right) \qquad (15.23)$$

in which the constant γ was called the low-frequency reduction factor. Clough and Penzien (1975) suggested the following modification:

$$S(\omega) = \left(\frac{1 + 4\zeta_g^2(\omega/\omega_g)^2}{[1-(\omega/\omega_g)^2]^2 + 4\zeta_g^2(\omega/\omega_g)^2}\right)\left(\frac{(\omega/\omega_1)^4}{[1-(\omega/\omega_1)^2]^2 + 4\zeta_f^2(\omega/\omega_1)^2}\right) \qquad (15.24)$$

where the constants ω_l and ζ_l are selected to give the desired filter characteristics. It is more flexible but with more parameters to be determined. Figure 15.2 gives a set of results for different values of γ and μ for Barstein's spectrum.

15.2.3 Linear response of an SDF system subjected to a stationary input

The response $y(t)$ of an SDF system subjected to random input $x(t)$ is random and may be evaluated from equations 15.11–15.12, from which other statistic estimates may be derived. For example, a maximum y_{max} with given probability P of exceedance may be written as

$$y_{max}(P) = r_{T_d,P}\,\sigma_y \qquad (15.25)$$

where T_d is the duration of the stationary response and $r_{T_d,P}$ a peak factor related to duration and probability of exceedance. In terms of spectrum, the relative displacement, relative velocity and absolute acceleration response

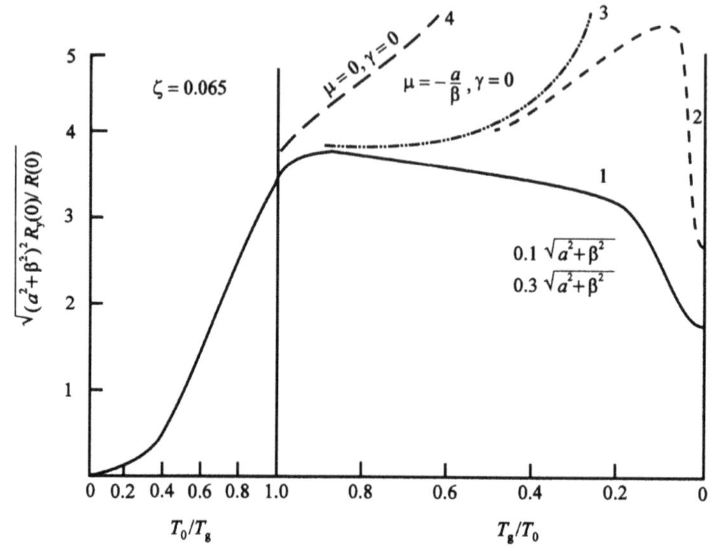

Figure 15.2 Effect of low-frequency reduction factor.

spectra SD, SV and SA are random. Taking the displacement spectrum SD of a stationary response as an example, the probability of nonexceedance is

$$\mathrm{FD}(\alpha)=\mathrm{P}[\mathrm{SD}\leqslant\alpha]=\exp(-\nu(\alpha)\,t)=P \tag{15.26}$$

Let

$$\mathrm{SD}_{T_d,P}=\alpha=r_{T_d,P}\sigma_y \tag{15.27a}$$

$$\overline{\mathrm{SD}}=E[\mathrm{SD}]=\bar{r}\sigma_y=r_{T_d,P=0.5}\sigma_y \tag{15.27b}$$

$$\tilde{\mathrm{SD}}=\mathrm{Var}[\mathrm{SD}]=\tilde{r}\sigma_y^2 \tag{15.27c}$$

$$\sigma_y=\frac{\pi S(\omega_0)}{2\zeta\omega_0^3} \tag{15.27d}$$

where ω_0 and ζ are respectively the natural frequency and damping ratio of the system.

1. For independent crossing:

$$\nu(r)=\frac{\omega_0}{\pi}\exp\left(-\frac{r^2}{2}\right) \tag{15.28}$$

$$FD(r)=\exp\left[-\frac{\omega_0 T_d}{\pi}\exp\left(\frac{-r^2}{2}\right)\right] \tag{15.29}$$

2. For two-state Markov crossing:

$$r_{T_d,P}=\left[-2\ln\left(\frac{\omega_0 T_d}{\pi\ln P}\right)\right]^{1/2} \tag{15.30}$$

$$\bar{r}=\left[2\ln\left(\frac{\omega_0 T_d}{\pi}\right)\right]^{1/2}+\frac{0.5772}{[2\ln(\omega_0 T_d/\pi)]^{1/2}} \quad (15.31)$$

$$\tilde{r}=\frac{\pi^2}{12\ln(\omega_0 T_d/\pi)} \quad (15.32)$$

$$\nu(r)=\frac{\omega_0}{\pi}\left(\frac{1-\exp(\pi^{1/2}\delta_e r/2)}{\exp(-r^2/2)-1}\right) \quad (15.33)$$

where

$$\delta_e=\delta^{1.2} \text{ and } \delta=2\left(\frac{\zeta}{\pi}\right)^{1/2} \quad (15.34)$$

$$\bar{r}=\left[2\ln\left(\frac{\omega_{oe} T_d}{\pi}\right)\right]^{1/2}+\frac{0.5772}{[2\ln(\omega_{oe} T_d/\pi)]^{1/2}} \quad (15.35)$$

$$\tilde{r}=\left(\frac{1.2}{[2\ln(\omega_{oe} T_d/\pi)]^{1/2}}+\frac{5.4}{13+[2\ln(\omega_{oe} T_d/\pi)]^{3.2}}\right)^2 \quad (15.36)$$

where

$$\begin{aligned}\omega_{oe}&=\max(1.05, 2\delta\omega_0) && (0<\delta<0.1)\\ &=(1.63\delta^{0.45}-0.38)\,\omega_0 && (0.1<\delta<0.69)\\ &=\omega_0 && (0.69<\delta<1.0)\end{aligned} \quad (15.37)$$

The two-state Markov assumption gives good results. A detailed discussion of the independent and two-state Markov crossing is given in Nigam (1991).

15.2.4 Nonstationary response of an SDF system

(a) Nonstationarity from initial conditions

When a stationary input motion is suddenly applied to an SDF system at rest, the response will be nonstationary at the beginning. The total energy per unit mass can be derived for a white noise input as

$$\frac{\overline{2E}}{m}=\overline{\omega_0^2 y^2}+\bar{\dot{y}}^2=\frac{2S_1}{\zeta\omega_0}\left[1-\frac{1}{\omega_1^2}\exp(-2\zeta\omega_0 t)\,(\omega_1^2+2\zeta\omega_0^2\sin^2(\omega_1 t))\right] \quad (15.38)$$

$$\sigma_y^2(t)=\overline{y^2}\frac{S_0}{4\zeta\omega_0^3}\left[1-\exp(-2\zeta\omega_0 t)\left(\frac{1}{1-\zeta^2}-\frac{\zeta^2}{1-\zeta^2}\cos(2\omega_1 t)\right.\right.$$
$$\left.\left.+\frac{\zeta}{(1-\zeta^2)^{1/2}}\sin(2\omega_1 t)\right)\right] \quad (15.39)$$

where S_0 is the white noise spectral density of the input acceleration; S_1 is related to S_0; $\omega_1=(1-\zeta^2)^{1/2}\omega_0$.

For a sufficiently long time t, the response approaches its stationary value

$$\sigma_y^2(\infty)=\frac{S_0}{4\zeta\omega_0^3} \tag{15.40}$$

(b) Stationarized process

For a stationarized earthquake motion

$$x(t)=f(t)\,n(t) \tag{15.41}$$

where $n(t)$ is a stationary process and the strength envelope function $f(t)$ is a deterministic function of time, the correlation function and variance of the response $y(t)$ are

$$R_y(t,t')=\iint_{-\infty}^{\infty} h(t'-\tau')\,h(t-\tau)\,f(\tau)\,f(\tau')\,R_n(\tau-\tau')\,\mathrm{d}\tau\,\mathrm{d}\tau' \tag{15.42}$$

$$\sigma_y^2=R_y(t=t')=\iint_{-\infty}^{\infty} h(t'-\tau')\,h(t-\tau)\,f(\tau)\,f(\tau')\,R_n(\tau-\tau')\,\mathrm{d}\tau\,\mathrm{d}\tau' \tag{15.43}$$

which can be written in the following form:

$$\sigma_y^2=\frac{1}{\pi}\int_{-\infty}^{\infty} S_n(\omega)|H(\mathrm{i}\omega,t)|^2\mathrm{d}\omega \tag{15.44}$$

where $S_n(\omega)$ is the power spectral density of the related stationary input $n(t)$, $H(\mathrm{i}\omega, t)$ is the instantaneous transfer function for $x(t)=f(t)\exp(\mathrm{i}\omega t)$ and

$$|H(\mathrm{i}\omega,t)|^2=\iint_{-\infty}^{\infty} f(\tau)\,f(\tau')\,h(t-\tau)\,h(t-\tau')\cos\omega(\tau-\tau')\,\mathrm{d}\tau\,\mathrm{d}\tau' \tag{15.45}$$

Bolotin (1960) studied the response of a stationarized input with $f(t)=\exp(-ct)$ and a related stationarized power spectrum given by equation 15.18 and obtained the following approximate result:

$$\sigma_y^2(t,t')=\frac{1}{2\omega_1^2}\exp(-2\zeta\omega_0 t)\left(\frac{\exp(-2\Omega_1 t)\,Y_1-Y_2}{\Omega_1}-\frac{\exp(-2\Omega_1 T)(\Omega_1 Y_1-\omega_1 Y_3)}{\omega_1^2+\Omega_1^2}\right.$$
$$\left.+\frac{(\Omega_1 Y_2+\omega_2 Y_4)\cos(2\omega_1 t)-(\omega_1 Y_2-\Omega_1 Y_4)\sin(2\omega_1 t)}{\omega_1^2+\Omega_1^2}\right) \tag{15.46}$$

where

$$Y_{1,2}=\int_0^t \exp[-(\alpha\pm\Omega_1)\,\tau]\cos(\beta\tau)\cos(\omega_1\tau)\,\mathrm{d}\tau \tag{15.47a}$$

$$Y_{1,2}=\int_0^t \exp[-(\alpha\pm\Omega_1)\,\tau]\cos(\beta\tau)\sin(\omega_1\tau)\,\mathrm{d}\tau \tag{15.47b}$$

$$\Omega_1 = \zeta\omega_0 - C \tag{15.47c}$$

A simplified approach was suggested (Zhou, Hu and Chen, C1965) to consider the effect of nonwhite spectra and nonstationarity. For a stationarized process, the instantaneous response is usually smaller than the corresponding stationary case and the difference is controlled by the strength envelope function $f(t)$. It is assumed in the suggested approach that the ratio

$$f = \frac{\sigma_y(t)}{\sigma_y(\text{stationary})}$$

depends primarily on $f(t)$ but not on the spectral shape. With this assumption, the ratio f of the nonstationary case can be evaluated from a much simpler case of white noise input:

$$\sigma^2_{ynw}(t) = \sigma^2_{ynw}\ (\text{stationary}) \frac{\sigma^2_{yw}(t)}{\sigma^2_{yw}(\text{stationary})} \tag{15.48}$$

where nw stands for nonwhite and w for white input. This relation may also be written finally in the following form:

$$\sigma^2_{ynw}(t) = 4\zeta\omega_0^2\sigma^2_{ynw}(\text{stationary}) \int_0^t f^2(\tau)\, h^2(t-\tau)\, \mathrm{d}\tau \tag{15.49}$$

Table 15.1 shows a numerical example for comparison of the above-mentioned approximation, taken from Hu and Zhou (c1965).

In the numerical example, parameters in equation 15.18 and others are those used by Bolotin, i.e. $\alpha = 0.7$ Hz, $\mu = 0$, $\omega_0 = 10$ Hz, $\zeta = 0.04$ and $c = 0.02\,\omega_0 = 0.20$ with envelope function $f(t) = \exp(-ct)$. It can seen from this comparison that spectral shape for different values of β/ω_0 does have a strong effect on either $\sigma_y^2(\text{max})$ or $\sigma_y^2(\text{stationary})$, as shown in the second and third rows, but does not have much effect on their ratio f, which is the assumption made in the approximation.

15.2.5 Nonlinear response of an SDF system (Nigam, 1991)

For an ideal elasto-plastic system, the elastic energy is transferred into plastic work at the beginning of an inelastic excursion, i.e.

Table 15.1 Comparison of the approximation in Equation 15.44

	Bolotin (nonwhite)				Hu and Zhou (nonwhite)
β/ω_0	$\frac{w_0^4\sigma_y^2(\text{max})}{R(0)}$	$\frac{w_0^4\sigma_y^2(\text{stationary})}{R(0)}$	f^2	f^2 (white)	$\frac{w_0^4\sigma_y^2(\text{stationary})}{R(0)}$
1.05	24.2	47.2	0.51	0.50	59.2
1.10	15.6	31.2	0.50	0.50	29.4
1.15	11.4	20.0	0.57	0.50	16.0

$$g_y|\delta| = \frac{1}{2} m \dot{y}'^2|_{y(t)=y_y} \tag{15.50}$$

or

$$|\delta| = \frac{1}{2\omega_0^2 y_y} \dot{y}'^2\Big|_{y(t)=y_y} \tag{15.51}$$

where $y' = y - y_{pn}$ is the relative displacement of the system after the nth inelastic excursion; y_{pn} is the permanent displacement after the nth excursion; g_y and y_y are respectively the yield strength and displacement.

For random input, $|\delta|$ is random; its mean is

$$|\bar{\delta}| = \frac{1}{2\omega_0^2} \sigma_{\dot{y}}^2 = \frac{\sigma_y}{2r_y} \tag{15.52}$$

where $r_y = y_y/\sigma_y$. The total energy dissipated due to yielding is

$$E = \sum_i |D_i| \tag{15.53}$$

where D_i is the energy dissipated in the ith inelastic excursion.

The mean and variance are respectively

$$[\bar{E}] = \left(\frac{\omega_0 T_d}{2\pi}\right)\left(\frac{\sigma_y}{r_y}\right)\exp\left(\frac{-r_y^2}{2}\right) \tag{15.54}$$

$$\mathrm{Var}[E] = \frac{3}{2}\left(\frac{\omega_0 T_d}{2\pi}\right)\left(\frac{\sigma_y^2}{r_y^2}\right)\exp\left(\frac{-r_y^2}{2}\right) \tag{15.55}$$

15.2.6 Linear random responses of an MDF system

Let $Q(t)$ be any response quantity of a linear MDF system subjected to earthquake motion. Through modal decomposition technique, $Q(t)$ can be expressed in its modal responses $Q_i(t)$ $(i = 1, 2, \ldots)$:

$$Q(t) = \sum_i Q_i(t) = \sum_i \alpha_i q_i(t)$$

where α_i is the ith modal value of the quantity Q, which depends on the ith modal function, the participation factor of the ith mode and other structural parameters, but is not related to ground motion; the modal response $Q_i(t)$ depends on both the ith modal parameters and ground motion.

If the ground acceleration $a_g(t)$ is stationary with zero mean and one-sided power spectrum $G_g(\omega)$, the stationary response $Q(t)$ then has zero mean and its one-sided spectrum is

$$G(\omega) = \sum_j \sum_k \alpha_j \alpha_k H_j(i\omega) H_k^*(i\omega) G_g(\omega) \tag{15.56}$$

where $H_j(i\omega)$ is the transfer function of the jth modal response of $Q(t)$.

Theoretically speaking, if a relationship such as that given in equation 5.10 between the response spectrum $S_a(T, \zeta)$ and the power spectral density

$G(\omega)$ is given, the following relation may be derived between the response spectrum of $Q(t)$ and its modal responses from equation 15.56 under certain assumptions as follows (Der Kiureghian, 1981):

$$S_{aQ}^2 = \sum_j \sum_k \alpha_j \alpha_k \rho_{0,jk} S_{aj} S_{ak} \tag{15.57}$$

where S_a stands for the acceleration response spectrum $S_a(\omega, \zeta)$ and

$$\rho_{0,jk} = \frac{\lambda_{0,jk}}{(\lambda_{0,jj}\lambda_{0,kk})^{1/2}} \tag{15.58}$$

where

$$\lambda_{0,jk} = \mathrm{Re}\left[\int_0^{\infty} H_j(i\omega)\, H_k^*(i\omega)\, G_g(\omega)\,\mathrm{d}\omega\right] \tag{15.59}$$

The cross term $\rho_{0,jk}(j \neq k)$ represents the correlation of the responses of the neighboring modes. Both ρ and λ are functions of the duration of strong motion, the frequency of the neighboring response modes and the damping ratio of the system. If the duration is long enough compared to the considered periods of the structure and if the input spectrum is smooth over a wide frequency range, the following approximation may be used:

$$\rho_{ij} = \frac{8(\zeta_i\zeta_j)^{1/2}(\zeta_i + \beta\zeta_j)\,\beta^{3/2}}{(1-\beta^2)^2 + 4\zeta_i\zeta_j\beta(1+\beta^2) + 4(\zeta_i^2 + \zeta_j^2)\beta^2} \tag{15.60}$$

where $\beta = \omega_i/\omega_j$. In the case of constant damping ζ,

$$\rho_{0,ij} = \frac{8\zeta^2(1+\beta)\,\beta^{3/2}}{(1-\beta^2)^2 + 4\zeta^2\beta(1+\beta)^2} \tag{15.61}$$

The above results were obtained by Der Kiureghian (1981) and named the CQC (complete quadratic combination) method which reduces to the SRSS (square-root of sum of squares) method for structures with well separated modal responses, i.e. with separated modal frequencies or small damping. In this case, the neighboring modal responses of the system will be independent and the cross term $\rho_{0,ij}(i \neq j)$ vanishes.

The CQC method is now commonly used in practice, such as in the seismic design codes of the SEAOC and API.

Newmark and Rosenblueth (1971) suggested a similar result, in which an equivalent damping ratio

$$\zeta_i' = \zeta_i + \frac{2}{\omega_i T_d} \tag{15.62}$$

is defined to consider the effect of duration, and then the following approximate equation is proposed:

$$Q^2 = \sum_i Q_i^2 + \sum_i \sum_j \frac{Q_i Q_j}{1 + \zeta_{ij}^2} \tag{15.63}$$

Table 15.2 Comparison of CQC and Newmark–Rosenblueth (N–B) results

ζ	ω_i/ω_j	N–B	CQC	N–B/CQC
0.02	0.25	0.001	0.001	1.95
	0.5	0.004	0.003	1.19
	1	1	1	1
0.05	0.25	0.007	0.004	1.95
	0.5	0.02	0.02	1.2
	1	1	1	1
0.10	0.25	0.027	0.014	1.93
	0.5	0.083	0.070	1.18
	1	1	1	1

where

$$\zeta_{ij}=\frac{\omega_i-\omega_j}{\zeta_i'\omega_i+\zeta_j'\omega_j} \tag{15.64}$$

Unfortunately, this good result has been unduly ignored in practical applications.

Table 15.2 gives a comparison of the CQC method and the Newmark-Rosenblueth (N–B) result (for long duration) for the structure considered by Wilson, Der Kiureghian and Bayo (1981). It can be seen that the N–B results are very good.

Similar to the case of an SDF system, the mean and variance of $Q(t)$ are

$$Q_{T_d,P}=r_{T_d,P}\sigma_Q \tag{15.65}$$

$$\bar{Q}=\bar{r}\sigma_Q \tag{15.66}$$

$$\text{var}[Q]=\tilde{r}\sigma_Q^2 \tag{15.67}$$

where $\bar{r}$ and $\tilde{r}$ are the same as those given in equation 15.31–15.32 or equations 15.36–15.37.

15.2.7 Reliability analysis of earthquake action

Reliability analysis of earthquake action consists of three stages: seismic hazard assessment, structural damage analysis and damage level selection. The first stage has been treated in Chapter 12, and the other two will be treated here.

(a) Damage criteria

Damage criteria may be classified into two categories as follows. The damage threshold defines damage as the overpassing of a limit or threshold; if the response is below a specified limit, the structure is considered safe. The boundary of 'safe' and 'failure' is defined clearly, although it may be random. The other category is the damage index, where damage is defined by an index varying from 0 to 1, with 0 for safe, 1 for failure or complete damage and intermediate values for partial damage. In the threshold damage criterion, damage is given with a probability when the threshold and/or other quantity is considered random, while in the index damage criterion, damage is defined by an index showing the intensity or extent of damage.

Let y be a group of quantities of structural response and strength or resistance and damage criterion be defined by a function of y, $D=f(y)$.

The threshold damage criterion D is of the following form:

$$D=y-y_0\geqslant 0 \tag{15.68}$$

where y_0 is the threshold of y. In the commonly used criteria, y is the internal force, stress or deformation and y_0 the threshold or allowable value of y.

The index damage criterion is of the following form:

$$D=Ny^b-C \tag{15.69}$$

where N, b and c are constants and D takes continuous values from 0 to 1.

Take low-cycle fatigue as an example, with N to represent the number of cycles of the response; N may only be a few cycles for a very intense earthquake or a few tens of cycles for a moderate earthquake, while ordinary fatigue considers very large N values such as a few tens of thousands.

The Palmgren–Miner criterion defines the damage index for repeated action of nonequal amplitudes as

$$D=\sum_i \frac{n_i}{N_i} \tag{15.70}$$

where n_i is the actual number of repetitions and N_i the critical number of repetitions corresponding to a certain amplitude of y_i.

Ang and his group (Ang *et al.*, 1984) defined the following low-cycle fatigue criterion for reinforced concrete structures:

$$D=\frac{y}{y_0}+\beta\frac{\varepsilon}{Q_y y_0} \tag{15.71}$$

where y_0 is the maximum deformation for complete damage once occurred, Q_y the yield limit strength, and ε the total energy dissipated during the repetition of deformation, and β a constant. Jiang and Sun (C1987) adopted a similar form for some masonry structures as follows:

$$D=\left[\frac{y}{y_0}+3.67\left(\frac{\varepsilon}{Q_y y_0}\right)^{1.12}\right]^{1/2} \tag{15.72}$$

(b) Dynamic reliability analysis

For nonstationary responses, the dynamic reliability is defined as

$$R_d(T_d)=P[\max D|T_d]=\exp\left(-\int_0^{T_d}\nu(t)\,dt\right) \tag{15.73a}$$

For stationary responses it is

$$R_d(T_d)=\exp(-\nu t) \tag{15.73b}$$

Threshold damage criterion. For the following specific threshold damage criterion,

$$y(t) < -a \quad \text{or} \quad y(t) > a \qquad (0 \leq t \leq T_d) \tag{15.74}$$

where a is positive, the foregoing equation yields for a stationary Gaussian process

$$R_d(a, T_d) = P[\max|y(t)| > a | T_d]$$
$$= \exp\left[-\frac{(\lambda_2/\lambda_0)^{1/2}}{\pi} T_d \exp\left(\frac{-r^2}{2}\right)\right] \tag{15.75}$$

where $r = \alpha/\sigma_y$, and λ_n for $n = 0$ or 2 is the spectral parameter.

If the clumping effect is considered, the two-state Markov assumption will give a better result:

$$R_d(a, T_d) = \exp\left[-\frac{(\lambda_2/\lambda_0)^{12}}{\pi} T_d \exp\left(\frac{-r^2}{2}\right) \cdot \left(\frac{1 - \exp[-(\pi/2)^{1/2} \delta_e^{1.2} r]}{1 - \exp(-r^2/2)}\right)\right] \tag{15.76}$$

For nonstationary responses, an equivalent stationary duration $T_{deq} < T_d$ may be defined by the following expression:

$$R_d(T_d) = \exp\left[-\int_0^{T_d} \nu(t)\,dt\right] = \exp[-\nu(T_d)\, T_{deq}] \tag{15.77}$$

Introducing

$$m = \frac{\sigma^2(T_d)}{\sigma^2(T_d/2)} \tag{15.78}$$

the following approximate value may be obtained:

$$\frac{T_{deq}}{T_d} = \exp[-2(m-1)] = \exp\left[-2\left(\frac{\sigma_y^2(T_d)}{\sigma_y^2(T_d/2)} - 1\right)\right] \tag{15.79}$$

The results for stationary response can then be used for nonstationary cases if T_{deq} is used in place of T_d.

Index damage criterion. For the index damage criterion, the damage is defined as

$$D \geq R \tag{15.80}$$

where D is the actual damage index during the period 0–T_d and R a specified limiting index. D is random because earthquake motion is random; R is also random because structural strength is of large uncertainty. Usually the random parameters D and R may be treated as independent, and the structural reliability is

$$R_{\mathrm{D}}(T_{\mathrm{d}}) = \iint_{D<R} f_D(y_1) f_R(y_2)\,\mathrm{d}y_1\,\mathrm{d}y_2$$
$$= \int_0^\infty \left[\int_0^{y_2} f_D(y_1)\,\mathrm{d}y_1\right] f_{\mathrm{R}}(y_2)\,\mathrm{d}y_2$$
$$= \int_0^\infty F_{\mathrm{D}}(y_2) f_{\mathrm{R}}(y_2)\,\mathrm{d}y_2 \tag{15.81}$$

where $F_D(y) = P[D < y]$ is the cumulative probability function of D. Since this function is difficult to determine analytically, it is usually assumed to follow the log-normal distribution for both D and R and the following simple result for safety can then be obtained:

$$R_{\mathrm{D}}(T_{\mathrm{d}}) = P[D < R] = P[D/R < 1] = P[\ln D - \ln R < 0] \tag{15.82}$$

Let $z = \ln D - \ln R$. Since both $\ln D$ and $\ln R$ are normally distributed, z is also normally distributed and

$$R_{\mathrm{D}}(T_{\mathrm{d}}) = P[z < 0] = P\left[\frac{z - m_z}{\sigma_z} < -\frac{m_z}{\sigma}\right] \tag{15.83}$$

where m_z and σ_z are respectively the mean and standard error of z, and the reduced variable $(z - m_z)/\sigma_z$ is also normally distributed; the final result is

$$R_{\mathrm{D}}(T_{\mathrm{d}}) = \Phi(\beta) \tag{15.84}$$

where $\Phi(\cdot)$ is the standard normal probability function and the reliability index is

$$\beta = -\frac{m_z}{\sigma_z} = \frac{m_{\ln R} - m_{\ln D}}{(\sigma_{\ln R^2} + \sigma_{\ln D^2})^{1/2}} \tag{15.85}$$

This result may also be expressed in terms of mean, standard error and coefficient of variation for both R and D:

$$\beta = \frac{\ln\{(m_R/m_D)[(1 + V_D^2)/(1 + V_R^2)]^{1/2}\}}{\{\ln[(1 + V_R^2)(1 + V_D^2)]\}^{1/2}} \tag{15.86}$$

where $V_R = \sigma_R/m_R$ and $V_D = \sigma_D/m_D$ are respectively the coefficients of variation of R and D.

For practical cases, the mean and variance of the strength R are derived from test data and those for force parameter D are derived under simplified assumptions. Take the accumulated energy dissipation

$$D = \int \mathrm{d}E$$

as an example. It is possible to derive an expression for the energy $S(y)$ dissipated in one cycle and the total dissipated energy will be

$$D=\sum_i S(y_i)\,n_i \tag{15.87}$$

where n_i is the number of repetitions of amplitude y_i. For stationary normal responses with a narrow-band spectrum, the mean and variance of D will be

$$m_D=\lambda_0 T_{\mathrm{d}}\int_0^\infty \frac{S(y)}{\sigma_y^2}\,y\exp\left(-\frac{y^2}{2\sigma_y^2}\right)\mathrm{d}y \tag{15.88}$$

$$\sigma_D^2=\left(\frac{(\lambda_2/\lambda_0)^{1/2}\,T_{\mathrm{d}}}{2\pi}\right)^2\int_0^\infty \frac{S^2(y)}{\sigma_y^2}\exp\left(-\frac{\zeta^2}{2\sigma_y^2}\right)\mathrm{d}y-m_y^2 \tag{15.89}$$

(c) Damage level selection

Damage level or design level selection is a problem of determining a level of risk that one feels is tolerable or a safety level that one is willing to pay for. Theoretically speaking, this problem may be studied through benefit analysis to obtain a reasonable balance between investment and possible future loss. But the practical situation is to select such a level compatible with past practice because past experience is considered acceptable in a global sense, and it is not easy to consider a sudden change in design level without very strong reasons.

In many codes and in practice, the damage level used is a function of the importance of the structure. For ordinary building structures, probability of damage is around 10% in 50 years, but may reach 1% in 100 years for important facilities in nuclear power plants for the present practice of design and load evaluation, in which many uncertainties are still not evaluated. Once these uncertainties are considered, the probability of damage may change.

16 Fuzzy set theory and applications

An earthquake is a complex phenomenon. Despite great advances made in the past few decades, earthquake engineering remains as much an art as a science. Progress is constrained by a scarcity of data, the relatively short history of scientific recording and the complexity of the underlying physical process. Expertise is often needed in handling incomplete knowledge, combining information from different sources and intepreting results from sophisticated models.

Uncertainties exist in earthquake source modeling, seismicity, attenuation, structural response, hazard mitigation, risk assessment and decision making. The random aspects of these uncertainties have been studied quite thoroughly using probabilistic methods. By contrast, fuzzy information (such as vague, imprecise or linguistic data) has been largely ignored. Fuzzy information is sometimes incorporated obliquely in the ways engineers use their judgement and experience, but the manner in which this is done is ad-hoc, ill-structured and often unaccountable.

Fuzziness, like randomness, is a kind of uncertainty, and a question often asked is: 'What is the difference between fuzziness and randomness?' Randomness is due to our inability to predict the occurrence of an event, so we resort to repetitive tests and observed data to obtain frequency information. In this setting each event (single or compound) is discrete, which means that when an event takes place, we can definitely recognize it. The uncertainty is related to the occurrence of the event, which is a matter of chance. As is well known, the theory of probability, which is based on set theory, has been developed to deal with randomness. However, some items of information are nonstatistical in nature and have an inherent inexactness that is difficult to define, e.g. the meaning of 'large' in 'large earthquakes'. In this case, the uncertainty is embedded in the vague concept 'large earthquake', whose denotation is not an ordinary set but a fuzzy one. In reality many phenomena are so diverse and complex that they defy description in discrete quantities or precise statements. Rather, it is better to use vague statements. The physicist Duhem argued that confidence in the truth of a vague assertion may be justified simply because of its vagueness, which makes the assertion compatible with a whole range of observed facts (Negoita, 1985). When we encounter the vague information 'large earthquake', we are not sure whether magnitude 6 fits that description, but we are certain that magnitude 8 does. Hence, the uncertainty is related to the degree of belongingness. Magnitude 8 belongs to 'large earthquake' more so than magnitude 6 which, in turn, has a higher degree of belongingness than magnitude 5 and so on.

Fuzzy set theory was established by Zadeh in the 1960s (Zadeh, 1965). In three decades, it has gained broad recognition and rapid progress has been achieved in theory and applications in various fields.

In this chapter, we shall first describe the mathematical model for fuzzy information, and then discuss how such information can be processed. Several applications will be presented along with the methodology. Section 16.1 presents the basic concepts of fuzzy sets. The processing of fuzzy information for deterministic systems as well as fuzzy systems is discussed in section 16.2. An important application, viz. fuzzy classification, is described in section 16.3.

16.1 Basic concepts

A fuzzy set is an extension of an ordinary set. We review in this section the concepts of set theory and extend these concepts to fuzzy sets.

16.1.1 Ordinary sets

A collection of objects being discussed is called a universe, denoted by X, Y etc. A collection of some objects in a universe is called a subset or simply a set, denoted by A, B etc. We shall use the statement 'a set A on X' to imply that A is a subset of X.

The following notations will be used throughout this chapter.

- $x \in A$ – element x belongs to set A.
- $y \notin A$ – element y does not belong to A.
- $A \subseteq B$ – set A is contained in set B, or equivalently, set B includes A. B includes A if and only if all elements of A are also elements of B. Also, it follows automatically that $A \subseteq X$ (where X is the universe).
- $A = B$ – sets A and B are the same set, because they have exactly the same elements.
- $A \subset B$ – A is strictly contained in B (excluding equality). This implies that there is at least one element in B which does not belong to A.
- Φ – empty set: an empty set includes no elements. Note that Φ is a subset of X, i.e. $\Phi \subseteq X$.
- X – whole set: a whole set includes all elements in the universe. Note that X is also a subset of X, i.e. $X \subseteq X$ which implies that X is a subset of itself.
- Card (A) – the cardinality of set A: the cardinality of a set is the total number of distinct elements in the set.

If a set A has finite cardinality (e.g. if $A = \{2, 3, 5\}$, Card $(A) = 3$), we call it a finite set. If a set has infinite elements such as $X = [x \mid -\infty < x < +\infty]$, we call it an infinite set.

The characteristic function χ can be used to define a set A as follows:

$$\chi_A(x) = \begin{cases} 1 & \text{if } x \in A \\ 0 & \text{if } x \notin A \end{cases} \tag{16.1}$$

This can be interpreted as a mapping from X to a two-value set $Y = \{0, 1\}$, as shown in Figure 16.1. This mapping is expressed as follows: χ: $X \rightarrow Y = \{0, 1\}$.

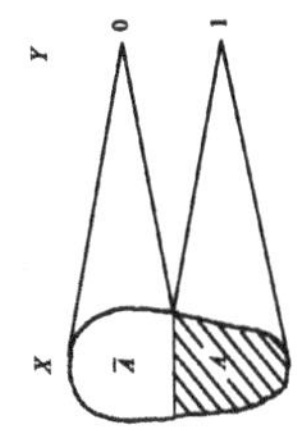

Figure 16.1 Mapping $\chi_A : X \rightarrow Y = \{0, 1\}$.

Using the characteristic function, a set A can be expressed as an ordered pair:

$$X = \{(x, \chi_A(x) | x \in X\} \tag{16.2}$$

For convenience of later development in fuzzy sets, we shall use the notation

$$A = \sum_{x \in X} \frac{\chi_A(x)}{x} = \frac{\chi_A(x_1)}{x_1} + \ldots + \frac{\chi_A(x_n)}{x_n} \tag{16.3}$$

for the finite sets and

$$A = \int_{x \in X} \frac{\chi_A(x)}{x} \tag{16.4}$$

for infinite sets. The notations +, Σ and ∫ stand for 'union' (instead of summation); the denominators are the elements in the universe and the numerators are their corresponding characteristic function values.

From a universe X, we can construct a set with all subsets of X as its elements. Such a set is called a power set, denoted by $P(X)$. For example, suppose that $X = \{a, b, c\}$, then its power set $P(X)$ is

$$P(X) = \{\Phi, \{a\}, \{b\}, \{c\}, \{a, b\}, \{b, c\}, \{a, c\}, \{a, b, c\}\},$$

In probability theory, we use terms such as 'sample space', 'single event', and 'compound event' (Benjamin and Cornell, 1970). The universe of set theory corresponds to the sample space; the element to a single event; and the subset to the compound event. The power set corresponds to the set of all events (single and compound).

The set-theoretic operations are defined as follows:

$$\text{union,} \quad A \cup B = \{x | x \in A \text{ or } x \in B\} \tag{16.5}$$

$$\text{intersection,} \quad A \cap B = \{x | x \in A \text{ and } x \in B\} \tag{16.6}$$

$$\text{complement,} \quad \overline{A} = \{x | x \notin A \text{ and } x \in X\} \tag{16.7}$$

$$\text{difference,} \quad A - B = \{x | x \in A \text{ and } x \notin B\} \tag{16.8}$$

and they can be visualized with the aid of the Venn diagrams in Figure 16.2.

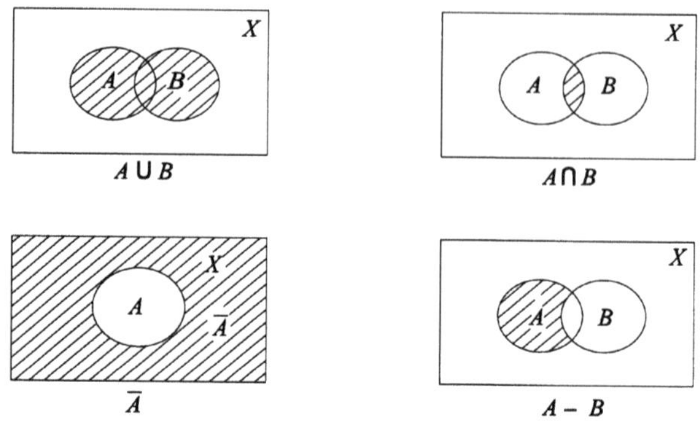

Figure 16.2 Venn diagrams.

Using the characteristic function, the set-theoretic operations described in equations 16.5–16.8 can be replaced by their corresponding function-theoretic operations. Thus, the characteristic functions of union, intersection, complement and difference of two sets A and B are given by

$$\chi_{A \cup B}(x) = \chi_A(x) \vee \chi_B(x) \tag{16.9}$$

$$\chi_{A \cap B}(x) = \chi_A(x) \wedge \chi_B(x) \tag{16.10}$$

$$\chi_{\bar{A}}(x) = 1 - \chi_A(x) \tag{16.11}$$

$$\chi_{A-B}(x) = \chi_A(x) \wedge \chi_{\bar{B}}(x) \tag{16.12}$$

where $\wedge$ and $\vee$ denote min and max operations, respectively.

It should be noted that for ordinary sets, we have the excluded-middle laws, i.e.

$$A \cup \bar{A} = X; \quad A \cap \bar{A} = \Phi \tag{16.13}$$

which imply that no element can belong to A and $\bar{A}$ at the same time.

16.1.2 Fuzzy sets

When we want to use ordinary sets to define linguistic definitions, we run into difficulty because of the ambiguity of our language. There are many classes of objects in which the transition from member (belongingness) to nonmember (nonbelongingness) is gradual rather than abrupt as dictated by the binary characteristic function χ. To model these classes, we use a fuzzy set A whose elements have different degrees of membership (belongingness). In place of the characteristic function $\chi_A(x) \in \{0, 1\}$ in ordinary sets, we

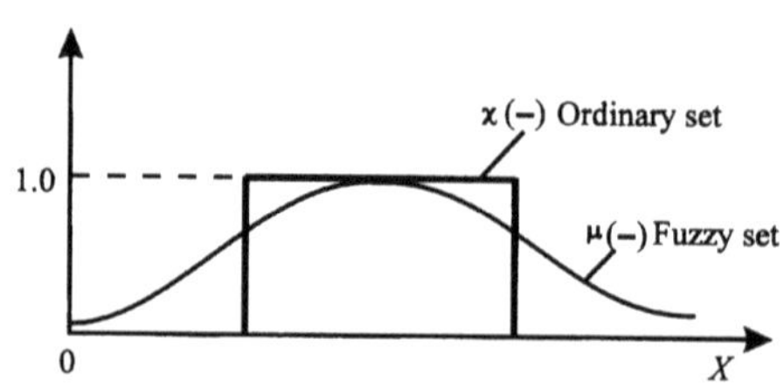

Figure 16.3 Characteristic function χ for an ordinary set and membership function μ for a fuzzy set.

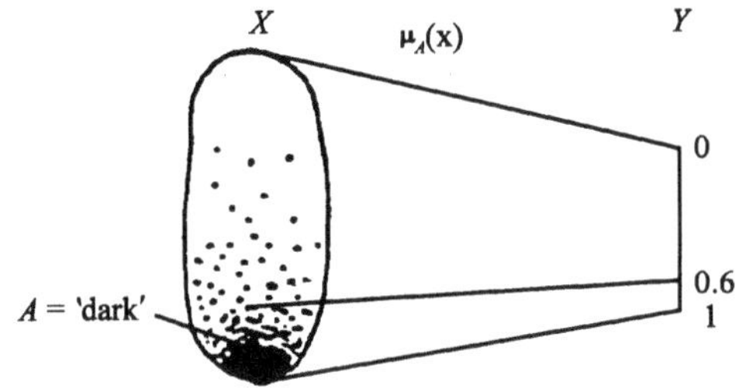

Figure 16.4 A mapping μ: $X \rightarrow [0, 1]$.

adopt the membership function $\mu_A(x) \in [0, 1]$ to express the grade of membership of x being A (Figure 16.3). The membership function is a mapping from X to [0, 1] as shown in Figure 16.4; that is μ: $X \rightarrow [0, 1]$.

Formally, a fuzzy set is expressed by ordered pairs

$$A = [(x, \mu_A(x)) | x \in X] \tag{16.14}$$

When X is finite, i.e. $A = \{x_1, x_2, \ldots x_n\}$, then

$$A = \sum_{x \in X} \frac{\mu_A(x)}{x} = \frac{\mu_A(x_1)}{x_1} + \ldots + \frac{\mu_A(x_n)}{x_n} \tag{16.15}$$

and when X is infinite

$$A = \int_{x \in X} \frac{\mu_A(x)}{x} \tag{16.16}$$

where the numerators are their membership values. Equations 16.14–16.16 correspond to equations 16.2–16.4 for ordinary sets.

A fuzzy set A in X is a subset of X, i.e.

$$A \subseteq X \rightarrow \mu_A(x) \leqslant \mu_x(x)$$

The empty set Φ is defined as

$$\text{Any } x \notin X,\ \mu_\Phi(x) \equiv 0$$

The whole set X is defined as

$$\text{Any } x \in X,\ \mu_X(x) \equiv 1$$

Finally, all fuzzy sets in X (including Φ, X and all ordinary sets) constitute the power set $F(X)$.

We use the function-theoretic operation to define the union, intersection and complement of two fuzzy sets A and B as follows:

$$\mu_{A \cup B}(x) = \mu_a(x) \vee \mu_B(x) \tag{16.17}$$

$$\mu_{A \cap B}(x) = \mu_A(x) \wedge \mu_B(x) \tag{16.18}$$

$$\mu_{\bar{A}}(x) = 1 - \mu_A(x) \tag{16.19}$$

The extended Venn diagram in Figure 16.5 illustrates these operations.

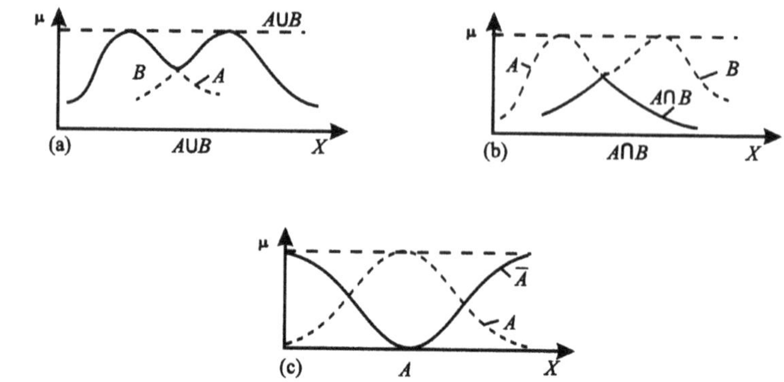

Figure 16.5 Extended Venn diagrams for fuzzy sets.

It is natural that the excluded-middle laws do not hold for fuzzy sets. Instead of equation 16.13, we have (Figure 16.5(c))

$$A \cup \bar{A} \neq X; \; A \cap \bar{A} \neq \Phi \tag{16.13a}$$

To illustrate, suppose that A and B are two fuzzy sets in X with their membership functions as given in Table 16.1. Results of the function-theoretic operations on A and B are shown in Table 16.2. These results agree with equation 16.13a, as seen in the bottom two rows of the table.

16.1.3 The α cut

As mentioned previously, the belongingness to a specific class changes gradually rather than abruptly. However, for decision- making purposes, a decision is usually made on the basis of the classification of whether an object belongs to a certain class or not. Thus, we need a discrete criterion to facilitate the classification. Usually, people will select a threshold level to divide objects into different classes. This concept leads to a formal definition of the α-cut.

Let $\alpha \in [0, 1]$. Then define an α-cut set of A as follows:

$$A_\alpha = [x \mid \mu_A(x) \geqslant \alpha] \tag{16.20}$$

Table 16.1 Membership values for fuzzy sets A and B

x^x	x_1	x_2	x_3	x_4	x_5	x_6
$\mu_A(I)$	0.1	0.6	0.8	0.9	0.7	0.1
$\mu_B(I)$	0.9	0.7	0.5	0.2	0.1	0.0

Table 16.2 Operations on fuzzy sets A and B

x^x	x_1	x_2	x_3	x_4	x_5	x_6
$\mu_{A \cup B}(I)$	0.9	0.7	0.8	0.9	0.7	0.1
$\mu_{A \cap B}(I)$	0.1	0.6	0.5	0.2	0.1	0.0
$\mu_{\bar{A}}(x)$	0.9	0.4	0.2	0.1	0.3	0.9
$\mu_{A \cap \bar{A}}(I)$	0.1	0.4	0.2	0.1	0.3	0.1
$\mu_{A \cup \bar{A}}(I)$	0.9	0.6	0.8	0.9	0.7	0.9

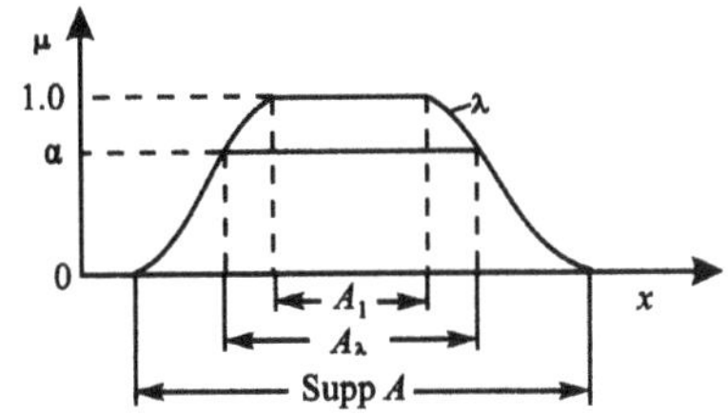

Figure 16.6 The α-cut.

An α-cut of a fuzzy set is an ordinary set. Any element $x \in A_\alpha$ implies that it belongs to A with a grade of membership greater than or equal to α (Figure 16.6). For instance, suppose that there are five shapes a, b, c, d and e as shown in Figure 16.7. Let a universe be $X = \{a, b, c, d, e\}$ and the shape 'round' be defined by a fuzzy set in X as follows:

$$A = \frac{1}{a} + \frac{0.9}{b} + \frac{0.4}{c} + \frac{0.2}{d} + \frac{0.0}{e}$$

When different values of α are chosen, different shapes will be considered as 'round'. For instance, when $\alpha = 1.0, 0.9, 0.4, 0.2, 0.0$, we have, respectively,

$$A_{1.0} = \{a\}$$

$$A_{0.9} = \{a, b\}$$

$$A_{0.4} = \{a, b, c\}$$

$$A_{0.2} = \{a, b, c, d\}$$

$$A_{0.0} = \{a, b, c, d, e\} = X$$

In this case, the α value indicates the degree of 'roundness', and if we select $\alpha = 0.4$ as a criterion for being 'round', we will have three elements falling into this category, i.e. $A_{0.4} = \{a, b, c\}$. The α-cut is a concept which is used to classify elements into sets without discrete boundaries. The value α is a kind of threshold level below which an element is not qualified to be a member for a specific class.

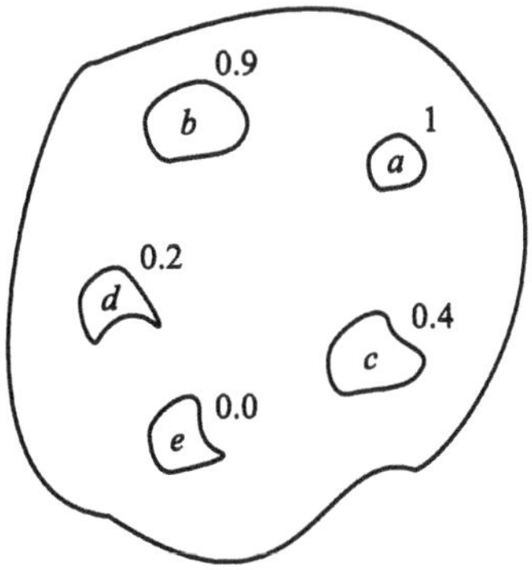

Figure 16.7 Fuzzy set for 'round'.

With the concept of an α-cut, define $A_{1.0}$ as the core of A. All elements in the core $A_{1.0}$ belong entirely to A. If $A_{1.0} \neq \Phi$, then A is called a normal fuzzy set. This implies that there is at least one element which has the full membership of A. Also, define

$$\text{Supp}A = A_{0^+} = \{x \mid \mu_A(x) > 0\} \tag{16.21}$$

which is called the support of A. The difference, Supp $A - A_{1.0}$, is the domain where all elements are neither full members nor nonmembers of A. It is the zone in which the grades of membership are in the open interval (0,1) (see Figure 16.6).

Any fuzzy set A on X may be decomposed as follows:

$$\mu_A(x) = \bigvee_{\alpha \in [0,1]} (\alpha \wedge \chi_A(x)) \tag{16.22}$$

which implies that any fuzzy set A can be imagined as a group of infinite α-cuts and the membership for an element x is the maximum α value among all α-cuts containing x. Equation 16.22 is called the decomposition theorem.

With the decrease in value from 1.0 to 0^+, A_α expands from the core $A_{1.0}$ to Supp $A(A_{0^+})$. The family of ordinary sets $\{A_\alpha \mid 0 < \alpha < 1.0\}$ forms a boundary where no element has full membership or nonmembership (Figure 16.6). This is the transition zone.

By the decomposition theorem, a fuzzy set can be defined by the domain of membership values such as

$$A = \sum_{\alpha_i} \frac{A_{\alpha_i}}{\alpha_i} \tag{16.23}$$

when discrete values of α are chosen, and

$$A = \int_{\alpha \in [0,1]} \frac{A_\alpha}{\alpha} \tag{16.24}$$

where α is continuous.

Equations 16.23–16.24 constitute a different way to express a fuzzy set and should be compared to equations 16.15–16.16 which define the fuzzy set in the universe X. The α-cut method of expression is very convenient when implementing fuzzy set computations as we shall show in the next section.

In most cases, the universe is the set of real numbers, i.e. $X = R = (-\infty, +\infty)$. The corresponding fuzzy sets are called fuzzy numbers. For a continuous fuzzy set A in R, for any $x < y < z \in R$, if the inequality

$$\mu_A(y) \geqslant \min(\mu_A(x), \mu_A(z)) \tag{16.25}$$

holds, then A is called a convex fuzzy set. In Figure 16.8(a), A is convex. However, in Figure 16.8(b), A is not because there are elements for which $\mu_A(y) < \min(\mu_A(x), \mu_A(z))$.

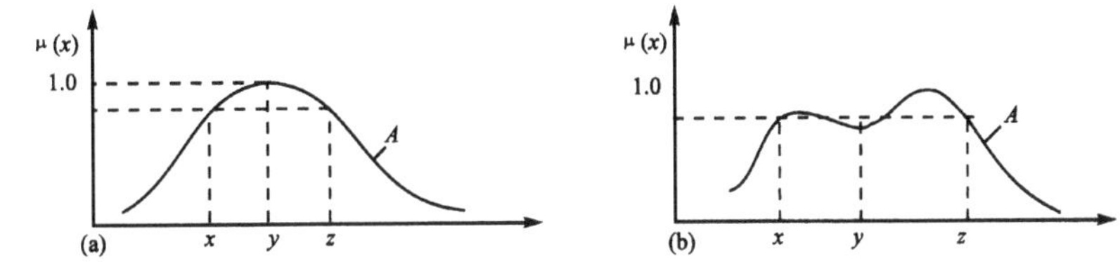

Figure 16.8 Convexity of fuzzy sets.

16.1.4 Fuzzy relations

Many engineering problems are concerned with the relations between different quantities, e.g. the relation between loading and deflection of a beam. We define a Cartesian product of two universes X and Y as

$$X \times Y = \{(x, y) | x \in X; y \in Y\} \tag{16.26}$$

which combines all $x \in X$ and $y \in Y$ in the form of ordered pairs as its elements. It is similar to a two-dimensional sample space.

With the Cartesian product, a fuzzy relation can be defined as

$$R = \{((x, y), \mu_R(x, y)) | (x, y) \in X \times Y\} \tag{16.27}$$

where $\mu_R(x, y)$ is the membership function which is a mapping from $X \times Y$ to [0, 1]. For instance, $\mu_R(x, y) = 0.8$ implies that the degree of relation between x and y is 0.8.

Referring to the ordered pair (x,y) as an element, the fuzzy relation R is a fuzzy subset of $X \times Y$. Hence, all operations for fuzzy sets can be transposed to the fuzzy relations. Thus, let R and S be fuzzy relations in $X \times Y$, then

$$\text{union} \quad \mu_{R \cup S}(x, y) = \max[\mu_R(x, y), \mu_S(x, y)] \tag{16.28}$$

$$\text{intersection} \quad \mu_{R \cap S}(x, y) = \min[\mu_R(x, y), \mu_S(x, y)] \tag{16.29}$$

$$\text{complement} \quad \mu_R(x, y) = 1.0 - \mu_R(x, y) \tag{16.30}$$

Similarly, the Cartesian product of two fuzzy sets A in X and B in Y can be defined as

$$R = A \times B = \{((x, y), \mu_R(x, y)) | (x, y) \in X \times Y\} \tag{16.31}$$

which is a special fuzzy relation of $X \times Y$. Its membership function is given by

$$\mu_R(x, y) = \mu_{A \times B}(x, y) = \min[\mu_A(x), \mu_B(y)] \tag{16.32}$$

Like the composition of discrete functions, two fuzzy relations can also be composed to become a new relation. Suppose that R is a relation in $X \times Y$, and S a relation in $Y \times Z$. Then a new relation in $X \times Z$ can be defined as:

$$T = R \circ S \tag{16.33}$$

where $\circ$ stands for the composition operation and the membership function for the new relation T is given by

$$\mu_T(x, z) = \bigvee_{y \in Y} (\mu_R(x, y), \mu_S(y, z)) \tag{16.34}$$

As an example, let two fuzzy relations R and S be given by

$$R = \begin{array}{c} \\ x_1 \\ x_2 \\ x_3 \end{array} \begin{array}{c} \begin{array}{ccccc} y_1 & y_2 & y_3 & y_4 & y_5 \end{array} \\ \begin{bmatrix} 0.1 & 0.2 & 0.0 & 1.0 & 0.7 \\ 0.3 & 0.5 & 0.1 & 0.2 & 1.0 \\ 0.8 & 0.0 & 1.0 & 0.4 & 0.3 \end{bmatrix} \end{array}$$

and

$$S = \begin{array}{c} \\ y_1 \\ y_2 \\ y_3 \\ y_4 \\ y_5 \end{array} \begin{array}{c} \begin{array}{cccc} z_1 & z_2 & z_3 & z_4 \end{array} \\ \begin{bmatrix} 0.9 & 0.0 & 0.3 & 0.4 \\ 0.2 & 1.0 & 0.8 & 0.0 \\ 0.8 & 0.0 & 0.7 & 1.0 \\ 0.4 & 0.2 & 0.3 & 0.0 \\ 0.0 & 1.0 & 0.0 & 0.8 \end{bmatrix} \end{array}$$

By equation 16.34, the value of the membership function for, say, (x_1, z_1) is

$$\mu_T(x_1, z_1) = \vee [(0.1 \wedge 0.9), (0.2 \wedge 0.2), (0.0 \wedge 0.8), (1.0 \wedge 0.4), (0.7 \wedge 0.0)]$$

$$= \vee [0.1, 0.2, 0.0, 0.4, 0.0] = 0.4$$

so that the fuzzy relation T is given by

$$T = \begin{array}{c} \\ x_1 \\ x_2 \\ x_3 \end{array} \begin{array}{c} \begin{array}{cccc} z_1 & z_2 & z_3 & z_4 \end{array} \\ \begin{bmatrix} 0.4 & 0.7 & 0.3 & 0.7 \\ 0.3 & 1.0 & 0.5 & 0.8 \\ 0.8 & 0.3 & 0.7 & 1.0 \end{bmatrix} \end{array}$$

16.2 Fuzzy information processing

In the last section, we discussed how fuzzy information can be modeled mathematically with fuzzy sets. We show now how items of fuzzy information can be processed.

Most engineering problems can be modeled as a system such as shown in Figure 16.9.

If the system is well-defined, then it can be represented by a function or a relation like

$$y = f(x) \tag{16.35}$$

or

$$R = \{[(x, y), \chi_R(x, y)] \mid x \in X; y \in Y\} \tag{16.36}$$

Figure 16.9 System model.

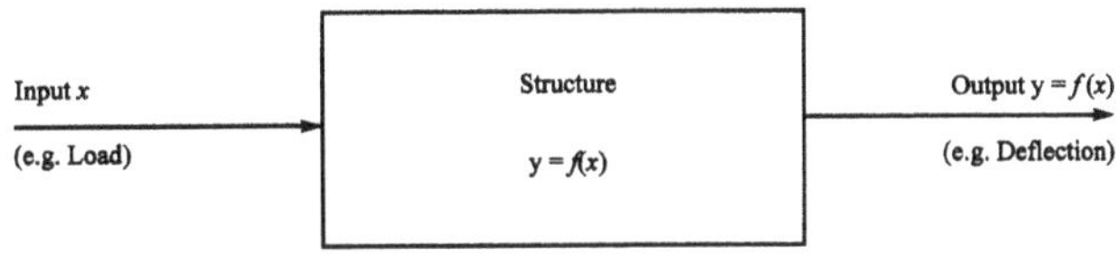

where

$$\chi_R(x, y) = \begin{cases} 1, & y = f(x) \\ 0, & \text{otherwise} \end{cases}$$

If the system is not well-defined, then it can be modelled as a fuzzy function or a fuzzy relation such as

$$y = \tilde{f}(x) \tag{16.37}$$

or

$$R = \{[(x, y), \mu_R(x, y)] | x \in X; y \in Y\} \tag{16.38}$$

where

$$\mu_R(x, y): X \times Y \rightarrow [0, 1]$$

We call the former a deterministic system, and the latter a fuzzy system.

For multiple variables, we can expand the well-defined function or relation as follows:

$$y = f(x_1, x_2, \ldots, x_n) \tag{16.39}$$

or

$$R = \{[(x_1, x_2, \ldots, x_n, y), \chi_R(x_1, x_2, \ldots, x_n, y)] | x_i \in X_i, i = 1, n; y \in Y\} \tag{16.40}$$

Correspondingly, when the system is not well defined, we have

$$y = \tilde{f}(x_1, x_2, \ldots x_n) \tag{16.41}$$

or

$$R = \{[x_1, x_2, \ldots, x_n, y), \mu_R(x_1, x_2, \ldots, x_n, y)] | x_i \in X_i, i = 1, n; y \in Y\} \tag{16.42}$$

In the following subsections, we shall address the issue of how the fuzzy input is passed through the system and how the output is obtained.

16.2.1 Extension principle

Zadeh (1965) established the extension principle, which is the basis for fuzzy information processing. For a deterministic system defined as $y = f(x)$, given the fuzzy input X, its image under this mapping (function) is

$$Y = f(X) \tag{16.43}$$

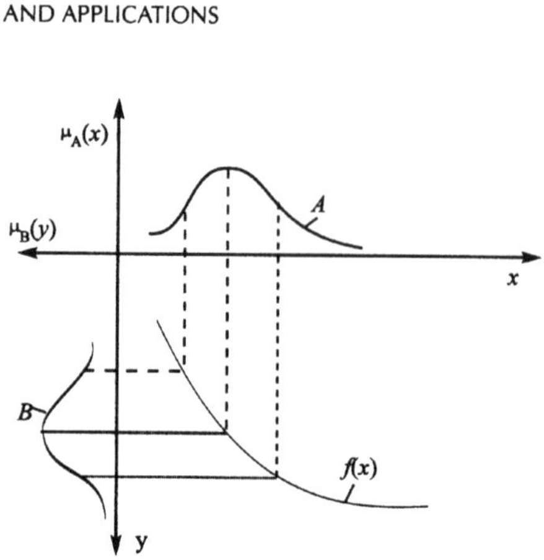

Figure 16.10 Mapping of a fuzzy set.

with the membership function for y as

$$\mu_Y(y) = \bigvee_{y=f(x)} [\mu_X(x)] \tag{16.44}$$

Figure 16.10 gives a graphic representation of this mapping.

For a deterministic system with multiple fuzzy inputs, defined as $y=f(x_1, x_2, \ldots, x_n)$, the output will be

$$Y=f(X_1, X_2, \ldots, X_n) \tag{16.45}$$

and the membership function for y is

$$\mu_Y(y) = \bigvee_{y=f(x_1,x_2,\ldots x_n)} [\mu_{X_1}(x_1) \wedge \ldots \wedge \mu_{X_n}(x_n)] \tag{16.46}$$

For a fuzzy system $y=\tilde{f}(x)$, even when the input x is discrete, or a single number (singleton), the output will in general be fuzzy because of the ill-defined nature of the system. The most general way to represent this kind of system is to use a fuzzy relation. Thus, for single input, the extension principle is given by

$$\mu_Y(y) = \bigvee_{y=f(x)} [\mu_R(x, y) \wedge \mu_X(x)] \tag{16.47}$$

and for multiple inputs,

$$\mu_Y(y) = \bigvee_{y=f(x_1,x_2,\ldots,x_n)} [\mu_R(x_1, x_2, \ldots, x_n, y) \wedge \mu_{X_1}(x_1) \wedge \ldots \wedge \mu_{X_n}(x_n)] \tag{16.48}$$

Hence, in theory, the extension principle gives the output from the system (deterministic or fuzzy) on the basis of deterministic or fuzzy input information. However, in practice, the implementation of the extension principle is very difficult except for the most trivial functional forms. In section 16.2.2, we shall discuss the implementation issues for a deterministic system, and in section 16.2.3, the implementation for fuzzy systems.

16.2.2 Fuzzy information processing in deterministic system – vertex method

Various methods have been developed to implement the extension principle. In particular, Schumucker (1984) used a discretization method to process fuzzy weighted averages to evaluate the reliability of a system. Baas and Kwakernaak (1977) introduced a non linear programming solution procedure which can obtain the closed-form solution for some simple cases. However, the procedure is complex and does not guarantee a solution. Other methods (Dubois and Prade, 1980; Mizumoto and Tanaka, 1979; Nakamura, 1984) have also been proposed, but it is fair to state that none produces results in a computationally simple manner while being devoid of anomalies.

The vertex method described in this section is based on the α-cut and interval analysis techniques (Dong and Shah, 1987; Dong and Wong, 1987). Using α-cuts, each fuzzy set is converted to a group of intervals with different α values as shown in equation 16.23. Then, intervals with the same α value from all fuzzy sets are processed by interval analysis, resulting in an interval function with the same α value. Hence, we begin the discussion with a review of interval functions.

Consider an interval function such as

$$Y=f(X_1, X_2, \ldots, X_n)$$

where X_i, $i=1, n$, are interval variables. These interval variables form an n-dimensional rectangle $X_1 \times X_2 \times \ldots \times X_n$ with 2^n vertices, and an example in three-dimensional space is given in Figure 16.11. The coordinates of all vertices of the n-dimensional rectangle are the combinations of n pairs of end points of the independent interval variables. Referring to the example in Figure 16.11, the coordinates of the eight vertices are:

$$(a_1, a_2, a_3); (a_1, a_2, b_3); (a_1, b_2, a_3); (a_1, b_2, b_3)$$

$$(b_1, a_2, a_3); (b_1, a_2, b_3); (b_1, b_2, a_3); (b_1, b_2, b_3)$$

The number of combinations of n-intervals is equal to 2^n. Use C_j to denote the jth combination (or the jth vertex) with $j=1, N$ where $N=2^n$. When

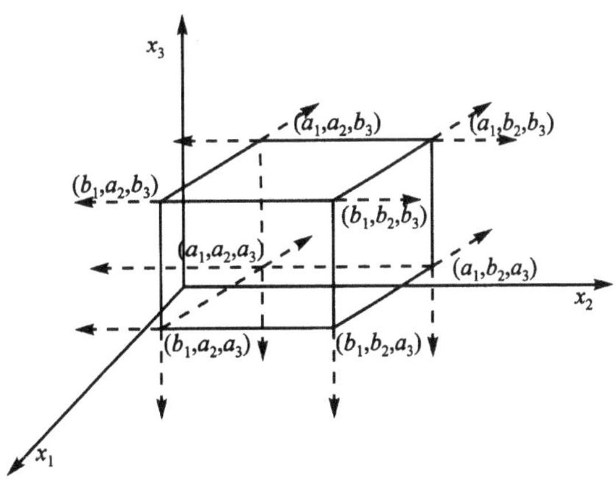

Figure 16.11 Three-dimensional rectangular region.

$Y=f(X_1, X_2, \ldots, X_n)$ is continuous in the n-dimensional rectanglar region, and if no extreme point exists in this region including the boundaries, the value of the interval function can be obtained by

$$Y=f(X_1, X_2, \ldots X_n)=[\min_j f(C_j), \max_j f(C_j)\}] \; (j=1, N=2^n) \quad (16.49)$$

where C_j is the ordinate of the jth vertex.

It should be emphasized that equation 16.49 is valid only when the continuity and no-extremum conditions are satisfied. Where there are extreme points of the function $Y=f(X_1, X_2, \ldots, X_n)$ within the n-dimensional region $X_1 \times X_2 \times \ldots \times X_n$, equation 16.15 must be modified to

$$Y=[\min_{j,k} (f(C_j), f(E_k)), \max_{j,k} (f(C_j), f(E_k))] \quad (16.50)$$

where E_k represents the coordinates of the kth extreme point.

To illustrate, let us look at a system modeled by a function as follows:

$$y=f(x_1, x_2)=x_1^2+x_2^2-4x_1+4$$

and the fuzzy inputs X_1 and X_2 are as defined in Figure 16.12. Let us find the output $Y=f(X_1, X_2)$.
Since

$$\frac{\partial y}{\partial x_1}=2x_1-4=0, \quad x_1=2,$$

$$\frac{\partial y}{\partial x_2}=2x_2=0, \quad x_2=0$$

there is a minimum point at (2,0). With $\alpha=0^+$, we have

$$X_{1,0^+}=[1, 4], X_{2,0^+}=[-1, 1]$$

$$C_1=[1, -1], C_2=[1, 1], C_3=[4, -1], C_4=[4, 1], E_1=[2, 0]$$

By equation 16.50, $f(C_1)=2, f(C_2)=2, f(C_3)=5, f(C_4)=5, f(E_1)=0$, and we obtain

$$Y_{0^+}=[\min(2, 2, 5, 5, 0), \max(2, 2, 5, 5, 0)]=[0, 5]$$

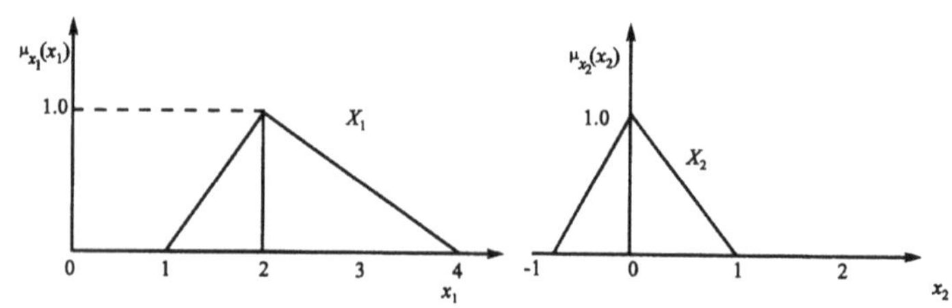

Figure 16.12 Fuzzy inputs of X_1 and X_2.

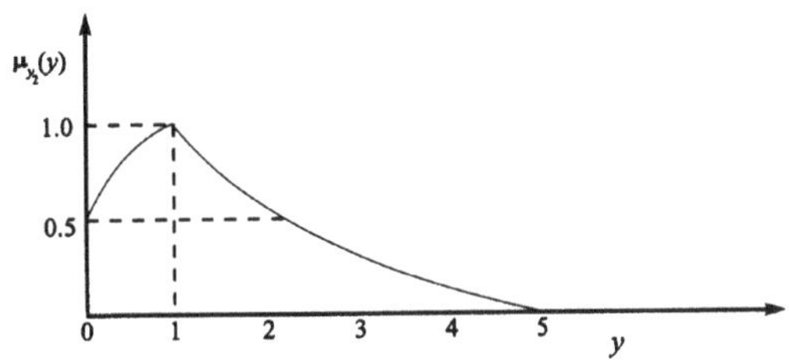

Figure 16.13 Extended Venn diagram for *Y*.

Note that if equation 16.49 is used, we obtain

$$Y_{0^+}=[\min(2, 2, 5, 5), \max(2, 2, 5, 5)]=[2, 5]$$

Similarly, when $\alpha=0.5$,

$$X_{1,0.5}=[2, 3.5],\ X_{2,0.5}=[-0.5, 0.5]$$

$$C_1=[2, -0.5],\ C_2=[2, 0.5],\ C_3=[3.5, -0.5],\ C_4=[3.5, 0.5]$$

$$E_1=[2, 0] \text{ on the boundary } x_1=2 \text{(line)}$$

we have $Y_{0.5}=[0, 2.5]$.

Finally, when $\alpha=1.0$,

$$X_{1,1}=3,\ X_{2,1}=0;$$

$$Y_1=f(3, 0)=3^2+0^2-4\cdot 3+4=1$$

The extended Venn diagram for Y is shown in Figure 16.13.

The vertex method has been applied to various problems. For example, Chen *et al.* (1988) used this method to find recurrence relationships from fuzzy earthquake magnitude data; Frangopol *et al.* (1987) used fuzzy boundaries for source areas and fuzzy attenuation to derive the fuzzy seismic risk maps. In general, for many deterministic systems with fuzzy input (information), the vertex method provides a powerful tool to process such information and obtain results.

16.2.3 Fuzzy information processing in fuzzy systems

When we deal with a fuzzy system, the computation becomes so intractable that discretization techniques are usually used to generate a numerical solution. To facilitate the operations to be described, we introduce the concepts of fuzzy vector and fuzzy matrix.

Define

$$\boldsymbol{a}=\{a_1 a_2 \ldots a_n\} \tag{16.51}$$

as a fuzzy vector* with elements $a_i \in [0, 1]$ for all i. The fuzzy vector a represents a fuzzy set A in a finite universe X, i.e.

* In this chapter, vectors will be denoted using curly brackets; matrices will be denoted using square brackets.

$$\mu_A(x_i)=a_i, \quad i=1, n \tag{16.52}$$

All the function-theoretic operations in equations 16.9–16.11 can be applied to fuzzy vectors as follows:

$$A\cup B\rightarrow a\vee b=\{a_i\vee b_i,\ i=1, n\} \tag{16.53}$$

$$A\cap B\rightarrow a\vee b=\{a_i\wedge b_i,\ i=1, n\} \tag{16.54}$$

$$\overline{A}\rightarrow \overline{a}=\{1-a_i,\ i=1, n\} \tag{16.55}$$

We can also define the fuzzy matrix which is nothing but the fuzzy relation

$$R=[r_{i,j}] \quad i=1, n;\ j=1, m \tag{16.56}$$

with

$$r_{i,j}=\mu_R(x_i, y_j)\in[0, 1] \tag{16.57}$$

The matrix represents a fuzzy relation in the two-dimensional universe $X\times Y$, where $X=\{x_1x_2\ldots x_n\}$ and $Y=\{y_1y_2\ldots y_m\}$. Let R and S be fuzzy relations in $X\times Y$. All function-theoretic operations for fuzzy relations in equations 16.28–16.30 can be expressed as follows:

$$R\cup S=[r_{i,j}\vee s_{i,j},\ i=1;\ n;\ j=1, m] \tag{16.58}$$

$$R\cap S=[r_{i,j}\wedge s_{i,j},\ i=1;\ n;\ j=1, m] \tag{16.59}$$

$$\overline{R}=[1-r_{i,j},\ i=1;\ n;\ j=1, m] \tag{16.60}$$

For the composition operation, let R be the relation in $X\times Y$ and S be the relation in $Y\times Z$, where $Z=\{z_1z_2\ldots z_1\}$; that is, $R=[r_{i,j}]$ and $S=[s_{j,k}]$. Then, the composition of R and S, denoted by T in $X\times Z$ is given by

$$T\circ S=[t_{i,k},\ i=1;\ n;\ k=1, l] \tag{16.61}$$

$$\mu_T(x_i, z_k)=t_{i,k}=\bigvee_{j=1,m}(r_{i,j}\wedge s_{j,k}) \tag{16.62}$$

where the notation $\circ$ stands for composition operation, defined in equations 16.33 and 16.34.

Note that a vector is a matrix which has only one row and n columns, i.e.

$$a=\{a_1a_2\ldots a_n\}$$

Hence, we can define the composition between a vector and a matrix as

$$b=a\circ R \tag{16.63}$$

where $\boldsymbol{b}=\{b_1b_2\ldots b_m\}$ and

$$b_j=\bigvee_{i=1,n}(a_i\wedge r_{i,j}) \tag{16.64}$$

For the Cartesian product of two fuzzy sets, we can also use the vector expression

$$R = a \times b = a^{\mathrm{T}} \circ b = [r_{i,j}]_{n \times m} \tag{16.65}$$

where

$$a^{\mathrm{T}} = \{a_1 a_2 \ldots a_n\}^{\mathrm{T}}$$
$$b = \{b_1 b_2 \ldots b_m\}$$

and

$$r_{i,j} = a_i \wedge b_j \tag{16.66}$$

As mentioned in section 16.2.1, for a fuzzy system modeled as a fuzzy relation

$$R = [\mu_R(x_i, y_j)] = [r_{i,j}]_{i=1,n;j=1,m}$$

if we know the fuzzy input A with fuzzy vector (membership function) $\boldsymbol{a} = \{a_1 a_2 \ldots a_n\}$, then its output (or image) can be obtained by exercising the extension principle or the composition of equations 16.63 and 16.64:

$$b = a \circ R$$

where $b = \{b_1 b_2 \ldots b_m\}$ and

$$b_j = \bigvee_{i=1,n} (a_i \wedge r_{i,j})$$

Thus, if the fuzzy system is defined and represented by a fuzzy relation R, then for any input A, we can use equations 16.63 and 16.64 to get fuzzy output B. The problem is how to identify a fuzzy system and to find fuzzy relations. This is to be addressed in the next subsection.

16.2.4 Identification of a fuzzy system

We now discuss the problem of how to identify a system based on input–output data pairs which are not precise but are linguistic or vague. Some examples of linguistic or vague statements are as follows.

- If the distance to the epicenter is 'short', then the predominant frequency of ground motion is 'high'.
- If the distance is 'medium', then the frequency is 'medium'.
- If the distance is 'long', then the frequency is 'low'.

It is not possible to derive from these statements a precise relation or function between 'distance' and 'frequency content' so that we can use it to predict the frequency of ground motion from distance or vice versa. However, this type of data does provide some information between distance and frequency.

Data such as these contain uncertainty which is not a matter of chance. It is judgemental knowledge derived from experience, and it is subjective. Hence, it is natural to model this type of information by a fuzzy system which represents the loose relation between the cause and effect (e.g. distance and frequency content in the examples). The basic task in the identification of a fuzzy system is to find the fuzzy relation which can represent the available data fairly well.

Thus, the problem of identification can be restated as follows. Given the data in the form of the following statements:

$$\left.\begin{array}{l}\text{if } x \text{ is } A_1, \text{ then } y \text{ is } B_1 \\ \text{else if } x \text{ is } A_2, \text{ then } y \text{ is } B_2 \\ \quad\vdots \\ \text{else if } x \text{ is } A_I, \text{ then } y \text{ is } B_I\end{array}\right\} \tag{16.67}$$

where $A_1, A_2, \ldots, A_I$ are fuzzy sets defined in the universe X and B_1, $B_2, \ldots B_I$ are fuzzy sets in the universe Y. Find a fuzzy relation R in $X \times Y$ which fits the above data most closely.

Mandani (1977; Mandani and Assilian, 1975) suggested that the relation can be obtained from the following equation:

$$R = \bigcup_{i=1,I} R_i = \bigcup_{i=1,I} [A_i \times B_i] \tag{16.68}$$

Mandani's method is simple, but has large 'fitting errors' as we shall show shortly. To improve accuracy of fit, extended regression methods have been proposed and used. One such technique is described in Pedrycz (1984) and Dong (1986) and is summarized below.

Using fuzzy vector and fuzzy matrix representation, the identification problem can be reformulated as follows. Given the data pairs

$$\{a_i, b_i, \quad i = 1, I\} \tag{16.69}$$

where $\boldsymbol{a}_i$ and $\boldsymbol{b}_i$ are fuzzy vectors of the universes X and Y such that

$$X = \{x_1 x_2 \ldots x_J\} \qquad Y = \{y_1 y_2 \ldots y_K\}$$

$$a_i = \{a_{i,1} a_{i,2} \ldots a_{i,J}\}, \qquad b_i = \{b_{i,1} b_{i,2} \ldots b_{i,K}\}$$

$$a_{i,j} = \mu_{A_i}(x_j), \qquad j = 1, J$$

$$b_{i,k} = \mu_{B_i}(y_k), \qquad k = 1, K$$

we want to find a relation matrix $R = [r_{j,k}] = [\mu_R(x_j, y_k)]$ which represents the I pairs of fuzzy data expressed by

$$\text{if } a_1, \text{ then } b_1$$

$$\text{else if } a_i, \text{ then } b_i$$

$$\text{else if } a_I, \text{ then } b_I$$

such that the error function Q (to be defined presently in equation 16.71) is minimized with respect to all elements of R, that is,

$$\frac{\partial Q}{\partial R}=0 \tag{16.70}$$

where 0 stands for the null matrix and $\partial Q/\partial R$ is a matrix with elements $\partial Q/\partial r_{j,k}$.

Using the vector representation, the error function can be written as

$$Q=\sum_{i=1}^{I}\sum_{k=1}^{K}[b'_{i,k}-b_{i,k}]^2 \tag{16.71}$$

with $b'_{i,k}=\mu_{B'_i}(y_k)$ the membership function of fuzzy set $B'_i=A_i\circ R$ and

$$b'_{i,k}=\max_j(a_{i,j}\wedge r_{j,k}) \tag{16.72}$$

Replacing equation 16.72 in equation 16.71, the error function becomes

$$Q=\sum_{i=1}^{I}\sum_{k=1}^{K}[\max_j(a_{i,j}\wedge r_{j,k})-b_{i,k}]^2 \tag{16.73}$$

Pedrycz suggested a modified Newton iteration approach to obtain the relation R. We recall here that the Newton iteration algorithm for finding the zeros of a function $f(x)$ is

$$x^{n+1}=x^n-\frac{f(x^n)}{f'(x^n)} \tag{16.74}$$

where $f'(x^n)$ is the first derivative of $f(x)$ at $x=x^n$. Hence, the Newton iteration scheme for finding the solution of equation 16.73 can be written in a format similar to equation 16.74. That is, if we write

$$F(R)=\frac{\partial Q}{\partial R}=\begin{bmatrix}\frac{\partial Q}{\partial r_{1,l}} & \frac{\partial Q}{\partial r_{1,2}} & \cdots & \frac{\partial Q}{\partial r_{1,J}}\\ & \vdots & & \\ & \frac{\partial Q}{\partial r_{s,t}} & & \\ & \vdots & & \\ \frac{\partial Q}{\partial r_{I,1}} & \frac{\partial Q}{\partial r_{I,2}} & \cdots & \frac{\partial Q}{\partial r_{I,J}}\end{bmatrix}=\left[\frac{\partial Q}{\partial r_{s,t}}\right]_{s=1,I;\,t=1,J} \tag{16.75}$$

then,

$$R^{n+1}=R^n-\alpha_n F(R^n) \tag{16.76}$$

where α_n is a particular value chosen such that a good rate of convergence can be obtained. A value for α_n is proposed by Pedrycz as

$$\alpha_n=\frac{1}{2I+n^{\beta}} \tag{16.77}$$

where n is the number of iterations and β is a constant selected explicitly in order to increase the rate of convergence and to avoid possible oscillations.

We illustrate the procedure with a hypothetical problem. Suppose that x stands for the distance to the epicenter and for simplicity we use a three-point discrete universe $X=[10, 50, 100]$ (miles). Define the fuzzy variable as

$$A_1 = \text{'short'} = \frac{1}{10} + \frac{0.5}{50} + \frac{0.0}{100}$$

$$A_2 = \text{'medium'} = \frac{0.1}{10} + \frac{1.0}{50} + \frac{0.1}{100}$$

$$A_3 = \text{'long'} = \frac{0.0}{10} + \frac{0.5}{50} + \frac{1}{100}$$

Similarly, y stands for the ground motion frequency, and $Y=[0.5, 1.0, 2.0]$(Hz). Define fuzzy variables as

$$B_1 = \text{'high'} = \frac{0.0}{0.5} + \frac{0.3}{1.0} + \frac{1.0}{2.0}$$

$$B_2 = \text{'medium'} = \frac{0.1}{0.5} + \frac{0.7}{1.0} + \frac{1.0}{2.0}$$

$$B_3 = \text{'low'} = \frac{1.0}{0.5} + \frac{0.5}{1.0} + \frac{0.0}{2.0}$$

First, we use Mandani's method. Hence, by equation 16.68, we have

$$R_1 = A_1 \times B_1 = \begin{bmatrix} 0.0 & 0.3 & 1.0 \\ 0.0 & 0.3 & 0.5 \\ 0.0 & 0.0 & 0.0 \end{bmatrix}$$

where the elements of R_1, such as $(r_{2,2})_1$, are calculated as

$$(r_{2,2})_1 = \mu_{R_1}(50, 1.0) = \min[\mu_{A_1}(50), \mu_{B_1}(1.0)] = \min[0.3, 0.5] = 0.3$$

Similarly, we have

$$R_2 = A_2 \times B_2 = \begin{bmatrix} 0.1 & 0.1 & 0.1 \\ 0.1 & 0.7 & 1.0 \\ 0.1 & 0.1 & 0.1 \end{bmatrix}$$

$$R_3 = A_3 \times B_3 = \begin{bmatrix} 0.0 & 0.0 & 0.0 \\ 0.5 & 0.5 & 0.0 \\ 1.0 & 0.5 & 0.0 \end{bmatrix}$$

Mandani's method (equation 16.68) then gives

$$R = \bigcup_i R_i = \begin{bmatrix} 0.1 & 0.3 & 1.0 \\ 0.5 & 0.7 & 1.0 \\ 1.0 & 0.5 & 0.1 \end{bmatrix}$$

where, for example,

$$r_{2,2}=\max[(r_{2,2})_1(r_{2,2})_2(r_{2,2})_3]=\max[0.3\ 0.7\ 0.5]=0.7$$

Note that the relation R defined by Equation 16.68 cannot in general recover the original data. That is, if we input again A_1, A_2, A_3 into R, then by the composition operation we obtain

$$B_1'=A_1\circ R=\frac{0.5}{0.5}+\frac{0.5}{1.0}+\frac{1.0}{2.0}\neq B_1$$

$$B_2'=A_2\circ R=\frac{0.5}{0.5}+\frac{0.7}{1.0}+\frac{1.0}{2.0}\neq B_2$$

$$B_3'=A_3\circ R=\frac{1.0}{0.5}+\frac{0.5}{1.0}+\frac{0.5}{2.0}\neq B_3$$

One measure of the 'error' associated with a given relation is the square error function defined as (equation 16.71)

$$Q=\sum_i d^2(B_i', B_i) \tag{16.78}$$

where $d(B_i', B_i)$ is the distance between two fuzzy sets B_i' and B_i. For this example, it is given by

$$d^2(B_1', B_1)=(0.5-0.0)^2+(0.5-0.3)^2+(1.0-1.0)^2=0.29$$

$$d^2(B_2', B_2)=(0.5-0.1)^2+(0.7-0.7)^2+(1.0-1.0)^2=0.16$$

$$d^2(B_3', B_3)=(1.0-1.0)^2+(0.5-0.5)^2+(0.5-0.0)^2=0.25$$

Hence,

$$Q=0.29+0.16+0.25=0.70$$

Using the extended regression method described earlier, the system matrix will be

$$R=\begin{bmatrix}0.0 & 0.47 & 1.0\\ 0.0 & 0.7 & 1.0\\ 1.0 & 0.5 & 0.44\end{bmatrix}$$

The corresponding error $Q=0.30$, which is much smaller than that obtained by Mandani's method.

16.3 Fuzzy classification

The basic purpose of all sciences is to find the inner structure of physical or social processes. The only way to achieve this goal is through abundant observations and experiments over a long period of time. On the basis of data obtained from observations and experimental measurements, scientists search for and predict the structure embedded in the data. They try to classify the sampled data and find the patterns for different classes.

In some circumstances classes do exist, and the difficulties with classification are due to variation in the features of the class members. For instance, in identifying handwriting, we know a handwriten letter of the alphabet should be one of 26 characters. However, due to differences in ways of writing, one might have difficulty in identifying a particular letter and might misclassify it. In this case, it seems reasonable to use a statistical approach and develop a Bayesian classifier based on a lot of data, thereby minimizing the probability of misclassification. In most cases though, the classes are subjective and human-oriented, i.e. it is people who subjectively postulate the number of classes and formulate a way to classify objects for some purpose, such as type of soil, intensity, different classes of age such as 'young', 'old' etc.

The approaches developed for classification so far are mostly discrete, or so-called hard-partition, i.e. each sample either belongs to one class with full membership or does not. There is no interim or ambiguity. In engineering practice, however, we often encounter borderline cases where classification is vague. For example, earthquake engineers assign an intensity after an earthquake from the qualitative and vague damage pattern and linguistic description of human reaction. With these pieces of fuzzy information, the boundaries of the different classes are not well defined. Such inexactness is rampant in real life and has stimulated the use of fuzzy set theory in classification, the subject of this section.

Generally speaking, a classification problem can be defined as follows. Consider a sample set $X=\{x_k|k=1, 2, \ldots, n\}$. For each sample x_k, there are m features, so x_k is a vector with m elements, i.e.

$$x_k=\{x_{k,1}x_{k,2}\ldots x_{k,m}\}=\{x_{k,j}|\, j=1, 2, \ldots m\}$$

The vector x_k can be referred to as a sample point in m-dimensional sample space and X is a point set with n points (samples) in the sample space. A geometric interpretation is usually helpful in visualizing the structure of the data, especially for two- and three-dimensional spaces.

A family of fuzzy sets $\{A_i, i=1, 2, \ldots, c\}$ is called a fuzzy-c-partition of X if

$$\mu_{ik}=\mu_{A_i}(\boldsymbol{x}_k)\in[0,1] \tag{16.79}$$

and

$$\sum_{i=1}^{c}\mu_{ik}=1, \qquad k=1, n \tag{16.80}$$

and

$$0<\sum_{k=1}^{n}\mu_{ik}<n \tag{16.81}$$

Let $\boldsymbol{U}=[\mu_{ik}]\in\boldsymbol{M}_{fc}$ be a fuzzy-c-partition. Define the fuzzy-c-partition space $\boldsymbol{M}_{fc}$ as follows:

$$\boldsymbol{M}_{fc}=\left[U|\mu_{ik}\in[0, 1];\ \sum_{i=1}^{c}\mu_{ik}=1;\ \sum_{i=1}^{n}\mu_{ik}<n\right] \tag{16.82}$$

where the cardinality of M_{fc} is infinite.

In order to select the optimal fuzzy partition U^* from the partition space M_{fc}, we can use the least-squares objective function methods. Define the objective function

$$J_t(U, V)=\sum_{k=1}^{n}\sum_{i=1}^{c}(\mu_{ik})^t(d_{ik})^2 \tag{16.83}$$

where $U=[\mu_{ik}]\in M_{fc}$ is a fuzzy-c-partition, and d_{ik} is the Euclidean distance given by

$$d_{ik}=\left[\sum_{i=1}^{m}(x_{kj}-v_{ij})^2\right]^{1/2} \tag{16.84}$$

and $V=[v_i|i=1, c]$ is the set of class centers with $v_i=\{v_{i1}v_{i2}\ldots v_{im}\}$ and

$$v_{ij}=\left(\sum_{k=1}^{n}\mu_{ik}^t\right)\frac{x_{kj}}{\mu_{ik}^t} \tag{16.85}$$

where t is the weighting factor, $t\in[1, \infty]$.

The optimal fuzzy partition U^* is

$$J_t(U^*, V^*)=\min_{U\in M_{fc}}[J_t(U, V)] \tag{16.86}$$

The optimization involves finding $U=[\mu_{ik}]$ such that it minimizes the function $J_t(U, V)$. It should be noted that V is also a function of U (from Eq. 16.82). To express $J_t(U, V(U))$ explicitly as a function of U and to find the minima over U is a complicated process. It is much easier to use the sequential iteration approach, i.e. first fix $V^{(0)}$, find $U^{(1)}=[\mu_{ik}^{(0)}]$ which minimizes $J_t(U, V^{(0)})$, then update $V^{(0)}$ to $V^{(1)}$ based on $U^{(1)}$ and repeat the iterative process. It can be shown that when V is fixed, U can be determined as follows (Bezdek, 1981). Let

$$I_k=[i|1\leq i\leq c,\ d_{ik}=0] \text{ and } \bar{I}_k=[1, 2, \ldots, c]-I_k \tag{16.87}$$

When $I_k=\Phi$ (no class center is coincident with the sample point x_k), then

$$\mu_{ik}=\left[\sum_{i'=1}^{c}\left(\frac{d_{ik}}{d_{i'k}}\right)^{2/t-1}\right]^{-1} \tag{16.88}$$

and when $I_k\neq\Phi$, then

$$\mu_{ik}=0 \quad \text{for all } i\in\bar{I}_k \quad \text{and} \quad \sum_{i\in I_k}\mu_{ik}=1 \tag{16.89}$$

This is the case where all class centers $i\in I_k$ are coincident with x_k, so x_k uniformly belongs to these classes.

The general steps of this method are as follows.

1. Fix $c(1\leq c<n)$ and select t. Initiate $U^{(0)}=[\mu_{ik}^{(0)}]_{c\times n}\in M_{fc}$. Then at iteration h, $h=0, 1, 2, \ldots$.

2. Calculate the c fuzzy class centers $V^{(h)}=[v_i^{(h)},\ i=1, c]$ using equation 16.85 on the basis of $U^{(h)}$.
3. Update $U^{(h)}$. The new membership values must be calculated by

$$\mu_{ik}^{(h+1)}=\left[\sum_{f=1}^{c}\left(\frac{d_{ik}^{(h)}}{d_{fk}^{(h)}}\right)^{2/t-1}\right]^{-1} \quad \text{if } I_k=\Phi$$

Table 16.3 Past damage record and classification

Sample no.	Past classification	Fraction of buildings with damage level i: x_1	x_2	x_3	x_4	x_5[a]	New classification
1	VI	0.13	0.52	0.28	0.07	0	VI
2	VI	0.12	0.46	0.33	0.09	0	VI
3	VI	0.19	0.47	0.32	0.02	0	VI
4	VI	0.15	0.50	0.30	0.05	0	VI
5	VI	0.17	0.48	0.32	0.03	0	VI
6	VI	0.14	0.49	0.31	0.06	0	VI
7	VI	0.20	0.45	0.33	0.02	0	VI
8	VI	0.09	0.52	0.27	0.12	0	VI
9	VI	0.14	0.55	0.26	0.05	0	VI
10	VI	0.18	0.44	0.31	0.07	0	VI
11	VI	0.17	0.47	0.31	0.05	0	VI
12	VI	0.14	0.53	0.28	0.05	0	VI
13	VII	0.06	0.34	0.40	0.16	0.04	VII
14	VII	0.06	0.38	0.38	0.17	0.10	VII
15	VII	0.09	0.31	0.42	0.12	0.06	VII
16	VII	0.07	0.37	0.40	0.09	0.07	VII
17	VII	0.08	0.31	0.42	0.16	0.03	VII
18	VII	0.05	0.32	0.43	0.18	0.02	VII
19	VII	0.09	0.32	0.40	0.14	0.05	VII
20	VII	0.10	0.28	0.42	0.17	0.03	VII
21	VII	0.11	0.32	0.41	0.10	0.06	VII
22	VII	0.04	0.36	0.42	0.11	0.07	VII
23	VII	0.07	0.29	0.44	0.13	0.07	VII
24	VII	0.05	0.26	0.50	0.17	0.02	VII
25	VII	0.10	0.39	0.36	0.10	0.05	VII
26	VIII	0	0.16	0.28	0.40	0.16	VIII
27	VIII	0	0.18	0.27	0.38	0.17	VIII
28	VIII	0	0.18	0.24	0.42	0.16	VIII
29	VIII	0	0.14	0.30	0.41	0.15	VIII
30	VIII	0	0.15	0.31	0.38	0.16	VIII
31	VIII	0	0.14	0.30	0.41	0.10	VIII
32	VIII	0	0.14	0.26	0.41	0.19	VIII
33	VIII	0	0.18	0.29	0.42	0.11	VIII
34	VIII	0	0.15	0.28	0.43	0.14	VIII
35	VIII	0	0.15	0.31	0.38	0.16	VIII
36	IX	0	0.06	0.12	0.22	0.60	X
37	IX	0	0.02	0.04	0.09	0.85	XI
38	IX	0	0.10	0.18	0.46	0.26	IX
39	IX	0	0.04	0.12	0.22	0.62	X
40	X	0	0.01	0.06	0.12	0.81	XI
41	X	0	0.03	0.04	0.20	0.73	X
42	X	0	0.01	0.03	0.18	0.78	X
43	XI	0	0	0	0.01	0.99	XI
44	XI	0	0	0	0.03	0.97	XI

[a] x_5 is a dependent feature of the other four and consequently neglected in the analysis.

or

$$\mu_{ik}^{(h+1)} = 0 \text{ for all } i \in \bar{I}_k \text{ and } \sum_{i \in I_k} \mu_{ik}^{(h+1)} = 1$$

4. If $\max_{i,k} [\mu_{ik}^{(h+1)} - \mu_{ik}^{(h)}] \leq \varepsilon$, stop; otherwise set $h = h+1$ and return to step 2.

This algorithum has been applied to the MM intensity classification (Boissonnade *et al.*, 1984). Using past earthquake records, there are 44

Table 16.4 Fuzzy partition for 44 regions (t = 1.75)

VI	VII	VIII	IX	X	XI
0.987	0.010	0.001	0.001	0.000	0.000
0.941	0.052	0.004	0.002	0.001	0.000
0.973	0.022	0.002	0.001	0.001	0.000
0.999	0.000	0.000	0.000	0.000	0.000
0.991	0.007	0.001	0.000	0.000	0.000
0.999	0.001	0.000	0.000	0.000	0.000
0.941	0.048	0.005	0.003	0.001	0.001
0.899	0.077	0.012	0.007	0.003	0.002
0.966	0.025	0.004	0.003	0.001	0.001
0.960	0.033	0.003	0.002	0.001	0.000
0.995	0.004	0.000	0.000	0.000	0.000
0.987	0.010	0.002	0.001	0.000	0.000
0.006	0.991	0.002	0.001	0.000	0.000
0.085	0.890	0.016	0.007	0.002	0.001
0.005	0.992	0.002	0.001	0.000	0.000
0.073	0.911	0.009	0.004	0.002	0.001
0.003	0.994	0.001	0.001	0.000	0.000
0.015	0.974	0.007	0.003	0.001	0.000
0.002	0.996	0.001	0.000	0.000	0.000
0.019	0.965	0.010	0.004	0.001	0.001
0.026	0.965	0.005	0.002	0.001	0.000
0.032	0.956	0.007	0.003	0.001	0.001
0.011	0.981	0.005	0.002	0.001	0.000
0.050	0.885	0.041	0.016	0.005	0.002
0.268	0.706	0.015	0.007	0.003	0.001
0.000	0.000	0.998	0.002	0.000	0.000
0.001	0.004	0.965	0.029	0.001	0.000
0.002	0.005	0.925	0.067	0.001	0.000
0.000	0.001	0.994	0.005	0.000	0.000
0.001	0.003	0.978	0.017	0.001	0.000
0.004	0.011	0.912	0.069	0.003	0.001
0.001	0.004	0.900	0.093	0.001	0.000
0.002	0.005	0.971	0.021	0.001	0.000
0.000	0.001	0.991	0.007	0.000	0.000
0.001	0.003	0.978	0.017	0.001	0.000
0.004	0.005	0.011	0.019	0.935	0.026
0.001	0.002	0.002	0.003	0.091	0.901
0.000	0.000	0.000	1.000	0.000	0.000
0.002	0.003	0.006	0.010	0.963	0.017
0.003	0.004	0.006	0.008	0.427	0.552
0.001	0.001	0.002	0.002	0.972	0.023
0.002	0.003	0.005	0.008	0.793	0.188
0.001	0.001	0.001	0.002	0.017	0.978
0.000	0.000	0.000	0.001	0.008	0.991

regions with various building damage information (Table 16.3). We use the above procedure to obtain a fuzzy classification. Initialize the $U^{(0)}$ by a random number generation routine and make $U^{(0)} \in M_{fc}$. Select $\varepsilon = 0.001$ and $t = 1.75$ and the classification is made, with results shown in Table 16.4. The intensity assigned to each region is the intensity with the largest membership value. The results of fuzzy classification are also included in Table 16.3 for comparison. Note that regions 36, 37, 39 and 40 have a different classification than before. In particular, the intensity for region 37 jumps two levels from its previous value as a result of using fuzzy classification.

Symbols

A	area, maximum acceleration
A_a	effective acceleration
A_v	acceleration corresponding to effective velocity
A_n	distribution coefficient of story shear at nth floor
a	acceleration, maximum acceleration, constant
$\boldsymbol{a}$	vector of initial condition and parameters
a_{max}	maximum acceleration
B	width, cost of repair
b	length, constant
C	structural coefficient
C_ζ	damping structural coefficient
C_μ	ductility structural coefficient
C_e	eccentricity structural coefficient
C_s	coefficient of stiffness uniformity
C_{es}	eccentricity-stiffness coefficient
C_d	amplification factor of structural displacement
C_D	coefficient of concentrated force at top of structure
$C_V = \sigma/m$	coefficient of variance
C_R	response modification coefficient
C_T	coefficient of natural period
C_{cr}	critical damping constant
c	damping constant
COV	covariance
D	fault distance, damageability, dead load
d	denominator, maximum displacement, damping force, displacement, length
d_s	depth of sand below surface (m)
d_{max}	maximum displacement
d_w	water level (m)
E	elastic modulus, energy
$E[\ ]$	expectation, assemblage expectation
EPA	effective peak acceleration
EPV	effective peak velocity
e	eccentricity
e_S	static eccentricity
e_D	accidental eccentricity
F	external force
F_t	force concentrated at top of structure
$F(\omega)$	Fourier complex spectrum
$F(x)$	probability of not exceeding x
$f(x)$	function of x, probability density
$f(x\|A)$	probability density of x giving A
$f = \omega/(2\pi)$	engineering frequency (Hz)
G	shear modulus, static load, restoring force
$G(\omega)$	one-sided power spectral density
g	gravitational acceleration ($g = 98\ \mathrm{cm/s^2}$)
$g(x)$	function of x
H	horizontal force, height, total height of building, depth below surface

H_n	top level of *n*th story, *n* from bottom
$\overline{H}$	height of center of gravity
$H(i\omega)$	transfer function
h	focal depth (km), depth of soil layer (m), story height (m), height
h_n	height of *n*th story
$h(t)$	impulse response function
I	earthquake intensity, importance factor
I_0	epicentral intensity
I	unit matrix
Im[ ]	imaginary part
i	damage index
J	target function
j	integer variable, overturning moment reduction
j_g	site type variable
j_v	component variable
K	stiffness, lateral stress factor, constant
k	stiffness, earthquake coefficient, coefficient
k_1	earthquake coefficient of first level
k_2	earthquake coefficient of second level
L	fault rupture length (km); length, live load
l, ℓ	length, wavelength
M	magnitude, mass, bending moment, moment
M_L	local magnitude
M_s	surface wave magnitude
M_u	upper limit magnitude
M_w	moment magnitude
M_0	lower limit magnitude
m	mass, magnitude, mean, no. of times,
m_b	body wave magnitude
N	number of times, earthquake number, number of cycles, number of stories, axial force
N_c	critical number of blows
n	number of times, story number from bottom, ratio of moduli of reinforcement and concrete
$P(x)$	$P(x) = 1 - F(x)$ = probability of exceeding x
$P[S]$	probability of occurrence of S
$P[S\|A]$	probability of occurrence of S giving A
PI	plastic index
P_c	percentage clay particle content
P	probability, reinforcing ratio, external force
$p(x)$	probability density function
PGA	peak ground acceleration
PGV	peak ground velocity
PSV	pseudo-velocity spectrum
Q	restoring force, quality factor, structural response
Q_u	reserved ultimate horizontal capacity
Q_y	yield shear
Q_{un}	Q_u of *n*th story
q	modal response
R	focal distance (km), energy center distance, radius, reduction coefficient of structural response, allowable value of structural capacity or strength or response

$R_{xy}(t_1, t_2)$	cross-correlation function
$R_t = \beta/\beta_{max}$	shape function of response spectrum
Re[ ]	real part
r	radius, ratio
r_s	stiffness ratio, reciprocal of story rotation or its average
$r_e = e/r_i$	eccentricity ratio (ratio of eccentricity to radius of gyration r_i)
$r(t)$	enveloping process
S	spectral density, state of structural response, site coefficient, site type
$S(T, \zeta)$	response spectrum (usually with $\zeta = 0.05$)
$S_a(T)$	(absolute) acceleration spectrum
$S_v(T)$	(relative) velocity spectrum
$S_d(T)$	(relative) displacement spectrum
s	second
T	period, structure life (years), duration
T_0	corner period of response spectrum (s)
T_d	duration of motion (s)
T_g	predominant period of ground motion
T_s	predominant period of site
T_j	period of jth mode
t	time
u	relative displacement, offset
V	shear, vertical force, maximum velocity, wave velocity
V_0	base shear
V_{eo}	elastic base shear
V_n	shear at nth floor
V_{en}	elastic shear at nth floor
VAR[ ]	variance
v	velocity, wave velocity, displacement
v_{max}	maximum velocity
v_P	longitudinal wave velocity (km/s)
v_S	transverse wave velocity (km/s)
v_f	fault rupture velocity (km/s)
W	weight, power, energy
W_n	weight of building from nth story up
w	displacement, weighting function
w_n	weight of nth story
X, Y, Z	coordinate axes, position variable
x, y, z	coordinate axes, position variable
y	response variable
y^0	vector of initial condition
Y	function
Z	seismic zoning coefficient
α	angle, curvature, base shear coefficient, longitudinal wave velocity, $\alpha(T) = k\beta(T)$,
$\boldsymbol{\alpha}$	parameter vector
α_n	shear coefficient at nth story
β	angle, coefficient, transverse wave velocity
$\beta(T)$	$(= S_a(T)/a_{max})$ amplification spectrum
$\Gamma(\,)$	Gamma function
γ	shear deformation, modal participating factor
Δ	epicentral distance (km), displacement, increment
Δ_y	yield limit deformation

δ	logarithm decrement, story drift
ε	axial deformation
ε_b	bending deformation
$\zeta = C/C_{cr}$	damping ratio, coefficient of yield strength
θ	Fourier transformation of $\theta(t)$
θ	angle
λ	Lamé constant, wavelength, spectral moment
μ	Lamé constant, $\mu = G$, ductility factor $\mu = \Delta_{max}/\Delta_y$, central moment
ν	Poisson's ratio, annual rate of earthquake
ρ	mass density
σ	stress, standard deviation
$\sigma(\tau)$	$= R(\tau)/R(0)$ nondimensional correlation function
τ	shear stress, time variable, time lag, time increment
Φ	mode shape matrix
ϕ	modal shape, function, phase angle, amplification factor of spectrum, coefficient
ψ	$\psi_\mu = \phi_\mu\phi_e$ amplification factor of spectrum
ω	$(= 2\pi f)$ circular frequency (rad.)
ω_0	natural frequency of a single-degree-of-freedom (SDF) system
ω_D	damped natural frequency
ω_j	jth natural frequency, spectral parameters
Ω	frequency variable, spectral parameter $\Omega = \omega_2$
log	common logarithm
ln	natural logarithm

Subscripts and superscripts

a	acceleration
b	bending
D	damping
d	dynamic, displacement, duration
E	earthquake
e	elastic, equivalent, eccentricity
f	failure
G	dead load
g	ground, site, soil, ground motion
h	horizontal
i, j, k	sequence number
L	live load
l	load
M	magnitude
m	maximum
n	story number from bottom, integer
p	probability
s	structure, site, snow, static
T	superscript indicates matrix transformation
t	total
u	ultimate
v	vertical, velocity
W	wind, wall
Y	yield, response

μ	ductility
cr	critical
0	base, fundamental, initial
•	over character indicates derivative with respect to time t
–	over character indicates average
~	over character indicates SSI variable
*	conjugate
′	distinguishes another quantity or variable

References

Abbreviations

ASCE: American Society of Civil Engineers
ASME: American Society of Mechanical Engineers
ASS: *Acta Seismologica Sinica*, China
ATC: Applied Technology Council
BERI: *Bulletin of Earthquake Research Institute*, Tokyo University
BSSA: *Bulletin of Seismological Society of America*
CCEE: Canadian Conference on Earthquake Engineering
ECEE: *European Conference on Earthquake Engineering*
EEEV: *Earthquake Engineering and Engineering Vibration*, China
EERI: Earthquake Engineering Research Institute, Stanford, CA, USA
EERL: Earthquake Engineering Research Laboratory, CIT, USA
EESD: *Earthquake Engineering and Structural Dynamics*
ERC: *Earthquake Research in China*, China
FEMA: Federal Emergency Management Agency
GEEESD: ASCE Geotechnical Engineering Division Specialty Conference on Earthquake Engineering and Soil Dynamics
ICMZ: International Conference on (Earthquake) Microzonation
ICSZ: International Conference on Seismic Zonation
IEM: Institute of Engineering Mechanics, Harbin, China
IIS: *Report of the Institute of Industrial Science, University of Tokyo*
ISMER: Workshop on Interpretation of Strong-Motion Earthquake Records Obtained in and/or near Buildings, UCLA
IWEES: International Workshop on Earthquake Engineering in Shanghai, Tong-Ji. University, Shanghai, China
JASCE: *Journal of American Society of Civil Engineers*
JEES: Japanese Earthquake Engineering Seminar
JGR: *Journal of Geophysical Research*
JNSEE: *Proceedings of Japan National Symposium of Earthquake Engineering*, Tokyo, Japan, Architectural Institute of Japan
JSCE: Japan Society of Civil Engineers
NEHRP: National Earthquake Hazard Reduction Program
PES: *Problems on Engineering Seismology*, Science Publishers, Moscow
SADRCS: US/PRC Workshop on Seismic Analysis and Design of Reinforced Concrete Structures
SMIRT: International Conference for Structural Mechanics in Reactor Technology
SoilDEE: International Conference on Soil Dynamics and Earthquake Engineering
TAIJ: *Transactions of the Architectural Institute of Japan*
UCB/EERC: Earthquake Engineering Research Center, University of California, Berkeley
UCEE: US/PRC Bilateral Workshop on Earthquake Engineering, Harbin, China
UCMZ: Joint US/PRC Microzonation Workshop, Harbin, China
UISRS: Structural Research Series, Civil Engineering Studies, University of Illinois
USCGS: US Coast and Geodetic Survey (now US Geological Survey, USGS)
USNCEE: US National Conference on Earthquake Engineering, Stanford University, Stanford, California, USA, EERI
USGSOF: US Geological Survey, Open-file Report

USGSP: US Geological Survey Professional Paper

WCEE: World Conference on Earthquake Engineering (1st: June 1956, Berkeley, CA, USA, EERI, 2nd: July 1960, Tokyo and Kyoto, Japan, Science Council of Japan. 3rd: Jan.–Feb. 1965, Auckland and Wellington, New Zealand, NZ Nat. Comm. on Earthquake Engineering. 4th: January 1964, Santiago, Chile. 5th: June 1973, Rome, Italy, Ministry of Public Works and Italian Assoc. of Earthquake Engineering. 6th: January 1977, New Delhi, India, Indian Soc. of Earthquake Technology. 7th: September 1980, Istanbul, Turkey. 8th: July 1984, San Francisco, CA, EERI, Prentice-Hall. 9th: August 1988, Tokyo and Kyoto, Japan, Science Council of Japan, Nissei Kogyo Co.)

References in Chinese

Bao, Z.-W. and Lai, J.-Y. (1981) Analysis of microtremor signal of buildings. *EEEV*, **1**(2).

CEER (Center of Earthquake Engineering Research, Taiwan University) and others (1986) Field investigation report of building damage in North Taiwan on November 15, 1986. *Journal of Structural Engineering*, **2**(1).

Chen, D.-S. (1965) Some data on maximum acceleration and acceleration spectrum of ground motion. *IEM*, *Collected Papers*, **2**.

Deng, Q.-D. *et al.* (1980) Principle and methodology of earthquake intensity zonation in China, *ASS* **2**(1).

Earthquake Loss Mitigation Dept, State Seismological Bureau, PR China (1992) *Earthquake Handbook*, Seismological Press.

Fu, C.Y. and Liu, Z.R. (1960) On macroseismic determination of depth of focus of earthquakes. *Science Record*, New Series, **4**(5).

Guo, H. (1982) Review of research on earthquake response of liquid-storage tanks in USA. *EEEV*, **1**(3).

Hu, Y. (1982) Inverse analysis of structural vibration, in *Collected Papers on Earthquake Engineering*, Science Press. (Also IEM 79–025, 1979.)

Hu, Y. (1983) A suggestion of removing the term "earthquake intensity" from seismic design code. *Journal of Building Structures*, no. 1.

Hu, Y. and Hu, Y. (1983) On the vertical earthquake responses of smokestacks, *EEEV*, **3**(2).

Hu, Y., Hu, Y. and Chen, X.-W. (1985) Earthquake damage analysis of brick smokestacks. *China Earthquakes*, **1**(2).

Hu Y. and Zhou, X.-Y. (1962) Elastic response of structure under stationary and stationarized earthquake ground motion. *IEM*, *Collected Papers*, **1**.

Hu, Y. *et al.* (1965) Fourier spectral analysis of microtremor of structures. *IEM*, *Collected Papers*, **2**.

Hu, Y. *et al.* (1980) Effects of site conditions on earthquake damage and ground motion. *EEEV*, trial issue no. 1.

Hu, Y. *et al.* (1982) Functional relations of bedrock motion parameters with earthquake magnitude and distance. *Acta Seismologica Sinica*, **4**(2).

Huo, J.-R. (1989) Study on the attenuation laws of strong earth quake ground motion near the source. Ph.D. Thesis, IEM.

Huo, J.R., Hu, Y.X. and Feng, Q.M. (1991) Study on envelope function of acceleration time history. *EEEV*, **11** (1).

Institute of Engineering Mechanics of Academia Sinica (ed.), (1979) *Haicheng Earthquake Damage*, Seismological Press.

Jiang, J.R. and Sun, J.L. (1987) Seismic damage model of brick structure. *EESD*, **7**(1).
Lanzhou Earthquake Research Institute and Earthquake Team of Ningxia Autonomy, State Seismological Bureau (1980) *Great Haiyuan Earthquake of 1920*, Seismological Press.
Li, S.-B. (1980) *Chinese Earthquakes*, Seismological Press.
Liao, Z. and Wang, F. (1983) Vertical seismic input to structures. *EEEV*, **3**(2).
Liu, H.-X. (1965) Review and suggestions on analytical method of earthquake loads in design code. *IEM*, *Collected Papers*, **2**.
Liu, H.-X. (ed.) (1985–86) *Damage of Tangshan Earthquake*, Seismological Press.
Liu, J. and He, L.-M. (1983) Analysis of seismic responses of chimneys to combined horizontal and vertical ground motion. *EEEV*, **3**(2).
Qian, P.-F. (1983a) On vertical seismic load. *EEEV*, **3**(2).
Qian, P.-F. (1983b) *Seismic Analysis of Structures*, Seismological Press.
Qiao, T.-P. and Liu, H.-S. (1984) Risk analysis of liquefiable site, in *Earthquake Resistance of Industrial Buildings and Foundations*, Seismological Press.
Shi, Z.-L. *et al.* (1991) Research on the methodology and principle of seismic zonation – Results of the trials in Northern China. *ASS*, **13**, 179–89.
Song, Y.-T. and Zhu, J.-C. (1983) Effect of ground motion duration on earthquake response of multistory structures. *EEEV*, **3**(4).
Sun, C.-S and Chen, B.-W. (1982) Seismic microzonation of Lanzhou, *EEEV*, **1**(3).
Wang, F. (1987) Probabilistic model of earthquake occurrence and their application in seismic hazard assessment. Ph.D. Thesis, IEM.
Wang, J.-J and Hu, Y.-X. (1991) A study on rotational components of surface ground motion. *EEEV*, **11**(2).
Wang, Z.K. (1973) Statistical prediction of earthquake mitigation. *Geological Science*, **4**.
Xie, Y.-S. (1977) *Earthquake and Earthquake Resistance*, Science Press.
Yin, Z.-Q. and Peng, K.-Z. (1962) Analysis of Microtremor Phenomena of Buildings. *IEM*, *Collected Papers*, **1**.
Yin, Z. *et al.* (1981) Elasto-plastic earthquake responses of multi-story frames structures. *EEEV*, Trial Issue, no. 2.
Yin, Z. *et al.* (1983) Studies on several problems of seismic design method. IEM 83–051.
Yu, S.-S and Shi, Z.-J (1980) Analysis of liquefaction potential of horizontal soil strata. *EEEV*, Trial Issue, no. 1.
Zhou, S.-J *et al.* (1983) Experimental study on the crack location in the chimney subjected to shock vibration. *EEEV*, **3**(2).
Zhou, X.-Y (1965) Effect of soil condition on earthquake load on buildings. *IEM*, *Collected Papers*, **2**.
Zhou, X.-Y, Hu, Y.-X and Chen, D.-S. (1965) Approximate method of evaluating the nonstationary response of elastic systems. *IEM*.

References in Japanese

Aizawa, S. *et al.* (1990) Experimental study of dual active mass damper, in *Proc. 8th JEES*, Vol. 2.
Architectural Institute of Japan (1983) Ground motion and site.
Ishimoto, M. (1931a) Observation of earthquake acceleration. *BERI*, **9**(2).
Ishimoto, M. (1931b) Accelerograph and its record. *BERI*, **9**(3).
Ishimoto, M. (1931c) Ground motion properties studied from earthquake accelerograms. *BERI*, **9**(4).

Ishimoto, M. (1932) On ground motion acceleration. *Architectural Journal*, no. 554.

Ishimoto, M. (1935) On earthquake ground motion. *Proc. JSCE*, **21**(6).

Iwasaki, T. *et al.* (1980) Prediction of level of site liquefaction during earthquake, *Soil and Foundation*, **29**(4).

Japan Association of Electricity (1987) Technical guidelines of seismic design of nuclear power plant. JEAG 4601.

Japan Road Association (1980) Highway bridge guidelines and commentaries, V. Seismic design.

Kanai, K., *et al.* (1954–68) Measurement of the microtremor, reports 1–12. *BERI*, **32–46**.

Kato, B. and Akiyama, H. (1976) Collapse conditions of multistory shear frames during earthquake. *TAIJ*, No. 244.

Kato, B. and Akiyama, H. (1978) Damage distribution of multi-story steel shear frames during earthquake. *TAIJ*, No. 270.

Shimizu, N. (1972) Studies on seismic design method of machine and pipeline system. *IIS*, **22** (1).

Takata, T. *et al.* (1965) Damage of bridges (Niigata earthquake). *Reports of Civil Engineering Institute*, No. 125.

References in western languages

Aki, K. (1984) Asperities, barriers, characteristic earthquake and strong motion prediction. *JGR*, **89** (B7).

Aki, K. (1988) Seismological synthesis of strong ground motion, in *Proc. 9th WCEE*, Vol. 8.

Aki, K. and Richards, P.G. (1980) *Quantitative Seismology*, Freeman, San Francisco.

Algermissen, S.T. and Perkins, D.M. (1976) A probabilistic estimate of maximum acceleration in rock in the contiguous United States. USGSOF 76–416.

Allen, C.R. *et al.* (1987) Seismic hazard evaluation of a highly active fault – the Xianshuihe Fault of Southeastern China, in *Proc. Int. Seminar on Seismic Zonation*, Guangzhou, China.

Ambraseys, N.N. (1969) Macroseismic observations for some recent earthquakes, in *Proc. 4th WCEE*, Santiago, Chile, Vol. 3.

Ambraseys, N.N. (1973) Dynamics and response of foundation materials in epicentral region of strong earthquakes, in *Proc. 5th WCEE*, Vol. 1.

Ambraseys, N.N. and Sarma, S. (1969) Liquefaction of soils by earthquakes. *BSSA*, **59**, (2).

Amin, M. and Ang, A.H.-S. (1968) Nonstationary stochastic Model of Earthquake Motions, *JASCE*, 94 (EM2).

Anagnos, T. and Kiremidjian, A.S. (1984) Stochastic time-predictable model for earthquake occurrence, *BSSA*, **74**, 2593–611.

Anderson, A.W. *et al.* (1952) Lateral forces of earthquake and wind. *JASCE*, **117**.

Anderson, J.G. and Quaas, R. (1988) Effect of magnitude on the character of strong ground motion: an example from the Guerrero, Mexico strong motion network. *Earthquake Spectra*, **4** (3).

Ang, A.H.-S. and Hu, Y.X. (1983) Development of probability bases for earthquake resistant design. C.E. Studies, Str. Res. Sec. No. 511, UILU-ENG-83-2016, University of Illinois, Urbana, IL.

Ang, A.H.-S. and Tang, W.H. (1975) *Probability Concepts in Engineering Planning and Design*, Vol. 1, *Basic Principles*, John Wiley.

Ang, A.H.-S. *et al.* (1984) Seismic damage of R.C. buildings and related ground motion parameter, in *Proc. ATC Workshop on Earthquake Ground Motion and Building Damage Potential*, San Francisco.

Anon. (1973) Effect of soil on earthquake vibrational intensity. *PES*, **15**.

API (1982) *API Recommended Practice for Planning, Design, and Constructing Fixed Offshore Platforms (API RP 24)*, American Petroleum Institute.

Aptikaev, F. and Kopnichev, J. (1980) Correlation between seismic vibration parameters and type of faulting, in *Proc. 7th WCEE*, Vol. 1.

Arias, A. (1970) A measure of earthquake intensity, in *Seismic Design for Nuclear Power Plants* (ed. R. Hansen), MIT Press, Cambridge, MA.

Ariman, T., Lee, B.J. and Chen, Q.Q. (1987) Failure of buried pipelines under ground deformation, in Proc. *3rd SoilDEE*, Princeton, NJ.

Ariman, T. and Muleski, G.E. (1981) A review of the response of buried pipelines under seismic excitations, PVP-34, *Lifeline Earthquake Engineering – Buried Pipelines, Seismic Risk, and Instrumentation*, ASME, pp. 1–29. (Also *EESD*, **9** (2).)

ATC (1978) Tentative provisions for the development of seismic regulations for buildings. ATC-3-06.

ATC (1979) Proceedings of a workshop on earthquake resistance of highway bridges. ATC 6–1.

Baas, S.J. and Kwakernaak, H. (1977) Rating and ranking of multiple-aspect alternatives using fuzzy sets. *Automatica*, **13**, 47–58.

Barstein, M.F. (1960) Application of probability methods for design: the effect of seismic forces on engineering structures, in *Proc. 2nd WCEE*, Vol. 2.

Basham, P.W. *et al.* (1982) New probabilistic strong seismic ground motion maps of Canada: a compilation of earthquake source zones, methods and results, Earth Physics Branch Open-File 82–33, Energy, Mines and Resources, Ottawa, Canada.

Battis, J. (1981) Regional modifications of acceleration attenuation functions. *BSSA*, **71** (4).

Bayo, E. and Wilson, E.L. (1983) Numerical techniques for the evaluation of soil-structure interaction effects in the time domain. UCB/EERC-83/04.

Beck, J.L. (1978) Determining models of structures from earthquake records. EERL 78–01.

Beck, J.L. (1982) System identification applied to strong motion records from earthquakes, in *Earthquake Ground Motion and Its Effects on Structures*, Applied Mechanics Division, Vol. 53, ASME.

Benjamin, J.R. and Cornell, C.A. (1970) *Probability, Statistics and Decision for Civil Engineers*, McGraw-Hill, New York.

Berger, E. *et al.* (1975) ALUSH – a computer program for seismic response analysis of axisymmetric soil-structure systems. UCB/EERC 75–31.

Bertero, V.V. *et al.* (1980) Lessons from structural damages observed in recent earthquakes, in *Proc. 7th WCEE*, Vol. **4**.

Bezdek, J.C. (1981) *Pattern Recognition with Fuzzy Objective Function*, Plenum Press.

Blume, J.A. *et al.* (1961) *Design of Multistory Reinforced Concrete Buildings for Earthquake Motions*, Portland Cement Association.

Boissonnade, A., Dong, W., Liu, S. and Shah, H. (1984) Use of pattern recognition and Bayesian classification for earthquake intensity and damage estimation. *Journal of Soil Dynamics and Earthquake Engineering*, **3** (3), 145–49.

Bollinger, G.A. (1977) Reinterpretation of the intensity data for the 1886 Charleston, South Carolina Earthquake, in *Studies Related to the Charleston, South Carolina Earthquake of 1886* – a Preliminary Report (ed. D.W. Rankin), USGSP 1028–B, pp. 17–32.

Bolotin, B.B. (1960) Application of statistical method for estimation of structural strength under earthquake action. *Engineering Papers*, **27**.

Bolt, B.A. (ed.) (1987) *Seismic Strong Motion Synthesis*, Academic Press.

Bolt, B.A. and Abrahamson, N.A. (1982) New attenuation relations for peak and expected accelerations of strong ground motion. *BSSA*, **72** (6).

Bolt, B.A. *et al.* (1982) Earthquake strong motion recorded by a large near-source array of digital seismographs. *EESD*, **10** (4).

Booker, J.R. *et al.* (1976) GADFLEA – a computer program for the analysis of pore pressure generation and dissipation during cyclic or earthquake loading. UBC/EERC 76–24.

Boore, D.M. and Joyner, W.B. (1984) Prediction of high frequency strong ground motion, in *Proc. IWEES*, Shanghai, China, Vol. 1.

Brankov, G. and Sachanski, S. (1980) Effects of Vranchea earthquake, March 1977 on the territory of Bulgaria, in *Proc. 7th WCEE*, Vol. 9.

Campell, K.W. (1981) Near-source attenuation of peak horizontal acceleration. *BSSA*, **71** (6).

Campell, K.W. (1985) Strong motion attenuation relations: a ten year prospective. *Earthquake Spectra*, **1**, (1).

Carder, D.S. (1936) *Vibration Observations, Earthquake Investigations in California 1934–1935*, USCGS Special Publ. No. 201.

Carder, D.S. (1966) *Earthquake Investigations in the Western US 1931–1964*, USCGS Publ. No. 41–2.

Cassaro, M.A. and Martinez-Romero, E. (eds) (1987) *The Mexico Earthquakes–1985*, ASCE, New York.

Chandra, U. (1979) Attenuation of intensities in the United States. *BSSA*, **64**(6).

Chandra, U. *et al.* (1979) Attenuation of intensities in Iran. *BSSA*, **69** (1).

Chen, D., Dong, W. and Shah, H. (1988) Earthquake recurrence relationships from fuzzy earthquake magnitudes. *Journal of Soil Dynamics and Earthquake Engineering*, **7** (3), 136–42.

Chinese Institute of Civil and Hydraulic Engineering (1981) *A Short Course in Earthquake Engineering*, Teipei, China.

Chopra, A.K. (1978) Earthquake resistant design of concrete gravity dams. *JASCE*, **104** (ST6).

Clough, R.W. and Hidalgo, P. (1973) Design of a shaking table test for a reinforced concrete frame structure, in *Proc. 5th WCEE*, Vol. 1.

Clough, R.W. and Penzien, J. (1975) *Dynamics of Structures*, McGraw-Hill.

Clough, R.W. *et al.* (1973) Earthquake simulator test of a three story steel frame structure, in *Proc. 5th WCEE*, Vol. 1.

Cluff, L.S., Patwardhan, A.S. and Coppersmith, K.J. (1980) Estimating the probability of occurrence of surface faulting earthquakes on the Wasatch Fault Zone, Utah. *BSSA*, **70**, 1463–78.

Cornell, C.A. (1968) Engineering seismic risk analysis. *BSSA*, **58** (5).

Cornell, C.A. and Vanmarcke, E.H. (1969) The major influences on seismic risk, in *Proc. 4th WCEE*, Vol. 1.

Cornell, C.A. and Winterstein, S.R. (1988) Temporal and magnitude dependence in earthquake occurrence models. *BSSA*, **78**, 1522–37.

Cornell, C.A. *et al.* (1993) Seismic hazard induced by mechanically interactive fault segments. *BSSA*, **83**(2).

Davis, R.O. and Berrill, J.B. (1982) Energy dissipation and seismic liquefaction in sands. *EESD*, **10** (1).

Der Kiureghian, A. (1981) A response spectrum method for random vibration analysis of MDF system. *EESD*, **9** (5).

Der Kiureghian, A. and Ang, A. H.-S. (1977) A fault-rupture model for seismic risk analysis. *BSSA*, **67** (4).

Dong, W. (1986) Applications of fuzzy set theory in structural and earthquake engineering. Ph.D. Thesis, Stanford University.

Dong, W. and Shah, H. (1987) Vertex method for computing function of fuzzy variables. *Journal of Fuzzy Sets and Systems*, **24** (1), 65–78.

Dong, W. and Wong, S. (1987) Fuzzy weighted averages and implementation of the extension principle. *Journal of Fuzzy Sets and Systems*, **21** (2), 183–99.

Donovan, N.C. and Bornstein, A.E. (1978) Uncertainties in seismic risk procedures. *JASCE*, **104** (GT7).

Dubois, D. and Prade, H. (1980) *Fuzzy Sets and Systems: Theory and Applications*, Academic Press, New York.

Duda, S.J. and Nuttli, O.W. (1974) Earthquake magnitude scale. *Geophysical Surveys*, **1** (4).

Durate, R.T. (1980) An investigation of the influence of large magnitude earthquakes on seismic risk, in *Proc. 7th WCEE*, Vol. 1.

Dutton, C.E. (1889) The Charleston earthquake of August 31, 1886, in *Ninth Annual Report of the U.S. Geology Survey to the Secretary of the Interior 1887–1888*, J. W. Powell, Director, Government Printing Office, Washington, DC, pp. 203–528.

Esteva, L. and Villaverde, R. (1973) Seismic risk, design spectra and structural reliability, in *Proc. 5th WCEE*, Vol. 2.

Eykhoff, P. (1974) *System Identification – Parameter and State Estimation*, John Wiley & Sons.

FEMA (1988) *NEHRP Recommended Provisions for Seismic Regulations for Buildings*, FEMA 95.

Frangopol, D. M., Ikejima, K. and Hong, K. (1987) Seismic hazard prediction using a probabilistic-fuzzy approach, in *Reliability and risk analysis in civil engineering, Proc. ICASP5*, University of British Columbia, Vancouver, Canada, May 25–29, pp. 783–90.

Fukutake, K. *et al.* (1990) Analysis of saturated dense sand-structure system and comparison with results from shaking table test. *EESD*, **19** (7).

Fuller, M.L. (1912) The New Madrid earthquake. *US Geological Survey Bulletin* No. 494, 119.

Gardner, J.K. and Knopoff, L. (1974) Is the sequence of earthquakes in Southern California, with aftershocks removed, Poissonian? *BSSA*, **64**, 1363–7.

Ghanaat, Y. and Clough, R.W. (1983) Analytical correlation of a dynamic brace buckling experiment. *EESD*, **11** (1).

Goto, H. and Toki, K. (1969) Structural response to nonstationary random excitation, in *Proc. 4th WCEE*, Vol. 1.

Gutenberg, B. and Richter, C.F. (1942) Earthquake magnitude, intensity, energy, and acceleration. *BSSA*, **32** (3).

Gutenberg, B. and Richter, C.F. (1956) Earthquake magnitude, intensity, energy, and acceleration (second paper). *BSSA*, **46** (2).

Hadjian, A.H. (1972) Scaling of earthquake accelerograms – a simplified approach. *JASCE*, **98** (ST2).

Haldar, A. and Tang, W.H. (1979) Probabilistic evaluation of liquefaction potential. *JASCE*, **105** (GT2).

Hanks, T.C. (1979a) b values and $\kappa^{-\gamma}$ seismic source models: implications for tectonic stress variations along active crustal fault zones and the estimation of high-frequency strong ground motion. *JGR*, **85** (B5).

Hanks, T.C. (1979b) Seismological aspects of strong motion seismology, in *Proc. 2nd USNCEE*, Stanford, CA, EERI.

Hanks, T.C. and Kanamori, H. (1979) A moment magnitude scale. *JGR*, **84** (B5).

Hardin, B.O. and Drnevich, V.P. (1972) Shear modulus and damping in soils. *JASCE*, **98** (SM6 and SM7).

Hart, G.C. and Yao, J.T.P. (1976) System identification in structural mechanics. *JASCE*, **102** (EM6).

Hartzell, S.H. (1978) Earthquake aftershocks as Green's functions. *Geophysics Letters*, 5 (1).

Hattori, S. (1980a) Seismic risk maps in Asian countries – China, Philippines, Indonesia and others, in *Proc. 7th WCEE*, Vol. 1.

Hattori, S., (1980b) Seismic risk maps in Turkey, Iran and Mediterranean areas, in *Proc. 7th WCEE*, Vol. 1.

Hays, W.W. *et al.* (1978) Preliminary ground response maps for the Salt Lake City, Utah area, in *Proc. 2nd ICMZ*, Vol. 1.

Hays, W. W. (ed.) (1981) Facing geologic and hydrologic hazards- earth-science considerations. USGSP 1240-B, p. 109.

He, Q.J. and Wei, L. (1981) On seismic strengthening of existing reinforced concrete frames, in *Proc. SADRCS.*

Hershberger, J. (1956) A comparison of earthquake accelerations with intensity ratings, *BSSA*, **46** (4).

Hisano, M., Nagashino, I. and Kawamura, S. (1990) Earthquake observation of a base-isolated building with slide bearings, in *Proc. 8th JEES.*

Hopper, M. G. (ed.) (1985) Estimation of earthquake effects associated with large earthquakes in the new Madrid seismic zone. USGSOF 85–457.

Housner, G.W. (1947) Characteristics of Strong Motion of Earthquakes, *BSSA*, **37** (1).

Housner, G.W. (1952) Spectrum intensities of strong-motion earthquakes, in *Proc. Symposium on Earthquakes and Blast Effects on Structures*, RESI.

Housner, G.W. (1965) Intensity of ground shaking near the causative fault, in *Proc. 3rd WCEE*, Vol. 1.

Housner, G.W. (1979) Earthquake and earthquake engineering, in *Proc. 3rd CCEE*, Vol. 1.

Housner, G.W. and Jennings, P.C. (1964) Generation of artificial earthquake, *JASCE*, **90** (EM1).

Housner, G.W. and Jennings, P.C. (1972) The San Fernando earthquake. *EESD*, **1** (1).

Housner, G.W. and Jennings, P.C. (1984) Historical view of earthquake engineering, in *Proc. 8th WCEE*, Post-Conf. Volume.

Hu, Y. (1980) Input identification from structural vibration responses. UCB/EERC-80/26.

Hu, Y. (1991) State-of-the-art of seismic hazard assessment in China, in *International Seminar on Earthquake Prediction and Hazard Mitigation Technology*, Tsukuba, Japan.

Hu, Y. and He, X. (1986) A new method of generation of response spectrum-compatible ground motion, in *Proc. 8th WCEE*, Vol. 1.

Hu, Y. and Huo, J. (1988) Regressional principle and relationship between earthquake intensity and acceleration, in *Proc. 9th WCEE*, Vol. II.

Hu, Y. and Sun, P.S. (1982) Seismic microzonation and site effects, in *Proc.* 3rd *ICMZ*, Seattle, Wash., Vol. 2.

Hu, Y. and Tao, X.X. (1984) Selection and estimation of earthquake motion for critical structure engineering sites, in *IWEES*, Vol. 1.

Hu, Y. and Zhang, M.Z. (1983) Attenuation of ground motion for regions with no ground motion data, in *Proc. 4th CCEE*, University of British Columbia, Vancouver, Canada.

Hu, Y. *et al.* (1989) Attenuation relationship of ground motion for North China partly based on data from Tangshan earthquake, in *Selected Papers from Chinese Journals of Structural Engineering* (ed. R.C.Y. Young), ASCE.

Husid, R.L. (1969) Analisis de terremotos: analisis general. *Revisita del IDIEM, Santiago, Chile*, **8** (1).

IAEE (1980–) Earthquake resistant regulations – a world list, International Association for Earthquake Engineering, Tokyo, Japan. (Published every four years.)

Ibanez, P. (1977) Review of analytical and experimental techniques for improving structural dynamic models. Rept 1149–1, Applied Nucleonics Co.

Idriss, J.M. and Kennedy, R.P. (1980) Analysis for soil-structure interaction effects for nuclear power plants. Report by the Ad Hoc Group on Soil-Structure Interaction of the Committee on Nuclear Structures and Materials of the Structural Division of ASCE.

Iemura, H. and Jennings, P.C. (1974) Hysteretic response of a nine-story reinforced building. *EESD*, **3**(2).

Irikura, K. (1988) Estimation of new-field ground motion using empirical Green's function, in *Proc. 9th WCEE*, Vol. 8.

Ishihara, I. (1981) Dynamic soil properties, in *Proc. UCMZ*, IEM.

Iwan, W.D. (1984) Characteristics of digital strong motion accelerographs, in *IWEES*, Vol. 1.

Iwan, W.D. and Gates, N.C. (1979) Estimating earthquake response of simple hysteretic structures. *JASCE*, **105** (EM3).

Iyengar, R.N. and Iyengar, K.T.S.R. (1969) A stationary random process model for earthquake accelerograms. *BSSA*, **59** (3).

Joyner, W.B. and Boore, D.M. (1981) Peak horizontal acceleration and velocity from strong motion records from the 1979 Imperial Valley, California, earthquake. *BSSA*, **71** (6).

Joyner, W.B. and Boore, D.M. (1982) Prediction of earthquake response spectra, *USGSOF*, 82–977.

Kagawa, T. and Kraft, L.M. Jr (1980a) Lateral load-deflection relationships of piles subjected to dynamic loadings. *Soil and Foundations*, **20** (4).

Kagawa, T. and Kraft, L.M. Jr (1980b) Seismic responses of flexible piles. *JASCE*, **106** (GT8).

Kagawa, T. and Kraft, L.M. Jr (1981) Dynamic characteristics of lateral load-deflection relationships of flexible piles. *EESD*, **9** (1).

Kanai, K. (1983) *Engineering Seismology*, University of Tokyo.

Kanai, K. and Suzuki, T. (1968) Expectancy of the maximum velocity amplitude of earthquake motions at bedrock *BERI*, **46** (3).

Kanai, K. *et al.* (1966) Observation of strong earthquake motions in Matsushiro area, part 1. *BERI*, **44** (3).

Katayama, T. (1980) Distribution of structure's period-dependent seismic risk in Japan, in *Proc. 7th WCEE*, Vol. 1.

Kaul, M.K. (1978) Stochastic characterization of earthquakes through their response spectrum. *EESD*, **6** (5).

Kaya, I. and McNiven, H.D. (1978) Investigation of nonlinear characteristics of a three-story steel frame using system identification. UCB/EERC-78/25.

Kelly, J.M. (ed.) (1981) Control of seismic response of piping systems and other structures by base isolation. UCB/EERC-81/01.

Kelly, J.M. (1986) Aseismic base isolation: review and bibliography. *SDEE*, **5**(4).

Kelly, J.M. and Beucke, K.E. (1983) A friction-damped base isolation system with fail-safe characteristics. *EESD*, **11** (1).

Kiremidjian, A. and Anagnos, T. (1984) Stochastic slip-predictable model for earthquake occurrence. *BSSA*, **74**, 739–55.

Kitagawa, Y. and Matsushima, Y. (1980) Study of dynamic ground characteristics in range of short and long periods, in *Proc. 7th WCEE*, Vol. 2.

Knopoff, L. (1971) A stochastic model for the occurrence of main-sequence earthquakes. *Reviews of Geophysics and Space Physics*, **9**, 175–88.

Kobayashi, H. and Kagami, H. (1972) A method of local seismic intensity zoning maps on the basis of subsoil conditions, in *Proc. 1st ICMZ*, Vol. 2.

Kubo, K. (1975) Fundamental concept of aseismic design of underground piping systems, in *Proc. 5th ECEE*, Vol. 2.

Kubo K. and Penzien, J. (1979) Analysis of three-dimensional strong ground motions along principle axes, San Fernando Earthquake. *EESD*, **7** (3).

Lai, S.S.P. and Biggs, J.M. (1980) Inelastic response spectra for seismic building design. *JASCE*, **106** (ST6). (Also *Proc. 7th WCEE*, Vol. 4.)

Li, G. H. (1982) Earthquake response of Luan River Bridge near Tanshan, in *Proc. UCEE*, Vol. 1.

Liao, Z.-P. and Wong, H.L. (1984) A unified transmitting boundary of discrete models, in *IWEES*, Vol. 1.

Liu, H.X. (1980) On the Chinese seismic intensity scale, in *Proc. 7th WCEE*.

Liu, H.-X. *et al.* (1982) Physical measure of earthquake intensity studied by statistical discrimination, in *UCEE*, Vol. 2.

Liu, S.C., Lagorio, H.J. and Chong, K.P. (1991) Application of active control technology, in *Proc. China–Japan–USA Trilateral Symposium/Workshop on Earthquake Engineering*, Harbin, China.

Loh, C.H. *et al.* (1982) Engineering analysis of SMART-1 array accelerograms. *EESD*, **10** (4).

Lomnitz, C. (1974) *Global Tectonics and Earthquake Risk*, Elsevier.

Lomnitz, C. and Rosenblueth, E. (eds) (1976) *Seismic Risk and Engineering Decisions*, Elsevier.

Luco, J.E. (1982) Linear soil-structure interaction: a review, in *Earthquake Ground Motion and Its Effects on Structures* (ed. S.K. Datta), ASME.

Luo, Q.F. and Hu, Y.X. (1989) An improved semi-empirical method for synthesizing near-field accelerogram of earthquake, in *Proc. Sino-Japan Seismological Research*, Beijing, China.

Lysmer, J. (1978) Analytic procedures in soil dynamics. UCB/EERC 78–29.

Lysmer, J. and Kuhlemeyer, R.L. (1969) Finite dynamic model for infinite media. *JASCE*, **95** (EM4).

Mahin, S.A. *et al.* (1989) Pseudodynamic test method – current status and future directions. *JASCE*, **115** (ST12).

Mandani, E. (1977) Applications of fuzzy logic to approximate reasoning using linguistic systems. *IEEE Transactions on Computers*, **C-26**, 1182–91.

Mandani, E. and Assilian, S. (1975) An experimental in linguistic synthesis with a fuzzy logic controller. *International Journal of Man–Machine Studies*, **7**, 1–3.

Martin, P.P. and Seed, H.B. (1982) One dimensional dynamic ground response analysis. *JASCE*, **108** (GT7).

Matthiesen, R.B. (1980) Building instrumentation programs, in *Proc. ISMER*.

McGuire, R.K. (1974) Seismic structural response risk analysis, incorporating peak response regressions on earthquake magnitude and distance. Res. Rept R74–51, MIT, Department of Civil Engineering.

McGuire, R.K. (1976) EQRISK–evaluation of earthquake risk to site. USGSOF 76–67.

McGuire, R.K. (1977) Seismic design spectra and mapping procedure using hazard analysis based directly on oscillator response. *EESD*, 5 (2).

McGuire, R.K. (1978) Seismic ground motion parameter relations. *JASCE*, **104**, (GT4).

McGuire, R.K. and Barnhard, T.P. (1979) The usefulness of ground duration in predicting the severity of seismic shaking, in *Proc. 2nd USNCEE*.

McGuire, R.K. and Barnhard, T.P. (1981) Effects of temporal variations in seismicity on seismic hazard, *BSSA*, **71**, 321–34.

McVerry, G.H. (1979) Frequency domain identification of structural models from earthquake records. EERL 79–02.

Medvedev, S.V. (1962) *Engineering Seismology* (in Russian), National Publisher of Literatures on Structure, Architecture and Architectural Materials, Moscow.

Mizumoto, M. and Tanaka, K. (1979) Some properties of fuzzy numbers, in *Advances in Fuzzy Sets Theory and Applications*, (ed. M. Gupta), North-Holland, Amsterdam.

Mogami, T. and Kubo, K. (1953) The behaviour of sand during vibration, in *Proc. 3rd Int. Conf. Soil Mech. and Foundation Eng.*, vol.1.

Mostaghel, N. and Habibagahi, K. (1979) Cyclic liquefaction strength of sands. *EESD*, **7** (3).

Nagahashi, S. (1980) Effects of ground motion duration upon earthquake response of structure, in Proc. 7th *WCEE*, Vol. 7.

Nakamura, Y. (1984) Extension of algebraic calculus on fuzzy numbers using α-level sets. Presented at Fuzzy Information Processing (NAFIP-84) Symposium, Kaual, Hawaii.

Nau, J.M. and Hall, W.J. (1982) An evaluation of scaling methods for earthquake response spectra. UISRS 499.

Negoita, C. (1985) *Expert Systems and Fuzzy Systems*, Benjamin Cummings Publishing Company.

Neumann, F. (1952) Some generalized concepts of earthquake motion, in *Proc. Symposium on Earthquake and Blast Effects on Structures*, EERI.

Neumann, F. (1954) *Earthquake Intensity and Related Ground Motion*, University of Washington Press.

Neumann, F. (1958) Damaging earthquake and blast vibrations. *Trends in Engineering, University of Washington*, No. 1.

Neumann, F. (1960) A broad formula for estimating earthquake forces on oscillators, in *Proc. 2nd WCEE*, Vol. 2.

Newmark, N.M. (1965) Effects of earthquakes on dams and embankments. *Geotechnique*, **15** (2).

Newmark, N.M. and Hall, W.J. (1969) Seismic design criteria for nuclear reactor facilities, in *Proc. 4th WCEE*, Vol. 2.

Newmark, N.M. and Hall, W.J. (1975) Pipeline design to resist large fault displacement, in *Proc. 1st USNCEE*.

Newmark, N.M. and Rosenblueth, E.A. (1971) *Fundamentals of Earthquake Engineering*, Prentice-Hall.

Nigam, N.C. (1991) Probabilistic approach to earthquake engineering – emergence of a practical theory. *Bulletin of the Indian Society of Earthquake Technology*, **28** (1).

Nuttli, O.W. (1973) The Mississippi Valley earthquakes of 1811 and 1812: intensities, ground motion and magnitudes. *BSSA* **63**, 227–48.

Nuttli, O.W. (1981) Evaluation of past studies and identification of needed studies of the effects of major earthquakes occurring in the new Madrid fault zone, US Federal Emergency Management Agency Region VII, Kansas City, Missouri, p. 28.

Ohsaki, Y. (1966) Niigata Earthquakes, 1964, building damage and soil condition. *Soil and Foundations*, **6**(2).

Ohsaki, Y. (1979) On the significance of phase content in earthquake ground motions. *EESD*, **7**(5).

Ohsaki, Y. (1982a) Some notes on Masing's law and nonlinear response of soil deposits, in Dr. Ohsaki's Collected Papers, Department of Architecture, University of Tokyo.

Ohsaki, Y. (1982b) Dynamic characteristics and one-dimensional linear amplification theories of soil deposits, in Dr. Ohsaki's Collected Papers, Department of Architecture, University of Tokyo.

Ohsaki, Y. (1982c) Dynamic nonlinear model and one-dimensional nonlinear response of soil deposits, in Dr. Ohsaki's Collected Papers, Department of Architecture, University of Tokyo.

Ohsaki, Y. *et al.* (1980) Analyses of seismic ground motion parameters including vertical components, in *Proc. 7th WCEE*, Vol. 2.

Okamoto, S. (1984) *Introduction to Earthquake Engineering*, University of Tokyo Press.

Oliveira, C.S. (1980) Seismic hazard analysis for zones of intermediate seismicity, in *Proc. 7th WCEE*, Vol. 1.

Omote, S. *et al.* (1988) A newly developed strong motion seismograph giving high-fidelity seismic data with operations completely under the command of the center station, in *Proc. 9th WCEE*, Vol. 2.

Park, Y.J. and Ang, A.H.-S. (1983) A mechanistic seismic damage model for reinforced concrete. *JASCE*, **111**(ST4).

Parker, W.H., Architect and Stewart, F.S., Builder (inspectors) (1886) Report of the Inspectors, Board of Inspectors (for the insurance companies), Stockdell, H.C. Chairman, J.A. Thomas and Hutson Lee, Committee, in *Yearbook – 1886, City of Charleston, South Carolina*, News and Courier Book Presses, Charleston, South Carolina.

Patwardhan, A.S., Kulkarni, R.B. and Tocher, D. (1980) A Semi-Markov Model for Characterizing Recurrence of great Earthquakes. *BSSA*, **70**, 323–47.

Pauschke, J.M. *et al.* (1981) A preliminary investigation of the dynamic response of the Imperial County Services Building during the October 15, 1979 Imperial Valley earthquake. Report no. 49, John A. Blume Earthquake Engineering Center, Stanford University.

Pedrycz, W. (1984) An identification algorithm in fuzzy relational systems. *Journal of Fuzzy Sets and Systems*, **13**, 153–67.

Penzien, J. and Watabe, M. (1975) Characteristics of 3- dimensional earthquake ground motions. *EESD*, **3**(3).

Pilkey, W.D. and Cohen, R. (eds) (1972) *System Identification of Vibrating Structures*, ASME, New York.

Psycharis, I.M. and Jennings, P.C. (1983) Rocking of slender rigid bodies allowed to uplift. *EESD*, **11**(1).

Rainer, J.H. (1979) Dynamic testing of civil engineering structures, in *Proc. 3rd CCEE*, Vol. 1, McGill University, Montreal, Canada, Canadian Committee for Earthquake Engineering.

Rea, D. and Penzien, J. (1973) Dynamic response of a 20ft × 20ft shaking table, in *Proc. 5th WCEE*, Vol. 2.

Riddel, R. and Newmark, N.M. (1979) Statistical analysis of the response of nonlinear systems subjected to earthquakes. UISRS 468.

Roesset, J.M. and Tassoulas, J.L. (1982) Nonlinear soil structure interaction: an overview, in *Earthquake Ground Motion and Its Effects on Structures* (ed. S.K. Datta), ASME.

Rosenblueth, E. (ed.) (1980) *Design of Earthquake Resistant Structures*, Pentech Press.

Savy, J.B., Shah, H.C. and Boore, D. (1980) Nonstationary risk model with geophysical input. *JASCE*, **106**, 145–63.

Schmucker, K. (1984) *Fuzzy Sets, Natural Language Computations and Risk Analysis*, Computer Science Press.

Schnabel, P.S. and Seed H.B. (1973) Accelerations in rock for earthquakes in the western United States. *BSSA*, **63**(2).

Schnabel, P. *et al.* (1972) Modification of seismograph records for effects of local soil conditions. *BSSA*, **62**(6).

Schneider, H.M. *et al.* (1982) Experimental behavior of a spatial piping system with shock arrestors and energy absorbers under seismic excitation. UCB/EERC-82/03.

Seed, H.B. (1966) A method for earthquake resistant design of earth dams. *JASCE*, **92**(SM1).

Seed, H.B. (1979) Considerations in the earthquake – resistant design of earth and rockfill dams, *Geotechnique*, **29**(3), 215–63.

Seed, H.B. (1981) Earthquake resistant design of earth dams, in *International Conference on Recent Advances in Geotechnical Earthquake Engineering and Soil Dynamics*, Vol. 3, University of Missouri, Rolla, Missouri, USA.

Seed, H.B. and Chan, C.K. (1966) Clay strength under earthquake loading conditions. *JASCE*, **92**(SM2).

Seed, H.B. and Idriss, I.M. (1971) Simplified procedure for evaluation potential. *JASCE*, **97**(SM9).

Seed, H.B. *et al.* (1969) Characteristics of rock motions during earthquakes. *JASCE*, **95**(SM5).

Seed, H.B. *et al.* (1976) Site dependent spectra for earthquake- resistant design. *BSSA*, **66**, (1).

Serff, N. *et al.* (1976) Earthquake induced deformations of earth dams. UCB/EERC 76–4.

Sherif, M.A. (1981) Microzonation with respect to site amplification and soil liquefaction, in *Proc. UCMZ*, Paper 2, pp. 1–64.

Shi, Z. (1987) Seismic zoning of North China, in *Proc. Int. Seminar on Seismic Zonation*, Guangzhou, China.

Shibata, H. *et al.* (1978) On ground motion records for engineering purpose other than horizontal ground acceleration record, in *Proc. 5th JNSEE*, 185–92.

Shih, C.F. (1981) Failure of liquid storage tanks due to earthquake excitation. EERL 81–04.

Shing, P.B. and Manivannan, T. (1990) On the accuracy of implicit algorithm for pseudodynamic tests. *EESD*, **19** (5).

Shinozuka, M. and Sato, Y. (1967) Simulation of nonstationary random process. *JASCE*, **98**, (EM1).

Soong, T.T., Masri, S.F. and Housner, G.W. (1991) An overview of active structural control under seismic loads. *Earthquake Spectra*, **7**(3).

Sozen, M.A. and Moehle, J.P. (1980) Effect of wall height on earthquake response of reinforced concrete multistory frame-wall structures, in *Proc. 7th WCEE* Vol. 6, pp. 439–46.

Stiemer, S.F. and Chow, F.L. (1984) Curved plate energy absorbers for earthquake resistant structures, in *Proc. 8th WCEE*, Vol. **5** pp. 967–74.

Street, R. (1982) A contribution to the documentation of the 1811–1812 Mississippi Valley earthquake sequence. *Earthquake Notes*, **53**(2), 39–47.

Street R. and Nuttli, O.W. (1984) The Central Mississippi River Valley earthquakes of 1811–1812, in *Missouri Academy of Sciences Meeting*, Cape Girardeau, Missouri, April 27–28, 1984, p. 34.

Sugimura, Y. (1977) Earthquake observation and dynamic analysis of a building supported on long piles, in *Proc. 6th WCEE*, Vol. 2.

Tajimi, H. (1960) A statistical method of determining the maximum response of a building structure during an earthquake, in *Proc. 2nd WCEE*, Vol. 2.

Takanashi, K. *et al.* (1980) Pseudo-dynamic tests on a 2-story steel frame by computer-load test apparatus hybrid system, in *Proc. 7th WCEE*, Vol. 7, pp. 225–32.

Takizawa, H. and Jennings, P.C. (1980) Collapse of a model for ductile reinforced concrete frames under extreme earthquake motions. *EESD*, **8**(2), 117–44.

Tanimoto, K. and Noda, T. (1976) Prediction of liquefaction occurrence of sandy deposits during earthquakes by a statistical method. *Proc. JSCE*, No. 256, 79–86.

Terzaghi, K. (1950) Mechanisms of landslides. The Geological Survey of America, Engineering Geology (Berkeley) volume.

Trifunac, M.D. (1976) Preliminary analysis of the peaks of strong earthquake ground motion – dependence of peaks on earthquake magnitude, epicentral distance and recording site conditions. *BSSA*, **66**(1).

Trifunac, M.D. and Brady, A.G. (1975) On the correlation of peak acceleration of strong motion with earthquake magnitude epicentral distance and site condition, in *Proc. USNCEE*, pp. 43–52.

Trifunac, M.D. and Brady, A.G. (1976) Correlations of peak acceleration, velocity and displacements with earthquake magnitude, distance and site conditions. *EESD*, **4** (5).

Udwadia, F.E. and Trifunac, M.D. (1974) Time and amplitude dependent response of structures. *EESD*, **2**(4).

Umemura, H. and Aoyama, H. (1969) Evaluation of inelastic seismic deflection of reinforced concrete frame based on the tests of members, *Proc. 4th WCEE*, Vol. 1 p. B2.

Unruh, J.F. and Kana, D.D. (1981) An iterative procedure for the generation of consistent power/response spectrum. *Nuclear Engineering and Design*, **66**(3).

USGS (1989) Lessons learned from the Loma Prieta, California, earthquake of October 17, 1989. US Geological Survey Circular 1045.

Vanmarcke, E.H. (1972) Properties of spectral moments with applications to random vibrations. *JASCE*, **98**(EM2), 425–46.

Vanmarcke, E.H.(1976) Structural response to earthquakes, in *Seismic Risk and Engineering Decisions* (ed. C. Lomnitz and E. Rosenblueth) Ch. 8, Elsevier, Amsterdam and New York.

Velera, J.E. and Donovan, N.C. (1977) Soil liquefaction procedure – a review. *JASCE*, **103**(GT6), 607–25.

Vere-Jones, D. (1966) A Markov model for aftershock occurrence. *Pure Applied Geophysics*, **64**, 31–42.

Wallace, J.W. and Moehle, J.P. (1989) The 1985 Chile earthquake: an evaluation of structural requirements for bearing wall buildings. UCB/EERC-89/05.

Wang, L.R.L. and O'Rourke, M.J. (1977) State of the art of buried lifeline earthquake engineering, in *The Current State of Knowledge of Lifeline Earthquake Engineering, Proc. Technical Council on Lifeline Earthquake Engineering Specialty Conference*, ASCE, pp. 252–66.

Whitman, R.V. (ed.) (1989) Workshop on ground motion parameters for seismic hazard mapping. Tech. Rept NCEER-89-0038.

Wiegel, R.L. (ed.) (1970) *Earthquake Engineering*, Prentice-Hall.

Wilson, E.L., Der Kiureghian, A. and Bayo, E. P. (1981) A replacement for the SRSS method in seismic analysis. *EESD*, **9**(2).

Xu, Z.X. and Weng, D.G. (1982), The effect of duration of strong ground motion on the collapse of structures, in *UCEE*, Vol. 1, p. A6–1–20.

Yamada, M. (1969) Low cycle fatigue fracture limits of various kinds of structural members subjected to alternately repeated plastic bending under axial compression as an evaluation basis or design criteria for aseismic capacity, in *Proc. 4th WCEE*, Vol. 1, p. B2.

Yamamura, N. and Tanaka, H. (1990) Response analysis of flexible MDF systems for multiple-support seismic excitations. *EESD*, **19**(3).

Yao, J.T.P. and Munse, W.H. (1962) *Low-cycle axial fatigue behavior of mild steel*, ASTM, Spec. Tech. Publ. 338, pp. 5–24.

Yegulalp, T.M. and Kuo, J.T. (1974) Statistical prediction of the occurrences of maximum-magnitude earthquakes. *BSSA*, **64**(2).

Yim, C.S. *et al.* (1980) Rocking response of rigid blocks to earthquakes. *EESD*, **8**(6).

Yin, Z.-Q. (1991) Study on prediction of earthquake disasters and their economic losses, in *Proc. China–Japan–USA Trilateral Symposium/Workshop on Earthquake Engineering*, Harbin, China.

Yin, Z.Q. *et al.* (1981) Aseismic problems of multistory R.C. frame with infill walls, in *Proc. SADRCS*, pp. 252–62.

Yin, Z.Q. *et al.* (1991) A method for prediction and classification of earthquake disasters, *ERC*, pt 3, no. 3.

Zadeh, L. (1965) Fuzzy sets. *Information and Control*, **8**, 338–53.

Zhou, X.Y. *et al.* (1984) The seismic–geological background and the earthquake response of typical structures in Beijing, in *Proc. 8th WCEE*, Vol. 2.

Index

Abrahamson, N.A. 89, 93
Acceleration 70
effective 70
record, *see* Ground motion record
rms (root mean square) 70, 79
saturation 79, 80, 89
time history, *see* Accelerogram
see also Ground Motion
Accelerogram 61, 67
Accelerograph 61, 62, 66, 67
analog 61, 62, 65
digital 66, 67
Accumulative damage, *see* Low-cycle fatigue
Aki, K. 100
Akiyama, H. 171, 172
Alpha-cut 362
Algermissen, S.T. 266
Allen, C.R. 244
Ambient vibration 225
Ambreseys, N.N. 50, 75, 79, 192, 193
Amin, M. 40
Ang, A.H-S. 14, 25, 40, 225, 248, 250, 340, 353
Annual rate of earthquake occurrence 244
Aoyama, H. 221
API (American Petroleum Association, *see also* Seismic design code 351
Aptikaev, F. 92, 93, 96
Arias, A. 70
Arias intensity 70
ARMA (autoregression moving average) 98
ATC (Applied Technology Council) 70, 281, 266, 267, 282–284
see also Seismic design code

b value 14
Baas, S.J. 369
Barstein, M.F. 26, 39, 40, 344, 345
Battis, J. 87, 88, 92, 93
Base isolation 214, 307–309
Base shear 311
Basham, P.W. 266
Benjamin, J.R. 358
Bezdek, J.C. 379
Blow count of standard penetration 197, 329, 332
Blume, J.A. 219
Boissonnade, A. 381
Bolotin, V.V. 26, 348
Bolt, B. 78, 88, 89, 93
Boore, D.M. 71, 92, 94
Bracing system 300
Brady, A.G. 93, 94
Brankov, G. 54
Brick structure 110
Bridge 110, 115, 209
abutment 281
design consideration 282
differential motion of supports 281
falling of superstructure 115, 209
pier 209
Brittle failure 13, 279
BSSC 322
Brune, J.N. 79
Tall building, *see* Highrise structure
Bursa City, Turkey 51

Campell, K.W. 88, 93
Cartesian product 365
Cassaro, M.A. 52
Chandra, U. 59, 87
Characteristic function 358–360
Chimney, *see* Smokestack
Chong, K.P. 306
Chopra, A.K. 294
Clay 192, 187, 188, 332
Clough, R.W. 148, 214, 345
Coherence function 32, 227
Construction detail 305
Constructional column 304
Cornell, C.A. 15, 246, 247, 340, 343, 358
Correlation function 29, 332
auto 29
cross 30, 323, 176
matrix 32
Crossing 36

Dam
concrete 292
Dou River 185, 206
earth 110, 205–207
Jadeite 295, 296
Koyna 203
Miyun 205
Damage, *see* Earthquake damage
Damper 215
active mass 215
Damping 147, 150, 151, 160, 228, 318
artificial 149
complex 160, 201

Damping *contd.*
equivalent 162
material 89
radiation (geometrical) 160, 182
ratio 311
Rayleigh 162, 205
theory of 160
viscous 160
Decision-making 362
Deformation response 169
Dena, Q.D. 57
Der Kiureghian, A. 14, 25, 248, 250, 340, 351, 352
Design, *see* Seismic design
Design earthquake, *see* Design ground motion
Disaster reduction 140, 141
Distance 56, 80
energy center 89
epicentral 56, 80, 89
fault 56, 80, 89
focal 80, 89
Dong, M. 369
Donovan, N.C. 92, 93, 194
Doppler effect 55
Drnevich, V.P. 191
Dual-parameter fatigue criterion, *see* Low-cycle fatigue criterion
Ductility 278, 283, 284, 299, 318
coefficient 167
Duda, S.J. 24, 79
Durate, R.T. 266
Duhamal integral 147
Duration 71, 73, 75, 115, 172
effect of 172
damage and 115
ground motion 71, 96
of shaking 73
Dynamic property 5, 214, 217–231
of brick wall 300–3
of joint (connection) 220
of material 217
of offshore platform 214
of part of structure 221
of soil 186
of structural member 219
of structure 151, 213, 224
natural frequency 151
of shear wall 214
Dynamic reliability 353
Dynamic test, *see* Structural test

Earthquake
activity, *see* Seismicity
belt 12, 102–4, 137
catalog 3, 103
in China 101–16
damage 5, 45–60, 105, 114, 352, 380
accumulative effect of 114
bridge 110, 111, 114, 115, 139, 281
building 54, 109, 254
concrete dam 29
criterion 311
earth structure 114, 115, 205, 255
experience from 109–111, 115, 141, 276
fatigue 353
index 106–8, 354
level 356
modern city 114, 115, 139–41
prediction 236, 237, 259
probability matrix 226
secondary disaster 138, 140, 141
selective 51, 53, 54
site effect on 52, 54, 60, 72
smokestack 111, 112, 114, 255, 279–81
storage tank 285–7
threshold, *see also* First passage 354
underground pipeline 288
disaster mitigation, *see* disaster reduction
disaster reduction 5, 140
engineering 4, 5
fault 106–8, 127, 133, 135
historical data of 3, 103, 118, 238
intensity 45–60, 236, 253, 262, 380, 381
abnormal 55, 106, 108, 109, 115
attenuation of 56, 57, 78, 262
elliptical 59
in China 58
in Japan 57
in USA 57
distribution 55, 106, 108, 109, 113
epicentral 56, 59
and ground motion 83, 87, 97, 265
rating 46
site 60
intensity scale 45
China (1957) 46, 48–50
China (1980) 84, 85
JMA (Japan) 75
MM (modified Mercalli, USA) 46, 47, 120
Rossi-Forel 207
instrumental data 239
in Japan 136–41
loss 236
magnitude, *see* magnitude
maximum credible 282
mechanism 13
microzonation 262, 267–72
prediction 5, 108, 113, 140

province 102–4, 264
resistant design, *see* Seismic design
response 4, 145–213
concrete dam 292
earth dam 205
earth structure 184–213
equivalent linear 201
method of analysis of 145, 147
equivalent static 145
history of 145
nonlinear 146, 164
numerical 146, 164
random, *see* statistical
response spectral 145
statistical 145
static theory of 145
underground pipeline 289
of site 184, 200, 201
of structure 145–213
vertical 158
with multiple inputs 172
response analysis, *see* response
source 56, 78
spectrum 79, 80, 97
in USA 117–41
zoning factor 320, 311
zonation 5, 260–7, 306
map 263, 330
of Canada 326
of China 262
of Soviet Union (former) 263
of USA 263, 265, 323
Earthquake case-histories
Alaska, USA, 1964 3, 127–30, 209
Armenia, former Soviet Union 39
California (San Fernando, USA), 1971 3, 130
Charleston, USA, 1886 120–4
Central California, USA, 1980 285
Chile 1939, 1977, 1983, 1985 3
Coalinga, USA, 1983 54
Dixie Valley-Fairview Valley, USA, 1954 59
El Centro, USA, 1940 301
Fukui, Japan, 1948 209
Gediz, Turkey, 1970 51
Guerrero, Mexico, 1957, *see* Mexico earthquake
Haicheng, China, 1975 3, 108–13, 192, 200, 209, 329, 331
Haiyuan, China, 1920 3, 105
Hebgen Lake, USA, 1959 59
Heyuan, China, 1962 293
Iran, 1972 3
Japan, 1933 3
Kanto, Japan, 1923 60, 72, 138–9, 315
Kashmir, India, 1905 3
Koyna, India, 1967 297
Loma Prieta, 1989 133–135
Managua, 1972 3
Mexico, 1957, 1962, 1985 3, 52, 53
Miyagi, Japan, 1978 140–1, 210
New Madrid, 1811–12 119–120
Nihonkai-Chubu, 1983 141
Niigata, Japan, 3, 139–40, 185, 196, 209
North Taiwan, China, 1986 115, 116
Pakistan, 1935 3
Peru, 1970 3
Philippines, 1990 3
San Fernando, USA, 1971 130–3, 146
San Francisco, USA, 1906, 1957 3, 60, 124–7
Tango, Japan, 1927 3
Tangshan, China, 1976 3, 55, 113–15, 185, 200, 205, 281, 329, 331
Tokaichi-oki, 1968 140
Tokyo, Japan, 1923 3
Tonghai, China, 1970 56, 106
Turkey, 1939 3
Urakawa, Japan, 1982 141
Vrancea, Romania, 1977 53, 54
Zhaotong, China, 1975 107
Eccentricity 320
Economic analysis 8, 277
Elephant-foot instability 285
Energy-absorbing boundary, *see* Transmitting boundary
Energy dissipation 161, 276, 278, 301, 312
Engineering seismology 6
Epicenter 56
distance 56
macroseismic 56
Equivalent linearization 201
Esteva, L. 93, 94
Evaluation and retrofitting 5
Experimental method, *see* Structural test
Extension principle 367

Fatigue, *see* Low-cycle fatigue
Fault 56, 109, 133
active 133, 290
blind (buried) 108
distance 56
rupture 105, 106, 110, 126
rupture length 56
type of 92
FEMA (Federal Emergency Management Agency) 266, 322
Field Act 319

Finite element method 204, 212
First passage 37, 354
Force-deformation curve, *see* Hysteretic curve
Fourier spectrum 32, 72
Fourier transformation fast (FFT) 155, 156
Frame 302, 320
Frangopol, D.M. 371
Freefield problem 177
Frequency-domain analysis 155, 157, 181
Fu, Cheng-Yi 57
Fuzzy 45, 46, 77, 357
  classification 377
  information 366
  set 357
  system 373
  vector 371

Gataldi, J. 46
Geology 238, 262
Goto, H. 41
Green's function 99
Ground, *see also* Site
  layered 200
  classification 60, 324, 327, 328, 331, 332
  failure 50, 55, 60, 133, 135, 141, 270
  damage of, *see* Ground failure
Ground motion 5, 7, 61–100, 282
  attenuation 87, 88–97, 262
  artifital 97–100
  characteristics of 67
  DBE (design-based earthquake) 292, 296
  design 282
  EPA (effective peak acceleration) 70, 235, 266, 323
  EPV (effective peak velocity) 70, 266, 323
  evaluation 87, 96
  factors affecting 52, 78, 91
  long period 51, 53, 54, 140
  MCE (maximum credible earthquake) 282, 283, 292, 295
  near field 90, 93
  observation 61, 62, 65
    array 65
    instrument, *see also* Accelerograph 61, 62
    measurement, *see* Observation
  OBE (operation-based earthquake) 292, 295
  random representation of 38–40
  relation of – and intensity 46, 83, 87, 96
  rotation 77
  saturation of 79
  simulation, *see* Artificial 77
  spatial correlation of 176
  spectrum 71, 72
  SSE (safe shotdown earthquake) 292
  saturation, *see* Acceleration saturation
  time history 38
  vertical 158
Ground motion record 65–70, 262, 276
  Acapulco (Mexico, 1965) 75
  Alameda Park (Mexico, 1962) 70, 75
  Chalome No. 2 (USA, 1966) 70, 75
  El Centro (USA, 1940, 1979) 68, 69, 73, 75, 274
  Guerrero (Mexico, 1985) 53
  Hachinohe (Japan, 1968, 1978) 69, 70, 75
  Kushiro (Japan, 1967) 75
  Leukas (Greece, 1973) 75
  Hueneme (USA,1957) 75
  on liquefied area 139
  Lima (Peru, 1966) 63, 70
  Long Beach (USA, 1933) 274
  Lulong (China, 1982) 64, 69
  Matsushiro (Japan, 1955, 1966) 75
  Melendy Ranch (USA, 1972) 75
  Mexico City (1962, 1964, 1965, 1985) 53, 68, 74, 75, 81
  Mororan (Japan, 1968) 75
  Niigata (Japan, 1964) 75
  Pacoima (USA, 1971) 69, 75, 133
  Parkfield (USA, 1966) 69, 75
  San Fernando (USA, 1971) 86, 133
  Seattle (USA, 1949) 75
  SMART-1 (Taiwan, China) 78, 116
  Tangshan (China, 1976) 81, 116
  Tianjin (China, 1976) 70, 74
  Tokaimura (Japan, 1964) 75
  Vrancea (Romania, 1977) 53, 54
Gumbel, *see* Extreme value
Gutenberg, B. 14, 22, 24, 56, 83, 104

Hall, W.J. 146, 166, 290
Hanks, T. 24, 79
Hardin, B.O. 191
Hartzell, S.H. 99
Hattori, S. 266
Hays, W.W. 269
He, Guang-Qian. 171
Hershberger, J. 83, 84
Highrise structure 300
Housner, G.W. 26, 38, 61, 79, 82, 84, 146, 274, 275, 287, 292, 306, 309
Hu, Yuxian 26, 41, 46, 75, 87, 88, 90, 92–6, 98, 100, 159, 192, 193,

225, 226, 230, 244, 280, 344, 345, 349

Identification 373, 374
structural, *see* Structural identification
Idriss, I.M. 177
IEM (Institute of Engineering Mechanics, China) 281
Impedance 128
Importance factor 311, 322
Incomplete knowledge 45, 357
Intensity, *see* Earthquake intensity
Irikura, K. 100
Ishimoto, M. 84, 274
Isoseismal (intensity contour) 51, 55, 56, 107, 108, 113, 114, 116, 119, 124, 134, 138–40
Iwan, W.D. 162
Iwasaki, T. 200

Jennings, P. 82, 146, 172, 214, 274
Joyner, W.B. 71, 92, 94

Kanai, K. 20, 39, 40, 85, 94, 225–7, 266, 268, 345
Kanamori, H. 24
Katayama, T. 94, 266
Kato, B. 171, 172
Kaul, M.K. 73
Kelly, J.M. 214, 306
Kiremidjian, A. 343
Knopoff, L. 340
Kopnichev, J. 92, 93, 96
Koyna dam 293
Kubo, K. 68
Kwakernaak, H. 369

Lagorio, H.J. 306
Landslide 50, 55, 114, 129, 130, 133, 135, 281
Land-use planning 262
Lateral resisting system 111, 300, 302, 304
Li, Guo-Hao 282
Li, Shan-Bang 57
Liao, Zgeng-Peng 183, 281
Lifeline 141
Liquefaction 111, 114, 121, 123, 139, 184, 192–200, 281, 329–32
and damage 55, 114, 193
countermeasure 333
evaluation 193–9
index 332
observation 186, 205
potential 332
risk 199
of silt 332
Liquid-storage tank 214
Liu, Hui-Xian 82, 84, 85, 281
Liu, S.C. 306
Liu, Zheng-Rong 57
Loh, C.H. 78
Lomnitz, C. 15, 262
Loss, *see* Earthquake loss
Low-cycle fatigue 75, 114–15, 215, 218, 219, 225, 302
criterion 353
Palmgren-Miner 353
Luco, J.E. 177
Lysmer, J. 177, 183

Mandani, E. 374
Manson-Coffin, *see* Low-cycle fatigue
Magnitude 21, 56, 357
fault rupture length relationship 24
frequency relationship 14, 104, 243, 340
characteristic 244
local 22
moment 24
recurrence relationship, *see* Frequency relationship
saturation 105
surface-wave 22
upper limit 15, 244
Mahin, S.A. 228
Mapping 358, 361, 368
Martinez-Roero, E. 52
Masing, *see* Nonlinear mathematical model
Masri, S.F. 306, 309
Mathematical model 211
beam model 211
Mindlin static model 211
nonlinear
bilinear 221
Clough 222
Masing 221
pinching 223
McGuire, R.K. 75, 85, 92–4, 96, 323, 340
McNiven, H.D. 232
McVerry, G.H. 231
Medvedev, S.V. 57, 60, 84, 267
Melendy Ranch 75
Membership function 361
Mexico City 52
Microtremor, *see* Ambient vibration
Microzonation, *see* Earthquake microzonation
Modal response combination 65, 152, 312, 351
Moehle, J.P. 54, 214
Moho discontinuity 9

Muleski, G.E. 288

Nagahashi, S. 75
Naito, T. 314
Natural frequency 151
Natural mode 151
Negoita, C. 357
NEHRP 266, 323
Neumann, F. 83, 85
Newmark, N.M. 84, 146, 148, 150, 166, 171, 207, 221, 275, 290, 351
Nonlinear (inelastic) 215, 224
  deformation 171
  response 146, 155, 164–72
  model, *see* Mathematical model
NSF 323
Nuclear power plant 291, 296–8, 307
Numerical integration 146, 154
  Gauss-Newton method 230
  Newmark method 148
  Wilson-theta method 148
Nuttli, O.W. 24, 79, 119, 120

Offshore platform 214
Ohsaki, Y. 99, 193, 196, 266, 297
Oliveira, C.S. 266
Overturning
  moment 332
  of rigid body 214

P-delta effect 158, 160, 325
Park, Y.J. 225
peak factor 345
Pedrycz, W. 374
Penzien, J. 68, 148, 345
Perkins, D.M. 266
PGA, *see* Ground motion
PGV, *see* Ground motion
Phase (phase angle) 98, 99, 156
  difference 99
  difference spectrum 99
Pile foundation 209
  analysis of 211
  damage of 209
  strong motion observation of 210
Pile-soil interaction (see soil-pile interaction)
Pinching 223
Pipeline 214
Plastic index 330
Plate tectonics 10, 11, 13
Pore pressure 187, 208
Potential source area 245
Potential source zone (see potential source area)
Power spectrum density 32, 73
Predominant period 71

Probability density function 27
  extreme value 15
  log-normal 28
  normal (Gauss) 28
  Poisson 242, 244
Probability of Exceedance 247
Probability theory 358

Qian, Bei-Feng 279
Qiao, T.P. 200
Quality factor 20

Ramberg-Osgood, *see* Nonlinear mathematical model
Random method 26–41, 339–56
  of earthquake ground motion 344
  of earthquake response 343
Random process 28, 36, 38, 39
  filtered 39
  Markov 339, 342
  non-Poisson 340
  non-stationary 40, 347
  Poisson 339
  renewal 340
  segmental Poisson 340
  semi-Markov 339, 342
  stationarized 348
  stationary 30, 38, 345
  two-state Markov 346, 347
  variable 26
  vibration 26
  Weibul (gamma) 346, 347
  white-noise 39
Rate of crossing 36
  rate of zero crossing 37
Region with few acceleration data 96
Regression analysis 88, 90
Reinforced concrete structure 305
Resonance effect 51, 52, 72
Response, *see* Earthquake response
Response spectrum 65, 73, 149, 235, 274
  acceleration 149
  analysis 154
  design 311, 324, 339
  effect of damping on 150
  low-frequency reduction factor of 345
  method 166
  microzonation, *see* Microzonation
  nonlinear 166–170
  pseudo-velocity 149
  relation between – and Fourier spectrum 73
  relation between – and power spectrum 73
  site 324, 327, 328
  theory of 150

Retrofitting, *see* Evaluation and retrofitting
Richter, C.F. 14, 22, 24, 56, 83, 262
Rigid body 214, 215
Rosset, J.M. 177
Rosenblueth, E. 15, 26, 146, 148, 171, 177, 304, 351

San Andreas fault 121
Sand liquefaction, *see* Liquefaction
Sano, R. 314
Sarma, S. 192, 193
Scattering problem 177
Schebalin, N.V. 57
Schnabel, P. 92, 93
Schumucker, K. 369
SEAOC 321, 322
  code (guidelines) 320, 321, 351
Seed, H.B. 92, 93, 97, 191, 195, 197, 198, 208, 324
Seismic coefficient, *see* Earthquake coefficient
Seismic design 235, 273–310
  bridge 133, 233, 281–4, 291
  building 279
  column 133
  code, *see* Seismic code
  concrete dam 292–6
  constructional measure of 300, 302–14
  critical structure 291–2
  concrete dam 292–6
  detailing 140, 233, 299, 305, 314
  highrise structure 279–81, 315, 316
  level 277, 292, 294
  lifeline 135
  load 311
  nuclear power plant 291, 292, 296–8
  offshore platform 291, 298–300
  probabilistic 333
  principle (general consideration) of 216, 273–310, 314
  of shear wall 301
  of smokestack 279
  of storage tank 285–8
  of smokestack 280
  theory of 273–6
  torsion 140
  two-level 276, 318, 333
  of underground pipeline 288–91
Seismic design code 5, 262, 311–35
  API 146, 169, 267, 288, 299
  ATC-3 266, 267, 282–4, 323, 326, 333
  bridge 282–4
  Canada 326
  China 75, 82, 197, 275, 314–19, 326–35
  Columbia 326
  Japan 284, 314–19
  Mexico 82
  offshore platform 298
  nuclear power plant 151, 169, 297, 322
  storage tank 288
  trend of 334
  UBC 263, 319–21
  underground pipeline 288
  USA 282, 314, 319–26
  see SEAOC
Seismic gap 52
Seismic hazard assessment (SHA) 4, 235–51, 262
  deterministic 246
  point source 247
  fault-rupture 250
Seismic hazard analysis, *see* Seismic hazard assessment
Seismic risk analysis 235, 251–60
Seismic wave, *see* Wave
Seismic zoning factor, *see* Earthquake zoning factor
Seismograph 21, 61, 62
  Wood-Anderson 21
Seismology 7, 9–25
  hypothesis 241
  in China 103
  in Japan 136
  periodicity of 103–5
Set 358
Shah, H. 369
Shear wall 53, 300, 301
Shibata, H. 77
Shimizu, N. 41
Shinozuka, M. 41
Silt 331
Site, *see* Ground
Site condition 52, 60, 81, 262, 270
Site effect 51–3, 311, 315, 317, 319, 324, 327
  on ground motion 51–3, 89, 92, 93, 140, 313, 315, 324, 327
  on intensity 52, 60, 72, 107, 108, 135
Site response 200
Site selection 278
Slope slide, *see* Landslide
Smokestack 111–13, 123, 279
Soil 186
  clay 187
  clayey 187
  dynamic property of 186, 188, 190
  sand 187
Soil-pile interaction 211

Soil-structure interaction 155, 177, 211
Soong, T.T. 306, 309
Sozen, M.A. 214
Spectrum 23, 35, 60, 71, 72, 80, 97
  approximate 75
  coherence 227
  intensity 71
  parameter 35
Stationarity 30
Stationary process 30, 160
Stiffness degradation 171, 301
Stiffness discontinuity, *see* Weak story
Stochastic, *see* Random
Strength enveloping function 40
Stress drop 13, 79
Strong motion, *see* Ground motion
Strong motion observation 5, 61
  array 65
  data of 231
  instrument of 61, 66
Structural coefficient 311
Structural control, *see* Vibration control
Structural identification 229
Structural model, *see* Mathematical model
Structural test 213–31
  of base-isolation system 214
  of building 214
  member 215
  method 216
    ambient vibration 225
    dynamic 217
    earthquake simulations 214, 217
    of dynamic property , 214, 217
    pseudo-dynamic 215, 217, 227
  pseudo-static 216
  shaking table 214, 217
  offshore platform 214, 215
  pipeline 214
  rigid body 214
  storage tank 214, 287
  to failure 227
Substructure method 178
Suehiro, K. 60
Sugimura, Y. 210

Tajimi, H. 26, 39, 40, 45
Takizawa, H. 172
Tanimoto, K. 199
Tectonic plate, *see* Plate tectonics
Terzaghi, K. 207
Thomson, W.T. 26
Time domain analysis 155
TMD (tuned mass damper) 310
Toki, K. 41
Torsion 215, 313
Transfer function 156
Transmitting boundary 182
Trifunac, M.D. 93, 94, 231

UBC (Uniform Building Code), *see* Seismic design code
Uncertainty 235, 250, 357
  fuzzy 357
  incomplete knowledge 357
  random 250, 357
Underground pipeline 288–91
Uplift 287
USGS (U.S. Geological Survey) 214, 263, 266, 269
Umemura, H. 221

Vanmarcke, E.H. 15, 35, 73
Variance 27, 28
Venn diagram 358, 360
Vibration control, *see also* Base isolation 306–10
Vulnerability 236, 251, 260
  bridge 110
  brick building 110
  earth dam 110

Wall 300
  brick 306
  infill 303
  shear 300, 301, 304
Wang, L.R.L. 288
Watabe, M. 68
Wave 16
  attenuation 20, 72
  body 17
  Love 18
  propagation 77
  Rayleigh 18
  reflection and refraction of 18
  shear 17
  surface 18
Weak story 171
Wei, Lian 171
Wente Brothers Winery 285–87
Whitman, R. 266
Wiegel, K.L. 71, 211
Wilson, E.L. 177, 182, 352
Wong, S. 369

Xie, Yu-Shou 46
Xinfengjian dam 293
Xu,Zhi-Xin 172
Yao, J.T.P. 218, 225
Yegulalp, T.M. 15
Yin, Zhi-Qian 171, 172, 214, 225, 259

Zadeh, L. 357, 367

For Product Safety Concerns and Information please contact our EU representative GPSR@taylorandfrancis.com
Taylor & Francis Verlag GmbH, Kaufingerstraße 24, 80331 München, Germany

www.ingramcontent.com/pod-product-compliance
Ingram Content Group UK Ltd.
Pitfield, Milton Keynes, MK11 3LW, UK
UKHW050755180926
475701UK00007B/1065

* 9 7 8 0 3 6 7 8 6 5 9 0 0 *